A Critique of Science
How Incoherent Leaders Purged Metaphysics of Mind and God

Orogee Dolsenhe

ISBN   978-1-105-05044-2

ALSO BY OROGEE DOLSENHE:

Coherent Nature: The Structure and Process of Consciousness

A Genuine Theory of Everything: Explaining Consciousness Through a Unification of Psychology, Biology, and Physics

In loving memory of my mother

For the knowledge-seekers of the past, present, and future

PREFACE

All scholars, it has been said, have some blind spot.
-Henry Holorenshaw

Laws alone can not secure freedom of expression; in order that every man present his views without penalty there must be spirit of tolerance in the entire population.
-Albert Einstein

Most people do not realize that anything is wrong with some of the maladaptive cultural practices in which they were raised. Consider the Indian caste system, which has endured for millennia with almost no form of centralized authority or large-scale force. Perhaps this is because it had become a norm of India's Hindu culture long ago. Any foreigner, however, who encounters this system for the first time finds it a rather "surprising or shocking" (Dumont, 1966) experience. Ever since becoming an independent nation, India has sought to abolish caste discrimination; however, despite the implementation of the Protection of Civil Right Act in 1955, this traditional religio-social culture continues to survive, especially in rural areas. And yet on the national level, it is fading away largely due to intercultural communication and comparison. Outsiders can recognize the ills of a cultural tradition that insiders neither question nor see. As they say, fish are always the last to discover water. This book attempts to address the unjust and counterproductive stratified system in one of the most unexpected places, one that considers itself to be the epitome of knowledge and wisdom: science.

Robert Louis Stevenson's *The Strange Case of Dr Jekyll and Mr Hyde* became so successful that its title became part of the English literary world. According to Webster, the definition of Jekyll and Hyde is that of "a person having a split personality, one side of which is good and the other side evil." Modern science also has a dual personality, for people generally see science as all Jekyll and no Hyde. This book presents the latter point of view in an attempt to correct the resulting flawed perception of modern science's quest for knowledge. Despite its split personality, science is the best path for people who want to pursue knowledge and resolve the many challenges facing them to pursue. This, then, reveals why we need to control Hyde without reducing Jekyll's goodness. In other words, science needs an intellectual watchdog who can point out its flaws and errors, as expressed by Michael J. Mahoney in *Scientist as Subject: The Psychological Imperative* (1976):

> Relative to the last century—or even the last decade—today's scientists know quite a bit about virtually everything on our planet—with one ironic exception. The exception, of course, is the scientist. He remains a mysteriously *unexamined* (emphasis added) inhabitant of our planet. In his relentless analysis of everything under (and including) the sun, he has remained curiously self-exempt from the scrutiny which is the reed of his vocation....The oversight is not benign. Our continued neglect of scientist could well be the most costly blunder in the history of empiricism. We can hardly hope to have much confidence in the product of science if we choose to remain ignorant of the limitations imposed by its human embodiment....He is not the paragon of objective reason; the saintly purveyor of truth. On the contrary, he is a thoroughly fallible human being—capable of bigotry, ambition, and political expedience. Far from his mythical image, he is probably the most passionate of professionals....The average scientist tends to be complacently confident about his rationality and expertise, his objectivity and insight... this complacency is naïve, unfounded and potentially disastrous to the pursuit of knowledge.

Labinger & Collins (2001) make a similar point:

> In the sciences resentment is especially easy to understand as traditionally there has been no equivalent to the theater or the music critique; the sciences have been taken to be too difficult or esoteric, to allow outside observers to have anything sensible to say to those who inhabit the "Republic of Science."

In general, people are intimidated by the mystique of science. I contend that science, despite its supposedly difficult-to-understand subject matter and its "esoteric" nature, must not be placed beyond the reach of examination and critique. Removing this aura will allow us to understand why this Hyde side of science has arisen and realize that other domains of society are facing similar problems. Human impulses play a powerful role in hijacking minds so that they can be led to embrace, over time, viewpoints that are dysfunctional, maladaptive, and even extremist and fundamentalist. This book traces the development and maintenance of a maladaptive culture among scientists to a set of human impulses. I was able to identify these seductive impulses because I recognized them in myself. The inward-looking nature of my own decision-making processes has led me to believe that the impulses and biases that I found within myself can be applied to everyone. My hope is that readers will recognize that their own impulses are no different from those experienced by scientists and others so that they can make better decisions and be better prepared for any possible mishaps.

A tragic irony is that science has developed a sophisticated system that allows its practitioners to justify their intolerance and discriminatory practices against those who do not subscribe to the "correct" viewpoints. In essence, the scientific community may be one of the most underdeveloped domains in the progress of social justice in the free world. Instead of moving forward seen in much of religious and political domains, ever since the 19th century it has been steadily marching backward. One of the recurring themes of this book is the human propensity to follow dysfunctional and incoherent leaders. Historical and sociological data, as well as a psychological analysis, are provided to explain how the practitioners of science developed a maladaptive culture and why they are unable to see it for what it really is. In order to make science's flawed practice recognizable, this book will compare it with other kinds of maladaptiveness in the world.

The main victim of this culture of intolerance and discrimination is metaphysics. The word *metaphysics* comes from the two Greek words: *meta* (beyond) and *phussika* (physics). In the case of science, the two metaphysical concepts are supernatural agency (i.e., God) and unobservable mind. One of modern science's main tenets, the necessity to "purge" science of metaphysics (Smocovitis, 1996), is dealt with extensively. This viewpoint, which is certainly not based on empirical evidence or even thoughtful examination, bears a remarkable resemblance to demagoguery. For those who have been in conflict regarding the issue of God because science's explanation is so intellectually and spiritually unsatisfying, this essay may provide some relief from the overwhelming pressure of science's mystique and dogma.

My purpose is not to defeat science or to be anti-science, but to address the Hyde side of science. Despite all of its flaws and errors, science represents the best of human intellect, knowledge, and wisdom.

Science can and must do better so that it can solve its own internal challenges as well as those facing the world. Science must try to discover its own blind spots and flaws as much as it focuses on those found within other fields. Above all, scientists should set an example of treating opposing and unconventional viewpoints with respect so that the rest of society can follow their example. This essay examines the self-exempted examiner without any reservation or restraint. I have done my best to make it accessible to most people by presenting it in readily understandable terms and concepts and as straightforwardly as possible. My candid assessment will no doubt cause some readers who believe in the prestige of science to be offended; however, it will also allow the ongoing debate to move forward.

Contents

Preface --- 5
Contents -- 9
Introduction --- 15
    The higher standard for the yardstick --------------------------- 18
    Comparing scientific puzzle and crossword puzzle -------------- 20
    The value system in knowledge-seeking --------------------------- 22
1. Psychology's Bartering away Its Soul for Status -------------- 31
    The predicament and insecurity of psychology ------------------- 33
    John B. Watson: A good PR man who created
        "smoke without fire" -- 34
    Intellectual hijacking -- 35
    B. F. Skinner: The most eminent psychologist
        who denied the existence of consciousness ------------- 37
    Self-deception and intellectual laziness ------------------------ 38
    Group pressure and the confidence of judgment ----------------- 40
    A cognitive revolution -- 42
    The preoccupation of experiments and the
        fragmentation of psychology ------------------------------ 45
    The premature commitment to computer-metaphor
        psychology -- 46
    The consciousness phobia --------------------------------------- 47
**2. The Indian Caste System: The Most Extreme Form
    of Social Stratification** --------------------------------------- 49
    The preoccupation of hierarchy and separation ----------------- 50
    The origin of the caste system --------------------------------- 53
    The reference of the caste system in the ancient texts --------- 54
    The need-based and the power-based stratification ------------- 55
    The zero-sum impulse and stratification
        in the tribal society -- 56
    The obsession with purity -------------------------------------- 59
    Howard Hughes: Obsessive compulsive disorder
        of the genius -- 60
    The caste system: Obsessive compulsive disorder
        of the society --- 62
3. The Rise and Fall of Communism ------------------------------- 65
    Plato and the reactionary impulse ------------------------------ 66
    The industrial revolution and Marx's

      communism doctrines----------------------------------69
   The expediency of the communist doctrines
      for Lenin and Stalin-------------------------------------71
   Universal suffrage and changed social conditions-------------73
   The audacity of the Xiaogang farmers---------------------------74
4. Hijacking of Minds--79
   The two separate system of decision-making------------------79
   The importance of emotions------------------------------------81
   Amygdala versus Brodmann area 10 (BA10)-------------------83
   Marx's personality and the background of
      the development of communism----------------------87
   The two-valued impulse---90
   Group polarization---93
   Visual neglect---95
   Intellectual neglect in value judgment----------------------------97
   Intellectual neglect in certainty judgment------------------------99
   The confounding of value judgment and certainty judgment-103
   The trapping mechanism and cultural hijacking---------------103
   Individualism and collectivism-----------------------------------106
   The French Revolution and the domination of extremists----108
   Obedience to the authority: Commonness of evil--------------111
   Hitler's personality and his rise to power-----------------------113
   Social hijacking in the holocaust and communism-------------117
5. The Tripartite Relationship: Conflict and Cooperation
   between State, Religion, and Science-----------------------121
   Did man make history or did history make man?--------------122
Knowledge-seeking in China--123
   Confucianism---124
   Mohism--125
   Taoism---126
   Chuang-tzu--127
   Materialism---127
   Legalism---128
   The establishment of the value system in China:
   The palace examination--130
Knowledge-seeking in the Islamic world-----------------------------132
   Multicultural contribution of the birth of modern science----133
   The neglect of experimentation in the ancient world---------134
   The scientific methodology of Islamic science-----------------136

The establishment of Madrasas and the shift
  of the value system in Islamic science----------------138
The conflict between Aristotelian and Islamic thought--------141
Ibn Sina--142
Al-Ghazali--143
The rise of universities and the changing of the value system------144
  Christianity's transformation from
    the persecuted to persecutor----------------------------145
  The East Roman Empire and the West Roman Empire--------146
  The regression of the Greek knowledge and the Dark Age---148
  The establishment of universities and
    the reconfiguration of the value system-----------------150
  The Aristotelian doctrine and reconciliation-------------------153
  Examining the conflict between science in the Galileo
    affair--155
  Spinoza's God: The view of the boldest scholar
    in the 17th century Europe-------------------------------162

**6. The Flipping of the Value System and the Inception of
the Science's Caste System in the 18th and 19th Century----165**
  Naturalism, atheism, and anti-religion---------------------------166
  The shift from supernaturalism to
    natural approach in astronomy---------------------------173
  Positivism and the beginning of the caste system
    in the 19th century---176
  The study of fossils and the seed of doubt----------------------179

**7. The First Phase of the Evolutionary Absurdity:
The Survival of the Slipperiest----------------------------------183**
  Henslow's role in Darwin's success------------------------------184
  The tautological issue of natural selection----------------------186
  "A special difficulty" and Darwin's delay-----------------------189
  The confusion between artificial selection and
    natural selection---191
  The ambiguity of the teleological implication
    of the concept of natural selection---------------------191
  From "my theory" to "the theory"-------------------------------193
  Darwin's accomplices and the birth of the Darwin industry--198
  Darwin's bulldog who didn't believe in natural selection-----202
  Intellectual neglect in Darwin worship------------------------203
  The role of interpersonal conflict in science------------------205

Maupertuis: A forgotten genius who was
    a precursor to Darwin and Mendel--------------------207
Religious intolerance---211
Thomas Huxley: Science's extremist-----------------------------213
St. George Mivart: An independent thinker caught in the
    middle---215
Group polarization between science and religion-------------219
The diversity of evolutionary thoughts-------------------------220

8. The Second Phase of the Evolutionary Absurdity: The Mystique of Incomprehensible Mathematics----------------223

The rediscovery of Mendel and saltationism-------------------224
The triumph of naturalism in biology----------------------------227
Incomprehensibility as a form of value--------------------------229
The mystique of mathematics and the meltdown of Wall
    Street--230
The synthesis, the compromise, or the intimidation by
    mathematics?---231
A group of blind men follow the blind leader------------------237

9. The Third Phase of the Evolutionary Absurdity: The Complexity of Life--241

The limitation of reductionism----------------------------------243
The search for non-human intelligence-------------------------246
Natural theology, intelligent design, and irreducible
    complexity---247
Curtailing the ridicule, charges of disloyalty, and
    warmongering--252
Systems biology: The irreducible systems----------------------253
Intangible complexity and the genetic code--------------------254
DNA's packaging--258
The origin of life--259
Fine tuning of the universe--------------------------------------262
The adaptive capacities without natural selection-------------264
Teleology and teleonomy: Science's verbal tap dancing------265
The Scopes monkey trial---266
The Dover trial---267
Common errors made during evolutionist-creationist
    debates--274
Three devious ways to reject supernaturalism-----------------275
Building bridges between science and religion----------------276

    The problem of unanimity--------278
    Academia's hostile and intolerant intellectual landscape-------280
    The ironies of multiverse--------283
    Relativity, quantum mechanics, and the fixation of physicists--------284

10. The Demise of Randomness as the Cause of Biological Complexity: Gaps in the Neo-Darwinism Paradigm-------291
    The gap with accidental evolution--------293
    The gap with evolutionary patterns--------294
    The gap with endosymbiosis--------295
    The gap with directed mutation--------295
    The gap with epigenetic change--------297
    The gap with parallel evolution--------298
    The gap with isomorphism--------299
    The gap with mathematic equations in morphology--------300
    The gap within--------300
    The gap with society--------301
    The evasive champ and the dawn of revolution--------301

11. The Demarcating Wall of Science--------307
    The discriminatory canon of science--------308
    Karl Popper's falsificationism--------311
    Crude stupidities and the streetlight effect--------314
    Michel de Montaigne and self-knowledge--------316
    Metacognition, metaexecutive, and metaknowledge--------320
    The mutual dependence principle in science--------325
    The risk management of decision-making by individuals and groups--------327
    The black swan and the expectancy effect--------330
    Gullibility and rigidity: The "not me" myth--------331

12. Kennedy's Failure and Success--------337
    The Bay of Pigs--------338
    The unheeded doubts--------343
    What went wrong?--------344
    Kennedy's adjustments after the Bay of Pigs--------345
    The Cuban missile crisis--------347
    Symptoms of groupthink and their significance in science---351

13. Preventing Maladaptive Attitudes and Practices--------355
    Signs of trouble and the critical period--------356
    Impulse control and the general checklist--------358

  Applying institutional pressure----------------------------------362
  Prevention versus treatment--------------------------------------365
References--369

INTRODUCTION

Civilizations die from suicide, not by murder.

-Arnold Toynbee

Francis Galton, Charles Darwin's cousin, in the 19th century coined the term "nature versus nurture" to discuss the two types of influences that humans all are subjected to. The main controversy involving two influences is: which one plays a greater role in shaping a person's traits and faculties. On the extreme nurture side, the human mind is considered as tabula rasa (blank slate) at birth and a person's makeup comes from experience. On the opposite extreme, an individual's innate qualities (or genetics) determine individual differences. The 19th century favored the hereditary approach and the 20th century preferred the environmental view (Pinker, 2004). A compromise between the two extremes is generally accepted in present time: behaviors and capacities emerge from an interaction of the two.

The capacity of human's cultural innovations allows us to adopt a variety of environments. We do not have to wait for evolution of thick insulation in order to live in cold region. Our ability to make clothing by using the fur of other animals allows us to live in extreme cold environments like the Arctic. In addition to making warm clothing, kayak, harpoons, oil lamps, dog sleds, igloo, etc. are some cultural inheritances that are essential to survive in the Arctic region (Sterelny, 2007). The Inuits obviously have the cultural inheritances that are highly adaptive to the hostile environment they live in.

Notwithstanding the capacity to develop and transmit cultural adaptability, humans also can paradoxically produce and maintain maladaptive cultural tradition (Edgerton, 1992; Sterelny, 2007). Edgerton (1992), for instance, provides many maladaptive cultural practices that go against the presumed primitive harmony. Perhaps the tragic history of Easter Island exemplifies the catastrophic potential of the human's cultural development (Sterelny, 2007; Diamond, 2005). The name "Easter Island" was given by a Dutch explorer, Jacob Roggeveen, who was the first European to visit the island on Easter Day on April 5, 1722. The island is not very big: approximately 66 square miles (about one tenth of Oahu, the Hawaiian island where Honolulu and the Pearl Harbor are located). It is one of the most isolated inhabited islands. The nearest mainland is 2,000 miles away and the nearest habitable island is Pitcairn and it is 1,400 miles away. When Roggeveen saw the island for the first time, he was impressed by two unusual things. First was that, despite the fertile volcanic soil, the mild tropical climate, the adequate rainfall, the island was almost completely barren land devoid of any tree. Second was hundreds of huge stone statues with varying size that typically ranged 15 to 20 feet tall (the tallest one was 70 feet) and weigh between 10 to 270 tons. Most of these statues were carved in a quarry and transported up to six miles. Roggeveen was astonished with such achievement since there were no lumber, wheels, metal tools, or draft animals. As Roggeveen was, other visitors were also perplexed by the massive stone edifices that seem to defy logic.

Modern archaeological and paleontological findings that began in 1955 reveal a completely different past. Pollen records show that Easter Island used to be a sub-tropical paradise with dense forest when the first Polynesians landed on the island. The first inhabitants trace back to 400 to 700 AD. The most common tree in the forest was a species of palm tree which can grow up to 82 feet and 6 feet in diameter. The palm tree would have provided a variety of functions. It could have been a good source of food such as nuts and sap which can be made into sugar, syrup, honey, and wine. It also could have been an important material for constructing large canoes to fish. The bones unearthed from the early garbage heaps from the period 900 to 1300 AD show fishes, porpoises, seals, seabirds, and land birds. Within a few centuries after the first settlement, pollen records indicate the degradation of the forest. The 15th century marked the end of not only palm trees but the forest as a whole (Diamond, 2005). In addition, virtually all other animals larger than insects became extinct. So what happened?

The main culprit of ecological catastrophe of Easter Island was the preoccupation of statue-building. Jared Diamond, a geography professor at UCLA, argues that the ecological catastrophe was self-induced in *Collapse*. He describes the highly hierarchical and centralized structure of clans of the island headed by chiefs who had religious and political authority. Under the leadership of clan leaders, they were building "moai," monolithic human stone figures, up to 887 of them. Most of them were built between 1250 to 1500 AD.

The large wood and stone carving of humanoid forms, which is called "tiki," is common among Polynesian cultures. The cultural tradition of statue building of the Easter Islander, however, evolved into an obsession. The statues became larger and larger which Diamond (2005) describes as an "escalating spiral of one-upmanship" as rival clans try to outdo each others. Diamond (2005) compares this with the Egyptian pharaohs who built ever-larger pyramids or Hollywood movie moguls building ever bigger mansions. In the moai-building craze, trees were cut down to make rollers and rope to transport the massive statues. Diamond (2005) asks the key questions: "Why didn't they look around, realize what they were doing, and stop before it was too late? What were they thinking when they cut down the last palm tree?" He conjectures any islander who tried to stop the progressive deforestation would have been overruled by the carvers, bureaucrats, and chiefs who had vested interests. Diamond (2005) points out that the environmental collapse of Easter Island is not unique. The similar fate of other cultures such as the Greenland Norse, the Anasazi, and Maya all point to the human's susceptibility for cultural maladaptation in massive scale.

The main problem of self-induced environmental catastrophe may be that people who are living in this culture do not realize what they are doing. The East Islanders were raised in a culture that was preoccupied with building stone statues. Any outsider would have been puzzled by the obsession of building moai. They would be shocked to discover that the whole forest was disappearing because of it. Outsiders can easily see the pathology because he/she is unhindered by the cultural desensitization and would ask questions: What's the big deal about the stone statues? Why can't someone stop the whole nonsense?

This might be due to what Diane Vaughn (1996), a sociologist whose specialization is organizational and cultural failures at Columbia University, calls the "incremental descent." In her book *The Challenger Launch Decision*, deviance from what was originally intended, which was the utmost importance of safety, was normalized. The insiders, the managers and engineers of NASA, have become desensitized by the gradual repetition of overlooking safety which then resulted to the Challenger's disaster. The difficulty of changing cultural maladaptiveness was further demonstrated by a loss of another shuttle, Columbia. Columbia disintegrated during its re-entry into the earth's atmosphere in 2003. Just as the elite group of people at NASA, the Easter Islanders became desensitized with the gradual disappearance of its forest. If an outsider like Roggeveen visited the island before the demise of its forest, he would have noticed and warned them of the gravity of the situation. The point is that NASA's management and the Easter Islanders could have all benefitted from outsider's input. However, an outsider's critical assessment is not usually welcomed.

The Higher Standard for the Yardstick

> The idea that science is infallible and beyond criticism, is a delusion, and even dangerous one.
> -Anthony Standen

People all over the world respect, appreciate, and trust scientific knowledge and expertise. The confidence in science is translated into people's reliance on scientific know-how in dealing with the problems they face. If a child is sick, parents would most likely go to a hospital and put the fate of their loved one on medical doctors trained in science. Similarly, the judicial system often calls on for the testimony of scientific experts to help determine whether the accused is guilty or innocent. An incarcerated person for many years can be freed by a DNA evidence provided by a genetic scientist. India's Prime Minister Jawaharlal Nehru (1960) expressed his unequivocal support for science:

> It is science alone that can solve the problem of hunger and poverty, of sanitation and illiteracy, of superstition and deadening custom and tradition, of vast resources running to waste, of a rich country inhabited by starving people... Who indeed could afford to ignore science today? At every turn we seek its aid...The future belongs to science and to those who make friends with science.

Perhaps due to its success, people may think that science has no fault. After all, science is widely recognized as a "self-corrective" process. A phrase popularized by a philosopher Charles Peirce (1896): "This marvelous self-correcting property of Reason belongs to every sort of science...." Carl Sagan (1995) echoes Peirce: "Science is a collective enterprise with the error-corrective machinery often running smoothly." Similarly, Robert E. Krebs (1999) confidently expresses: "Science is the only human endeavor that operates with a self-correcting system using controlled experiments." Having faith that science has a self-corrective capacity may lead to the illusion of its infallibility and self-righteousness which then may lead to complacency. However, a scientist, being human, is susceptible to the same type of errors and flaws in judgment as any average person. If anything, a scientist may face a higher level of conformity pressure since his/ her job security is tied to one's reputation. When it comes to major issues, as we shall see later, it will resist all forms of corrective effort; science is close to being correction-resistant rather than self-corrective.

If science represents a yard stick that measures all others, it must itself have the highest standard and continue to find ways to improve it. To illustrate this point, let us talk about the standard of the real yard stick. Actually, it is the "meter bar." Up until 19th century, every country had its own system of measurement. Some countries like England even had three different ones. On May 20, 1875, the delegates of 17 countries signed the new standard system called the "metric" system which comes from the French word for measure. The metric system has seven base units called the International System of Units: length (meter), mass (kilogram), time, (second), electric current (ampere), temperature (Kelvin), amount of substance (mole), and luminous intensity (candela). Let us just focus on the unit of length. The "meter" is the unit of length which came from Pierre-Simon Laplace's proposal, which is one ten-millionth of the distance from the North Pole to the Equator (A meter is a little longer than a yard, approximately 3.29 feet). The designers of the metric system made a bar made of alloy 90% platinum and 10% iridium because its expansion rate due to temperature was minimal. In addition, it has the durability to take a very fine line which can reduce the error due to the width of the lines. Thus, the international prototype meter bar was born. The bar was kept at an underground vault at the International Bureau of Weight and Measures at Sevres, near Paris. The main drawback of the meter bar was that one had to travel a long distance to have the calibration. In 1960, a new method of standard was adopted: rather than a bar of alloy, the wavelength of orange light emitted by the gas krypton 86 was used. With this new standard, any laboratory in the world could reproduce the exact length without having to travel afar. In 1984, it made another change. One meter is now defined as the distance traveled by light through a vacuum in one-$299,792,458^{th}$ of a second. The continuous improvement and the highest standard imposed on the metric system ensure the precision for scientists and engineers in the globalized world. Parts can be produced in different locations that can be shipped and assembled at a place halfway around the globe. Such thing would not be possible without the highest standard imposed on the meter bar, which is used to calibrate other devices.

During the Middle Ages in Europe, it was the church that had enormous prestige and authority to judge all others in society based on its established yardstick. Yet, the correctness of its own yardstick was hardly a concern for the hierarchies of the church. The self-exempted status of the church of its own standard inevitably meant the systematic corruption and abuse of power. Those who dare to criticize the church like John Hus and Giralamo Savonarola were burned for heresy. It took the Protestant Reformation for the church leaders to realize and engage in upgrading the standards of its own yardstick. This was known as the "Counter-Reformation." The important lesson learned from the Reformation nearly five centuries ago is this: Those who enjoy prestige and power to judge others may develop an illusion of self-righteousness and infallibility. Consequently, they are susceptible to complacency and fail to keep up with its own standard. The worst case scenario is that self-appointed guards often thwart others from efforts for improvement. Hart (1952) addresses a human flaw: "The more we want to change others, the less we feel the need to change ourselves…." This also applies to science as argued by Standen (1950):

> Since scientists have such overweening confidence in their own ability – in their collective ability, that is to say – it is small wonder that they make no attempt to teach what are the limitations of science, for they hardly recognize any.

Comparing Scientific Puzzle and Crossword Puzzle

> The liberty of choice is of special kind; it is not in any way similar to the liberty of a writer of fiction. Rather, it is similar to that of a man engaged in solving a well-designed word puzzle. He may, it is true, propose any word as the solution; but, there is only one word which really solves the puzzle in all its parts. It is a matter of faith that nature—as she is perceptible to our five

senses—take the character of such a well-formulated puzzle. The successes reaped up to now by science...give a certain encouragement to this faith.

-Albert Einstein

The investigation of the truth is one way hard, in another easy. An indication of this is found in the fact that no one is able to attain the truth adequately, while, on the other hand, no one fails entirely, but everyone says something true about the nature of things, and while individually they contribute little or nothing to the truth, by the union of all a considerable amount is amassed.

-Aristotle

Science is a puzzle-solving activity that is geared toward attempting to understand nature. Scientists are not the only ones who are engaged in puzzle-solving. We are all well-experienced in all sorts of puzzle-solving. A woman might try to figure out why her cake doesn't taste as good as her friend's. She may try out things like adding a different ingredient. Parents might try to understand the causes of their teenage son's sudden mood change. They might ask some questions to their son or inquire with his teachers.

Among many types of puzzles, perhaps a crossword puzzle shares more things in common with a scientific puzzle than any other types because: 1) They consist of many parts. 2) Some of those accessible parts are more certain than others. 3) Some parts have more options than others. 4) Some parts are more crucial than others. 5) Parts are interconnected and interrelated. 6) The more interconnection with other parts, the more confidence of guesswork. A crossword puzzle consists of many words: approximately half are across and the other half are down. And each word has a specific number of letters accompanied by a clue. Where do you start the puzzle? Most people start from the top left side which is similar from the layout of a typical book unless it is in Chinese or Arabic; thus, proving that we already have a bias in choosing something as simple as where to begin the puzzle. Solving a scientific puzzle also faces many other biases as we will discuss later. If we are to seek help in learning or improving our skills in solving crossword puzzles, two most common advices are: 1) do the easier ones first and 2) use a pencil. Most people, I suspect, however, would figure out those simple methods on their own very quickly. However, something as obvious as the two advices in solving crossword puzzles are not followed well in science, particularly in crucial issues as we shall see later. Let's presuppose you completed about the two thirds of puzzle and, let's say, you face a roadblock in approaching the bottom right side. A common way to deal with an impasse is to recheck the inserted words and try other options. This is where using a pencil rather than a pen becomes useful. Again, this kind of common sense is not practiced well in science.

The obvious difference between a crossword puzzle and a scientific puzzle is that the former is small enough for a single person to finish it in a relatively short time whereas the latter is so huge that not one person or not even a sizeable group of people can manage to solve all of it. Because each person has a limited time and mental capacity, one can only handle a small part of the whole. This situation reminds us of a cliché that originated in India and popularized in a poem by the American poet John Godfrey Saxe in the West. It is about the story of six blind men trying to figure out what kind of animal (an elephant) by applying their sense of touch. Depending on where they touched (leg, tail, trunk, ear, belly, and tusk), they had completely different views (a pillar, a rope, a snake, a fan, a huge wall, and a spear, respectively). There is an important ethical lesson from this famous fable that we often forget. We may have diverse perspectives that contain partial truths. This means that our own viewpoint may be incomplete just like the others. Rather than seeing people with different viewpoints as a source of competition that we have to win over, they should be considered as ways to complement our own limitation. Simply put, a golden rule of knowledge is that we should respect the viewpoints of one another. This simple guideline, however, is extraordinarily difficult to follow in the scientific community as we shall see later.

The Value System in Knowledge-Seeking

> Values merge affect and concept. Persons are not detached or indifferent to the world; they do not stop with a sheerly factual view of their experience. Explicitly or implicitly, they are continually regarding things as good or bad, pleasant or unpleasant, beautiful or ugly, appropriate or inappropriate, true or false, virtues or vices.

A Critique of Science

-Robin M. Williams, Jr.

The enemy is no longer across the border, it is within, it is part of oneself...
-Susantha Goonatilake

In 2008, someone bought a license plate for $14.3 million at an auction in Abu Dhabi in the United Arab Emirates. You might think that the license plate must be made out of gold and diamond-studded. No, it is just like any other license plate. The only unusual thing about the plate is its number, "1". If you think the outlandish high price of a low numbered license plate is an oil-rich Arab thing, think again. In Britain, an "F1" plate was sold at $870,000. In the United States, a Delaware plate with a number "6" was sold at $675,000. Why such ridiculously high prices for plain license plates? The beginning of the license plate numbering originated in Massachusetts in 1903 and other states followed. It started with number "1" which was given to the governor, "2" to the lieutenant governor, and so forth. In addition to the usual status symbols of big expensive cars that high officials drive, people in states such as Rhode Island, Illinois, Massachusetts, and Delaware also considered low-numbered license plate to be a source of envy and fascination. It began in the early 20th century and is maintained until today. In fact, considering the extravagant price they fetch, you might say low-numbered plates seem to have surpassed the appreciation of cars.

Most people who hear about the excessive prices of low-numbered license plates for the first time would be puzzled. However, it might be a surprise that over-priced things are a lot more common than we might think. We just have become so used to them that we don't even notice them to be absurd, even though they are no different from high priced low-numbered license plates. Take gold, for example. Gold is one of the most sought-after precious metals. One ounce costs over $1,400 in the beginning of 2011. Yet, besides the glitter and the resistance to corrosion, it is hardly useful to make any kind of tool because of its softness. The almost universal appeal for gold began a similar way as the low-numbered license plates. The high price gradually rose due to supply and demand.

Society's assigned worth for specific things may not have an intrinsic value, yet they become so due to their significance as part of our culture and tradition, and yet some items with a considerable amount of intrinsic value may not receive the merit it deserves. The majority of valuations in society come in a form without monetary assessment. We may appreciate certain acts like a soldier's bravery or philanthropy's as a generous contribution. Conversely, we might assign negative value toward corrupt politicians or criminals. A great deal of children's learning consists of figuring out what is valued and what is not, and children learn the differing standard for different groups of people as to what is cool and what is not cool. What or how they say might be cool among his/her peers, but objectionable to teachers or parents. The evaluation of whether something is good or bad can occur at the level of unconscious (Zajonc, 1980). Anyone who travels to another country will quickly find out the vastly different value system than one's own. This is often referred to as "culture shock." The sense of value is one of the most pervasive things and it is very difficult to detach them from most of our experiences. Learning a value system is an essential component of acquiring knowledge and language.

This brings a question in mind as to whether knowledge-seeking has anything to do with value. One might ask what value has anything to do with the pursuit of truth. David Hume (1740), a Scottish philosopher, in the 18th century argued that "is" should not be derived from "ought" in the pursuit of knowledge. The "is-ought" distinction was differently phrased by Max Weber, a German sociologist and political economist. He articulated that statements of fact and statements of value are two different things and any confusion of the two is impermissible. The notion of value-free is one of science's lofty ideals that contrast with value-saturated religion; facts are objective and should not be inferred from subjective value. For those who believe in the virtue of science, value is a contamination that needs to be rid of. Some, however, have argued that social, political, economic, and personal values permeate scientific work at all levels (see McMullin, 2000). The promotion of value-free science is one of many examples that scholars try to make of science as being different from other domains, and scientists who have been trained in scientific methodology are seen as the intellectual elites who rise above and beyond the flaws and deficiencies of religion. However, value is so pervasive that it is difficult to completely be rid of. For instance, the difficulty of value-free ideal was pointed out by Lacey (1999): "science is value free" in and of itself is paradoxically held as a statement of value-promotion. Lacey's point underscores the value-ladenness of our expression that we don't even realize. The real danger of the promotion of value-free science is that it might give a false sense of security from the value-caused problems. Douglas (2009) came to a conclusion that science should be value-saturated rather than value-free. She makes a valid point that the value-free ideal depends on the premise of science's isolation from society. She believes that science can be value-laden, yet preserve its integrity.

Let me give you an example. In 1982, Peters & Ceci revealed startling findings on how scientists evaluate their works. Peters & Ceci (1982) selected 12 already published research articles by investigators from prestigious and highly productive American psychology departments. By substituting with fictitious names and institutions, the articles were resubmitted to the journals that had originally refereed and published them eighteen to thirty two months earlier. Out of the 12 journals, only three were recognized as resubmissions. Here is the most surprising and disturbing outcome: Out of the remaining 9 articles, 8 were rejected. The reasons for rejection in many cases were "serious methodological flaws" (Peters & Ceci, 1982). Without the names of the prestigious authors and institutions attached, the very same articles that were accepted earlier were considered as flawed and unworthy of publication. If the recognized value of authors and institutions overshadows the original aim to judge an article's merit, then, there is a serious problem in science.

Imagine a tennis tournament where the past winners of the tournament are given an incredible amount of advantages over other players. Imagine the calls from the line judges are largely determined because of past distinctions rather than where the ball lands. Imagine the line judge's thought: the ball must have landed inside the line since the great player hit it. As ridiculous as it may sound, this sort of thing really happens in the scientific community. While other domains of society, not only in sports but also in the court system, have put a great deal of effort to overcome the errors associated with value placing, science is curiously exempted from correcting the value-caused errors. This type of finding by Peters & Ceci (1982) was actually discussed in 1920 by Edward L. Thorndike, a psychologist at Columbia University, under the name of the "halo effect." He found that commanding officers rate their solders as generally good or bad across all categories of measurement. Thorndike

(1920) sums his findings:

> The "halo effect" is an extension of an overall impression of a person (or one particular outstanding trait) to influence the total judgment of that person. The effect is to evaluate an individual high on many traits because of a belief that the individual is high on one trait.

Then, he explains the opposite of halo effect, the "devil effect":

> Similar to this is the "devil effect," whereby a person evaluates another as low on many traits because of a belief that the individual is low on one trait which is assumed to be critical.

In addressing a similar psychological tendency, Merton (1968) gave the name the "Matthew effect" from the Gospel of Matthew (25:29) which says: "For everyone who has will be given more and he will have an abundance. Whoever does not have, even what he has will be taken from him."

In order to understand how such failures might arise, we need to examine one of the most important social faculties—the value judgment of people. We constantly compare our value with others which is called "social comparison" (Festinger, 1954). Social comparison provides much of social hierarchies, one's relative status in comparison with others. One is expected to behave in specific manners depending where he/she stands in relation to others. Our predisposition for social hierarchy is so engrained in our brain that social ranking would spontaneously emerge in children as young as two years old (Boyce, 2004; Cummins, 2000). Caroline Zink at NIMH with her colleagues demonstrated how powerful social status is even in playing a simple game. In their study, 72 subjects played an interactive computer game for money. They were told that their assigned status was based on their playing skill. The truth was that their assigned level (two-star player) was predetermined. Their opponents were a higher ranked player (three-star player) or a lower ranked player (one-star player) which was simulated by a computer. Remarkably, the brain activity of the participants was highly influenced by their perceived status. The brain's region activated when facing a superior player was different from when they thought they were playing an inferior player.

The studies of social hierarchy by Zink (2008) may at least partially explain the findings by Peter & Ceci (1982). We don't simply process information as what they are. Even before reviewers start reading submitted papers, perceived pecking order determines what part of the brain would be activated. In this sense, scientific papers may be largely judged by their covers rather than their contents. It would be ironic that the scientific community lags far behind the ideal of racial justice eloquently addressed by Martin Luther King, Jr. The main theme of *I Have a Dream* speech could be represented by a single sentence: "I have a dream that my four children will one day live in a nation where they will not be judged by the color of their skin, but by the content of their character." Because of the pervasiveness of value placing, it is not something we can or should be rid of. Instead, we should try to manage it. It is inescapable that some issues and discoveries are more valuable than others. Managing value is like riding a wild and unruly horse. However, if you can tame it, the value system could actually promote progress. Simply put, if you have the right value system, the rest will follow. As you shall see, a quantum leap of knowledge usually is associated with a value-shift in some fundamental way. Conversely, an inappropriate value system can drag down not only knowledge-seeking but also other domains of society.

The study of Peters and Ceci (1982) provoked some strong reactions: "We have met the enemy and he is us" (Cicchetti, 1982). Ziman (1982) calls for a "radical reform." Just as any sporting events, scientific endeavors must have a neutral space whereby the merit of scientist's work should be based on the work itself rather than the status of persons or institutions. The crucial findings by Peters and Ceci (1982) should have been a wakeup call. Yet, almost three decades have passed without any significant effort to correct the flawed peer-review process. I would suppose those who are in power to make change are benefitting from the obvious flaw. This is, in a way, similar to the fate of a campaign finance reform program that has been attempted many times since the 19th century. The main cause of this stalled campaign finance reform program—albeit a watered down version that almost has no effect on changing the problem has passed—is that the incumbents in the Congress are reluctant to limit their own privileges because they might have to compete with opponents on equal footing.

Society's value system is typically maintained by two ways: carrot or stick. People who follow the cultural norm are rewarded and those who do not are punished. In facing the outlook of acquiring a carrot or a stick, a human's innate tendency is to pay more attention to negative prospects than positive ones. This is known as "automatic vigilance." Pratto & John (2005) explains:

> There is a fundamental asymmetry in people evaluations of gains and losses, of joy and pains, and of positive and negative events....They [people] assign relatively more value, importance, and weight to events that have negative rather than positive, implications for them.

Pratto & John (2005) gives an evolutionary reason for the innate inclination for negativity:

> There are good evolutionary reasons for this widespread and pronounced asymmetry in people's evaluative reactions. Events that may negatively affect the individual are typically of greater time urgency than are events that lead to desirable consequences. Averting danger to one's well-being, such as preventing loss of life or limb, often requires an immediate response.

Ito et al. (1998) reports evidence that support automatic vigilance. The brain activity of negative information generates a greater brain activity than positive ones. Politicians know negative ads carry a more powerful punch than positive ones (Snsolabehere & Iyengar, 1995). There is no reason to think science is exempt from the power of automatic vigilance. The prospect of a negative reaction from peers would play a greater role in following the status quo even if one may not agree with it. This is especially problematic in the subjects that have been stigmatized. The separation anxiety can become the powerful motivation to maintain the prevailing paradigm in science. The label "unscientific" is something virtually no one is willing to risk. The "attention-grabbing power" of negativity is something that any society must be aware of. Hitler's rise to power was largely through negative portrayal of Jews. An unfortunate reality whether for the country or for the scientific community is that few individuals with exceptional communication skills can hijack an entire

A Critique of Science

society by effective use of negative information. Demagogue can triumph over reason, evidence, and justice. Yes, even in science, as we shall see.

Orogee Dolsenhe

Chapter One
Psychology's Bartering away
Its Soul for Status

For the greatest enemy of the truth is very often not the lie, deliberate, contrived and dishonest, but the myth, persistent, pervasive and unrealistic.

-John F. Kennedy

At any given moment there is an orthodoxy, a body of ideas, which it is assumed that all right-thinking people will accept without question. It is not exactly forbidden to say this, that or the other, but it is "not done" to say it, just as in mid-Victorian times it was "not done" to mention trousers in the presence of a lady. Anyone who challenges the prevailing orthodoxy finds himself silenced with surprising effectiveness.

-George Orwell

As mentioned earlier, it usually takes an outsider to recognize the profound flaw and injustice of the Indian caste system. Similarly, it may take an outsider to see the defects and unfairness in science that has been ongoing for more than 150 years. Let's start with how psychology bartered away its soul for a favored status in the stratified system of contemporary science. I choose to start with psychology because its profound flaw has been recognized and at least partially corrected. Its blunder is, however, only part of a much bigger systemic mess.

From the moment we wake up until we fall asleep, we are continuously conscious of something: the sound of alarm clock, the feeling of the toothbrush's bristles rubbing against our teeth and gums, the smell of minty toothpaste, seeing oneself in the bathroom mirror, or the hot sensation of hot water running down our body. These are some of the conscious experiences associated with morning rituals. The mental events taking place in my mind cannot be seen by others, cannot be measured, photographed, or compared with those of another person. And yet they are the underlying sources that drive and enable a person's observable behavior. Although they may not be observable, the importance of understanding another person's private conscious experience is recognized by virtually all of us. Tensions between husband

and wife or between friends may be reduced by accommodating the other person's concerns and emotions. Attempting to look into a person's inner experiences is indispensable for therapists and psychiatrists, both of whom seek to comprehend the person's or animal's inner world, which is not directly accessible. The prevalence of our consciousness can also be found in *Roget's Thesaurus,* which reveals that almost two-thirds of our vocabulary is devoted to expressing the content of one's consciousness (Baars, 2003).

But during the first half of the 20th century, psychologists removed consciousness from all scientific investigations. "Behaviorism," which would be the dominant school for almost 50 years, focuses, as the name suggests, on studying observable behavior and ignores what goes on in the mind of a person or an animal. John B. Watson (1913), who is sometimes called the father of behaviorism, is usually credited with founding this school with his "Psychology as the behaviorist views it," published in 1913. Known as "The Behaviorist Manifesto," its main point can be summed up as follows:

> Psychology as the behaviorist views it is a purely objective experimental branch of natural science. Its theoretical goal is the prediction and control of behavior....The time seems to have come when psychology must discard all reference to consciousness; when it need no longer delude itself into thinking that it is making mental states the object of observation.

Rejecting the consciousness is quite ironic, since psychology began as the academic discipline designed to study the consciousness. In fact, the word *psychology* is derived from two Greek words: *psyche* (mind or soul) and *logia* (study). Furthermore, consciousness is clearly pervasive in our perceptions, emotions, memories, and other experiences. Yet according to Baars (1986), behaviorism's rejection of consciousness came to dominate the academy totally:

> The chairmen of all the important departments would tell you that they were behaviorists....The power, the honors, the authority, the textbooks, the money, everything in psychology was owned by the behavioristic school . . . those of us who wanted to be scientific psychologists couldn't really oppose it.

You just wouldn't get a job.

Nowadays, people look back and puzzle over why this approach was adopted (e.g., Searle, 1993). How could someone not be baffled over the wholesale dismissal of consciousness by those very scientists whose supposed specialty was to investigate the introspective mind? What prompted Watson to propound such a thesis? Moreover, why did so many scientists accept it? In order to understand why this happened, we must examine the intellectual climate that gave birth to it.

The Predicament and Insecurity of Psychology

> Psychology may be the only science whose central issue that inspired the origin of the field is almost totally neglected.
> -Ernest Keen

> ...when the individual presents himself before others, his performance will tend to incorporate and exemplify the officially accredited values of the society, more so, in fact, than does his behavior as a whole.
> -Erving Goffman

Psychology's raison d'être (reason for existence in French) is to offer an explanation as to why people decide and act as they do by peering into the introspective events taking place inside their minds. This approach is particularly helpful in determining the cause of misbehavior. Let's now try to put ourselves in the shoes of the behaviorists so that we might understand how and why they decided to abandon the very introspective entity that psychologists were supposed to be investigating.

Psychology faced a particularly difficult challenge when it hoped to be accepted as "science" because much of its investigation falls beyond observable behavior. Kant even suggested that it could never be considered a true science because it lacked the mathematics seen in physics. Psychologists desperately wanted to demonstrate that their field could be a science. William James (1892), who is referred as the father of American psychology, even made a plea for psychology to be classified as a natural science. As established science is firmly grounded in experimentation and is able to replicate the results of its experiments in the laboratory (Staats, 1983), both of which are not the case with psychology, its practitioners developed "physics envy" and thus tried to apply that discipline's methods and aims to itself (Leahey, 1991). In essence, *being seen as doing science* was far more important than *actually doing science* for psychology during these years. Preoccupied with joining the elite science club, psychologists were willing to barter away their field's soul. The original aim of studying the mind was replaced with physics-like approaches, such as measuring the motion of particles, and gradually psychologists reduced their discipline to observing the bodily movement of animals and humans.

Daston (1982) points out that "social scientists of every stripe are notoriously preoccupied with establishing the scientific credentials of their disciplines....such science-envy has been assumed to be at best parochial." Abra (1988) makes a similar point: "Psychology's attempts to imitate and impress the natural sciences even at the expense of what should have been some major concerns was a tad childish." This problem still remains, for "psychology's devotees, students and professionals alike, still seem afflicted with massive inferiority complexes that make them feel they must apologize, especially to 'real scientists,' for what they do" (Abra, 1988).

John B. Watson: A Good PR Man Who Created "Smoke without Fire"

Watson perhaps played the most significant role in this odd episode. On the eve of his rise, psychology still was a relatively young science in disarray, for its practitioners had yet to agree on a definition for consciousness (Wozniak, 1997). Biology and physiology had already been accepted into the science club, and the question was whether psychology should become a subdiscipline under them or an independent discipline. Hence, the crucial factor in convincing psychologists to accept behaviorism was the climate of insecurity that surrounded the field. At a time when many other disciplines were eager to join science, Watson used his rhetorical skills and energy and eventually overcame reason. He did not introduce a novel or sophisticated concept of a new psychology, for even before he appeared on the scene psychology was moving away from mentalism and toward behaviorism. In fact, at the 1910 APA (American Psychological Association) meeting held two years before Watson's rise, Angell stated that the study of behavior was overshadowing the study of consciousness (Leahey, 1987). The concept of consciousness was steadily being replaced by observable motor responses and Costall (2006) points out that Watson was not alone in expressing doubt of intropsectionism. All Watson did was to take the existing intellectual landscape, which was already in flux, and simply articulate the behavioristic idea with "angry rhetoric" and create the "illusion of novelty" (Leahey, 1992). Three years after his "manifesto" was published, the nominating committee chose him as the APA's president; its members ratified the appointment.

His skill in public relations, which allowed him to create "smoke without fire" (Leahey, 1992), is reflected in his rapid rise in an advertising agency after the Johns Hopkins University dismissed him for having an affair with his research assistant. In less than two years, he became one of the agency's vice presidents and secured an income that was far higher than his professor's salary had been. Even without an academic position, his support among his peers remained strong. In 1943, a group of eminent psychologists rated Watson's manifesto as the most important article ever published in the psychology's top journal: *Psychological Review*. This may be one of the greatest insults to the discipline of psychology.

Intellectual Hijacking

Some politicians have a knack for self-promotion by dramatizing events that can induce emotional reactions. For instance, in the late 1940s and 1950s Senator Joe McCarthy (R-WI) sensationalized the threat of communism. One of the country's most visible and popular politicians, he was followed by microphones and television cameras whenever and wherever he spoke. Even Robert F. Kennedy briefly joined his anti-communist crusade. Based on his unsubstantiated claim that large numbers of communist spies and sympathizers had penetrated the government and other institutions, thousands of Americans lost their jobs, had their careers ruined, and were even imprisoned. Hollywood became a particular target. In fact, those of its screenwriters, directors, and producers who were on the so-called Hollywood Blacklist were forced to use fake names to secure any work at all. Just by mentioning the fear-rousing word of "communism," he was able to hijack the nation.

Due to the every-present fear that this campaign evoked, virtually no one challenged McCarthy. One exception was Edward R.

Murrow, a highly respected newscaster at CBS. Despite the objection of William Paley, his boss and friend, Murrow aired a 30-minute report entitled *A Report on Senator Joseph McCarthy* that exposed the senator's contradictions by means of his own speeches and claims. Perhaps the most memorable part came at the end:

> We must not confuse dissent with disloyalty. We must remember always that accusation is not proof and that conviction depends upon evidence and due process of law. We will not walk in fear, one of another. We will not be driven by fear into an age of *unreason* (emphasis added), if we dig deep in our history and our doctrine, and remember that we are not descended from fearful men.

With the help of Murrow and others, McCarthy's reign of terror came to an end. Soon after this damning report, his fellow senators and the media began to ignore him.

Something similar happened in the psychological community when it was dominated by behaviorism. Instead of "communism," here the dreaded term was "unscientific." For instance, Watson tried to dismiss James' idea as "unscientific" (Watson, 1925). Within the context of such an insecurity-infected field, the prospect of being labeled "unscientific" was – and still is – terrifying. Neither McCarthy nor Watson created the sentiments; instead, they rode on what was already there. The main reason they stood out was because they had exceptional communication skills, could shout longer and louder, and managed to elevate themselves as champions of worthy causes. Watson, a psychologist with a knack for publicity, was preoccupied with gaining as much scientific status for his doctrine of behaviorism as he possibly could. And so he sacrificed his intellectual integrity in order to enhance his reputation. Being *seen as doing science* was more important to him than *actually doing science*. The subsequent hijacking was so complete that no psychology textbook even listed "consciousness" in its index during behaviorism's domination. Here is a sad irony: McCarthyism lasted for several years, but the reign of behaviorism lasted for several decades. Scientists, who take pride in their logical reasoning, seem to find it harder to admit their mistake and undertake the necessary corrections than do politicians. Furthermore, as we will discuss later, the subsequent correction was only partial, for the lingering effect of consciousness-phobia still remains. Where, we might ask, were the Murrows of psychology during this sorry episode?

B. F. Skinner: The Most Eminent Psychologist Who Denied the Existence of Consciousness

> In some societies, such as the former USSR or China under Mao, hypocrisy has had an extreme or pathological character: Everybody knew that everybody's enthusiasm for fulfilling the plan or hatred of the class enemy was entirely faked, and yet one would lose one's job or be expelled from the party if one failed to conform.
>
> -Jon Elster

Behaviorism, which already had a radical view, would become even more radicalized thanks to B. F. Skinner's "radical behaviorism" approach. An excerpt from his *Particulars of My Life* illustrates this quite well: "But I prefer the position of *radical behaviorism* in which the existence of subjective entities is denied." Watson's exclusion of consciousness from science was followed by Skinner's view that consciousness does not even exist, based on his contention that mind and ideas are "non-existent entities" that were "invented" for the sole purpose of providing "spurious" explanations (Skinner, 1953). Thus, just as many political or religious doctrines become extreme or fanatical over time, an already excessively radical behaviorism became even more so. Alan Wallace, in his presentation at Google, calls this new theory "bizarre" and wonders how so many psychologists could buy into it. Baars (2003), who argues that Skinner's rejection of consciousness is like a physicist who boasts writing books without using the word "gravity," went on to say: "I cannot comprehend that attitude on purely intellectual grounds. I suspect that future historians will share my bafflement" (Baars, 2003).

This brings to mind an argument made almost four centuries ago by René Descartes, a French mathematician and philosopher. After asking what one could know for sure, he realized that the answer was one's own mind. His famous saying "cogito ergo sum" (I think, therefore I am) calls attention to the centrality of consciousness in our existence. Given this, did Skinner really believe what he taught? After all, he was chosen as one of the 20th century's most eminent psychologist based upon the frequency of journal citations, the frequency of introductory psychology textbook citations, and the survey responses of 1,724 psychologists (Haggbloom et al., 2002).

But why such an eminent scientist deny consciousness? Perhaps an examination of his life and personality will provide an answer to this apparent paradox. After receiving has B.A. in English (Hamilton College), he toiled for a year to write a novel at his parents' home. During this self-proclaimed "Dark Year," which was very difficult for him, his father lost his job as a lawyer with a Pennsylvania coal company. His father's subsequent (and very difficult) attempt to establish a private law practice put enormous pressure on Skinner, who was still struggling with his novel, to do something. During this time, his chance encounter with Bertrand Russell's (1927) book *An Outline of Philosophy* led him to Watson's behaviorism. He applied to Harvard University's psychology graduate program and soon began to shine as a researcher and maker of important contributions to the field.

Underneath Skinner's public façade was a contradictory inner dimension that can be glimpsed in his autobiography. A segment of a short story that he sent to Robert Frost, a recipient of four Pulitzer Prizes, describes the "conscious inner struggle" between a man (Will) and his wife (Elsa) (Baars, 2003). Baars (2003) fittingly calls this the "double life" that reveals a striking division between the private and public person and portrays Skinner as someone with "lifelong talent for publicity and his ability to dramatize himself." Perhaps his penchant for dramatization made him take the extreme position, for, after all, he wanted do outdo all others when it came to negating consciousness.

Self-Deception and Intellectual Laziness

> There are two cardinal sins from which all others spring: Impatience and laziness.
>
> -Kafka

> There is as much difference between us and ourselves as there is between us and others.
>
> -Michel de Montaigne

A Critique of Science

Skinner's double life prompts an important question: Was he aware this self-contradiction? Although we may never be sure, we can gain an inference from a rather perplexing attitude: he showed no embarrassment when revealing that he did not bother to examine his critics' arguments (Mahoney, 1979), even though some of them were so serious that they might have contributed to behaviorism's eventual downfall.

This brings up the issue of "self-deception," which has long been recognized for its importance and seriousness in human reasoning. Many people have commented upon this phenomenon: "This self-deceit, this fatal weakness of mankind, is the source of half of the disorder of human life" (Adam Smith, *The Theory of Moral Sentiments*); "Nothing is so difficult as not deceiving oneself" (Ludwig Wittgenstein, *Culture and Value*); and "...he is most deceived who is self-deceived" (Michael Mahoney [1976] in *Scientist as Subject*). The fundamental problem here is how can someone deceive himself or herself? Imagine that you are playing poker while pretending to be two different people. Can you bluff the "other" you? The "intending deceiver" and the "intended victim" are the same person (Mele, 1997). In order for self-deception to occur, one side has to pull it off without getting caught. Some people have approached the paradox by positing a division in the mind (e.g., Davidson, 1985; Pears, 1984). For them, such deception is possible if one suffers from the rare condition known as "split personality disorder." Mele (1997, 2001) rejects this position and attempts to explain the person's inclination for bias by presenting the following example: Sam wants to believe that his wife Sally would never have an affair. Despite the many signs of her infidelity, he continues to believe that she is a faithful wife. The desire to believe this is so strong that he ignores any and all apparent inconsistencies.

The term "self-deception" may be a misnomer. It is possible that the word may have originated as a metaphor and was, at some point, taken literally. It might be interesting to see if other languages have equivalent words. I suggest that the main cause of a double life stems from "intellectual laziness" rather than deceiving oneself. Intellectual laziness can become a habit and settle into person's thought habits, as was the case with Skinner denying his own consciousness. As mentioned earlier, he didn't bother to examine his critics' arguments.

Orogee Dolsenhe

Group Pressure and the Confidence of Judgment

Have you experienced a situation in which you were fairly sure about something until another person said that he was absolutely sure you were wrong? You begin to doubt your own certainty, only to find out later on that you were actually right. Such a thing, of course, can happen the other way around. Many social psychologists have demonstrated that caving into social pressure is very common. In Solomon Asch's (1951) classic experiment, participants were asked to judge a simple perceptual problem concerning the length of lines. There was nothing ambiguous or difficult to judge. Each participant was asked to choose which line among the three comparison lines matched with the standard lines. What the subject did not know, however, was that the other participants (usually six to eight) were cooperating with the experimenter. The experimenter would ask participants, one after the other, to give an answer. The accomplices, who answered aloud before the subject, all gave unanimous and incorrect answers, thereby placing the subject in the situation of having to assert his or her own perceptual evidence against the group's unanimous – and incorrect – opinion. The surprise was that 76 percent (37 out of 50) of the subjects went along with the false answers at least once in 18 trials.

The post-experiment interviews of subjects reveal some interesting results. The 37 subjects who yielded to group pressure can be categorized into three groups: (1) *Distortion of perception*. A small number of these subjects indicated that they actually saw the majority's choice as correct. This kind of perceptual error rising from expectation has been observed by Jerome Bruner and Leo Postman (1949). They presented subjects with an incongruous playing card, like a black heart instead of a normal red heart. Some subjects, even without the application of any pressure, actually saw the incongruous card as red; (2) *Distortion of judgment*. Most subjects fall into this category, for they realized that their perceptions were different but yielded because they assumed their judgment might be wrong; and (3) *Distortion of action*. These people did not suffer from any distorted perception or judgment; rather, they simply conformed to the majority out of the need to be similar. Asch's experiment shows just how fragile the human mind is to social pressure.

Even some of the 13 subject who maintained their independence could be divided into three categories: (1) *independence based on confidence,* (2) *independence and withdrawn,* and (3) *considerable tension and doubt.* In other words, only a small number of subjects could withstand the majority's pressure with confidence and vigor. Keep in mind that the task consisted of no more than a group of strangers engaged in making a simple perceptual comparison of the length of different lines. What if there had been additional pressures, as would be the case with students surrounded by prominent professors?

Let's now put ourselves in the shoes of those students during the heyday of behaviorism. Suppose you took your first psychology course in the 1950s at Harvard. You did not know much about psychology, but have always been interested in how the mind works. The introductory psychology course covered many diverse topics and you have become acquainted with many fascinating ideas and theories. During the third week, the professor began to talk about behaviorism, the discipline's prevailing view at that time. The issues were exciting, but for some reason you were somewhat bothered by your professor's contention that science only deals with observable facts and thus such unobservable entities like consciousness should not be discussed. You have always thought that mind was very a crucial part of one's behavior. But you soon realized that all of your classmates seem to embrace behaviorism rather enthusiastically. On the one hand, you feel a deep reservation about rejecting consciousness, but on the other hand you feel insecure and are afraid to voice your dissent. Above all, the label "unscientific" intimidates you so much that you keep your doubts about behaviorism to yourself. Like your fellow classmates, you eventually began to express your endorsement of this strange doctrine and ignore your earlier doubts. Time passed, and your subsequent courses gradually strengthen your commitment to behaviorism. The following year, you actually took a course taught by B. F. Skinner, one of the most prominent psychologists alive. This celebrated professor tells you that consciousness actually does not exist. A part of you said that is absurd, whereas another part said he is, after all, the top psychologist and therefore could be right. Again, you ignore your own doubts. During the third and fourth years, you became more and more dedicated to his radical position.

The psychology students who find themselves in such a situation would be far more susceptible to the pressure to conform than the subjects in Asch's experiment would be. First, these students would be exposed to years of dogmatic teachings – indoctrination in the guise of education. Second, the ideologies would be taught by professors whom students hold in high regard. Third, the students, particularly those aspiring to become psychologists, would have felt extra pressure to conform, since behaviorism stigmatizes alternative approach as "unscientific." This is the magic word that injects insecurity and fear into the minds of all inspiring students.

The point I am making is that it would have been particularly difficult for the average student, or even a professor for that matter, to defy the party line during the heyday of behaviorism. While some might have really believed in it, most of them would probably have gone along and kept their misgivings to themselves. A small number of psychologists, however, openly challenged behaviorism. Among them was Edward L. Tolman (1932, 1951), who emphasized the importance of unobservable *purpose* and *expectancy* in learning. His independent mind and willingness to jeopardize his reputation and career can be also seen in his defiance of McCarthy's anti-communism fervor. Targeted for dismissal by the University of California for refusing to sign a loyalty oath, he justified his action on the ground that doing so would violate his right to academic freedom. In 1955 the California Supreme Court overturned the oath and ordered the reinstatement of those who had refused to sign. Thus, Tolman had to endure being labeled as "disloyal" and "unscientific" while he remained true to himself. His personality seems to be the polar opposite of Skinner's, for he had no double life. His courage in politics as well as in psychology was finally recognized: the Education and Psychology faculty building at Berkeley was named "Tolman Hall" in his honor.

Here is the final irony: behaviorism was driven by two incoherent thinkers who were good PR men with knack for self-promotion through dramatization, men who were concerned more with their images than anything else. Their preoccupation with their public image parallels psychology's own obsession to be accepted as a real science. In essence, the psychological community was infected by their preoccupation.

A Cognitive Revolution

To reiterate, at a time of identity crisis among psychologists as to whether their field should become a subdiscipline of biology or physiology or an independent branch of science in its own right, Watson

A Critique of Science

was able to capitalize on this uncertainty by invoking the his own particular ideology: behaviorism. Ostracizing alternative views as "unscientific," he became a powerful demagogue like Senator McCarthy. Behaviorism's rapid rise and nearly five decades of domination, which underscores science's susceptibility to rhetoric over reason, lasted far longer than McCarthyism did. It was eventually overthrown by what is known as the "cognitive revolution."

Discontent had always existed underneath the straightjacket of behaviorism. It finally began to break out in 1948 at the Hixon Symposium on "Cerebral Mechanism in Behavior" held at the California Institute of Technology. These discussions were far broader than the implied official topic; in fact, two main trains of thought that came out of this event would set the groundwork for the cognitive revolution. One was the expansion of the already long list of behaviorism's inadequacies and defects. Karl Lashley's presentation in particular exposed this school's shortcomings by means of explaining the mind's complexity in relation to language and playing musical instruments. The second was a paper that introduced an analogy between the computer and the human brain. John von Neumann, a mathematician and a physicist, along with Warren McCulloch, a neurophysiologist, were instrumental in launching a new era of psychology based on how a computer processed information. Equating the human mind to a human invention is not new, for telephone switchboards, tape recorders, blank tablets, and conveyor belts are just some of the metaphors that have been used to describe mind. But the computer metaphor was different, for it generated a certain mystique. During World War II, the computer demonstrated its extraordinary capacity by breaking German codes and performing other amazing feats. Science fiction novels and movies also portrayed talking and thinking computers with superhuman capacities. The aura of this new invention's magical power generated an amount of attraction that had never been seen before in such metaphors.

About a decade after the symposium, two figures would become most conspicuous in the advent of the cognitive revolution: Noam Chomsky and George Miller. Chomsky (1959) was a young linguist at MIT when he reviewed Skinner's account of language. Skinner's (1957) book *Verbal Behavior* essentially reduced human language to the same type of learning found in the behavior of lower organisms, such as the pecking of pigeons or the maze running of rats motivated by the reward of a food pellet. According to him, a child's acquisition of language is based on receiving rewards for those of its utterances that meet the relevant features of a particular language. Chomsky, who has long been fascinated by language's intricate structure and creative role, felt that Skinner's behavioristic approach to the linguistic complexity was too simplistic and erroneous. He cited the child's mastering of grammar without constant parental training as indicating a remarkable innate capacity that could not be accounted by Skinner's reward-based mechanism. His unusually long 32-page review exposed the grave inadequacies and flaws of Skinner's behavioristic explanation. Ironically, the review would become more important and influential than the book itself, a truly rare event in any field. Despite Chomsky's vital points, Skinner and his followers did not bother to respond them until the reign of behaviorism was over.

George Miller's role in the cognitive revolution was rather different. Described as the "single most effective leader" in the emergence of the cognitive revolution (Baars, 1986), his strength was his lucid writing style, consummate lecture presentations, personal charm, and the ability to maintain a relationship with others (Baars, 1986). In this sense, he was more like the politically savvy Dwight Eisenhower than the tactically brilliant George Patton. His article entitled "The Magical Number Seven, Plus or Minus Two: Some Limits on Our Capacity for Processing Information," which appeared in psychology's top journal *Psychological Review* in 1956, was chosen as the most influential paper ever in a 1994 centennial issue of the journal. Neither a report of groundbreaking experiments nor of a novel theoretical proposal, it instead was a collection of many studies showing that a built-in limitation of capacity does not vary from one sensory function to another. His interpretation of the diverse data led him to state that a person's mental system has a central cognitive processing mechanism that has limitations similar to those of a computer's central processing unit (Gardner, 1985). Although applying the computer metaphor to the human mind had begun a decade earlier, Miller's article signaled the full legitimacy of the information-processing approach in psychology. The floodgates were opened, and a complete new approach to investigating the mind would soon sweep away behaviorism's supremacy.

The cognitive revolution met with only limited resistance from behaviorists for three reasons: (1) the strict adherence to experimental methodology, (2) the usage of the computer metaphor, and (3) the cautious choosing of terms that did not directly offend them. Previously "unscientific" concepts like "purpose" and "cognitive map" could be discussed by concealing them behind experimental procedures and results, computer metaphors, and terms that did not explicitly challenge behavioristic dogma. While these maneuvers might have enabled a smoother transition by lessening tensions and clashes with behavioristic loyalists, they failed to correct the school's fundamental problems. The very factors that facilitated this overthrow actually created the intellectual climate that would still impede psychology's ultimate goal: the study of mind. We examine these side effects in the following sections.

The Preoccupation of Experiments and the Fragmentation of Psychology

One of major characteristics of the presently dominant school of cognitive psychology is the large number of experimental reports. There are two reasons for this: (1) the preoccupation with conducting experiments to demonstrate psychology's worthiness as an empirical science, as well as the rigorous adherence to experiment so that they would not go against behaviorism's canon, and (2) the relative ease and low cost of conducting experiments in memory, perception, attention, language, and imagery. For example, a typical experiment required no more than a slide projector, a small room with a couple of chairs, and the continuous availability of college students as the pool from which subjects could be drawn. One drawback of this, however, was that the abundance of experimental results was not accompanied by theoretical advancement. What is lacking is not a sufficient number of theories, but the consolidation and unification of the experimental facts and diverse concepts that have been produced. Most researchers are more interested in branching off into more and more specialized fields and formulating their own custom-made theories to explain particular experimental results. Watkins (1990) calls this "an era of cheap theories." No other discipline is as segmented as psychology, because no other discipline is so conscious of its image. Staats (1983, 1985) calls this present fragmentary condition "psychology's crisis."

One might ask why this is a problem. In order to see the importance of unification, we need to look at a scientific discipline that is not infected with insecurity. For example, consider physics. Physicists have made truly impressive advancements by consolidating seemingly disparate phenomena. For example, Newton's unification of terrestrial and celestial motion (viz., his universal law of gravitation) combined Kepler's model of the moon's movement around Earth with Galileo's model of falling object on Earth's surface. The unification of three seemingly completely different phenomena – electricity, magnetism, and light – by H. C. Oersted, Michael Faraday, and James Maxwell into an electromagnetic force is one of the greatest discoveries in physics. Einstein's genius can be summed up with his theory that unified light with time, time with space, energy with matter, matter with space, and space with gravitation. If Einstein were preoccupied with experimental proof, the wide scope of his research would have been difficult to achieve.

The Premature Commitment to Computer-Metaphor Psychology
Explaining something unfamiliar or intangible by comparing it with a familiar entity is quite common in science. For instance, Rutherford used the analogy of the solar system for the atom's structure: negatively charged electrons orbiting a positively charged nucleus. Metaphor is particularly helpful when discussing unexplained mental process. Based upon the technology in use at the time in question, memory has been compared to a wax tablet, a telephone switchboard, a tape recorder, a conveyor belt, a hologram, and so on (Roediger, 1980). The computer metaphor, however, has become more than a metaphor in that many people now take it literally largely because of the computer's continuing mystique. Many of psychology's key terms have been borrowed from computer science, such as encode, store, retrieve, and decode. This explanation, however, has one very basic problem: How does the mind convert the incoming information into the form that neurons can handle and then decode it back to the mode in which it can be understood by people. If the human brain is a computer, the rules of communication, such as those associated with Morse code, a coding agent, and a decoding agent should exist. Unfortunately, none of these have ever been found in the brain. This should raise a serious doubt as to the legitimacy of a computer-based approach and open it up to alternative views.

The computer's mystique might have helped to overcome the

straightjacket of behaviorism, but there is a danger that it might have become a dogma itself and thus curtail alternative approaches. If this is true, one tyrant has been replaced by another. Roediger (1980) have cautioned:

> In 30 years, the computer-based information processing approach that currently reigns may seem as invalid a metaphor to the human mind as the wax-tablet or telephone-switchboard models do today.

The dirty secret of neuroscience is that its practitioners do not know exactly what the brain is doing, despite all of their seemingly groundbreaking discoveries. The structure of a neuron and the processes between neurons are fairly well understood. Studying patients with brain injuries has revealed many fascinating features of the human mind. Many non-invasive neuroimaging techniques like CAT (computed axial tomography) scans, PET (positron emission tomography) scans, and fMRI (functional magnetic resonance imaging) have made it possible for scientists to study the structure and function of a living human brain. Yet the mystery of what it is actually doing in order to accomplish our mental functions still remains. Despite this uncertainty, specialists are making a hasty and premature commitment to an understanding of the brain based upon a computer analogy. In sum, computer-metaphor psychology replaced behaviorism not because of any empirical evidence or theoretical persuasiveness, but because of the computer's mystique or value system.

The Consciousness Phobia

> As late as 1995, a world-famous cognitive psychologist proudly informed me at a cocktail party that he had written several books on cognition without once using the word "consciousness."
> -John F. Kihlstrom

During a half century of behaviorism, the word "consciousness" was banned from psychology for being categorized as the unobservable entity. The fear of being called "unscientific" trumped what was declared to be psychology's primary mission: studying the mind, namely, consciousness. The resulting timidity and insecurity still plague psychology today, decades after the supposed cognitive revolution. Tulving (1994), who laments the use of such behavioristically safe terms as memory "monitoring," mnemonic "behavior," memory "search," tip-of-the tongue "states," and feeling-of-knowing "experience," wonders: "Do they use these expressions deliberately in order to avoid the big bad 'C' word? Or do they take it for granted that when they refer to monitoring, and judgments, and experience, everyone understands that they are indeed talking about *conscious* (emphasis added) monitoring, conscious judgment, *conscious* (emphasis added) experience?" This phenomenon would never happen in physics. If psychologists, who are supposedly responsible for exploring the nature of consciousness, are afraid to even use the actual word "consciousness," then how can they ever hope to investigate it? And if they cannot, then, who will investigate it?

This phobia shifted dramatically when Francis Crick, co-discoverer of the double-helix of the DNA molecule, and his associate Christof Koch began promoting the study of consciousness as an important part of science in the 1990s. Horgan (1997) makes a poignant point: "Only Nixon could go to China. And only Francis Crick could make consciousness a legitimate subject for science." Indeed, since Crick's effort a flurry of books, journals, and conferences have been devoted to this previously stigmatized subject. Given this reality, psychologists should ask why the words of a Nobel molecular biologist are required to legitimate the investigation of the consciousness, something that they should have been doing all along? Are psychology's practitioners so insecure that they need the science's own Pope to tell them which subjects are sanctioned and which ones are not? As we shall see, the dysfunctional state of this field is only the tip of iceberg that plagues the entire scientific edifice. In this book, I hope to expose this iceberg for what it is: a hindrance to progress.

Chapter Two
The Indian Caste System:
The Most Extreme Form of Social Stratification

> Erroneous beliefs may have an astonishing power to survive, for thousands of years, in defiance of experience, and without the aid of any conspiracy.
>
> -Karl Popper

> Virtue can only flourish among equals.
>
> -Mary Wollstonecraft

> Real equality is immensely difficult to achieve, it needs continual revision and monitoring of distribution.
>
> -Mary Douglas

The psychology's recent history discussed in the last chapter should demand an explanation why scientists, the epitome of human intellect, would fall into such collective failure in decision-making. One of the main forces acting upon the psychological community was the status anxiety. As the university community began to form a stratification in the latter part of 19th century, psychology as many other disciplines was preoccupied with getting into the upper class called "science." In this chapter, we will compare with perhaps the most extreme form of stratification, the Indian caste system. I certainly do not imply that science's stratification is as bad as the Indian caste system. Instead, the same basic impulse is involved in the formation of both stratifications.

Other than slavery, the caste system may be considered as one of the most unjust social institutions. What is astounding about the Indian caste system is that it endured more than two thousand years of surviving anti-Hindu movements, foreign invasions, and the introduction of relatively egalitarian religions like Islam and Christianity (Kuran, 1995). It is even more remarkable that it has endured with almost no form of centralized reinforcing of authority or little use of force. An immediate reaction to the caste system from foreign visitors was surprise and shock, and some of them sought an explanation of its existence (Dumont, 1966).

The classical religious texts of Hinduism categorizes the population into four categories called *varnas*: the Brahmans (the priests and religious officials), the Kshatriyas (warriors and rulers), the Vaisyas (farmers, merchants, traders, and craftsmen), and the Sudras (laborers and servants). The first three castes have the status of "twice-born" which refers to the initiation ritual upon maturity. Then, the initiated one is considered as reborn as a caste member and Hindu. The Sudras are "once-born" and are excluded from the ritual and not even permitted to hear the Vedas, Hindu text. Then, there is a class that is outside of the four categories called the Untouchables who are even below the Sudras. Their occupations are considered to be polluted such as latrine cleaning, killing animals, and processing animal parts like hides. While the four categories are called *varna*, *jati* refers to a much smaller group or subcaste. Then, the English word the "caste" is used to denote both *varna* and *jati*.

The caste system in India has been described as a "state of mind" (Dumont, 1966). Human's state of mind determines what is to be valued and what is not. It is the basis for the formation and maintenance of ideology, culture, or dogma. The caste system has three major characteristics of the state of mind: separation, hierarchy, and impurity. They are more like preoccupations that everyday lives of all Hindu people are subjected to. They are the constant pressures that put the entire society into a peculiar form.

The Preoccupation of Hierarchy and Separation

A main characteristic of the caste system is "hierarchy" which ranks the groups as relatively superior or inferior to one another (Dumont, 1966). The metaphor of human body is used to illustrate the interrelationship among the four varnas. The Brahmans are at the very top of the ladder of the caste system and they do not have to struggle to maintain it for everyone recognizes it. Lower castes know that even poor Brahmans have a status they do not (Milner, 1994). The high status that Brahmans possess, however, is neither power nor authority (Dumont, 1966). Their formal assignment of political and economic power to others is the key to the maintenance of the high status (Milner, 1994). The Kshatriyas, the warriors and rulers class, are the ones with the political power. The caste system in India is unusual in that the political ruling class fall below the religious class. It is likely that there had been a long-drawn out conflict between the two classes and the Brahmans somehow had become victorious over the Kshatriyas (Cox, 1970). The sacred literature recalls the once dominant Kshatriyas which has been the obstacles in the path of Brahmanic power (Bouglé, 1908). The Brahman-Kshatriyas relationship is vital because the former depends on protection provided by the latter and the latter needs to legitimize their political power (Milner, 1994). The Vaisyas are essentially a working class like merchants, farmers, herders, and artisans. Paradoxically, however, they have the smallest percentage among the four varnas (Milner, 1994). They do not have a monopoly on the working professions, however. Many of those with trading and commercial professions belong to other varnas (Milner, 1994). The Sudras support the other three varnas. The Laws of Manu, one of the standard books in Hindu canon, explicitly states: "A Sudra, whether bought or unbought, may be compelled to do servile work; for he was created by the Self-existent to be the slave of a Brahman. A Sudra, though emancipated by his master, is not released from servitude; since that is innate in him, who can set him free from it?"

Whereas the classes of the four varnas are stable, subcaste (jati) can change over time. Subcastes may divide due to migration, change of occupation, the adoption of new religious practice, and internal group dissention (Cox, 1970). In fact, it is rare that a caste practicing the occupation which its name designates (Bouglé, 1908). Nesfield (1885) points out that the number of castes in India is constantly changing. Caste distinctions can be based on very slight cultural differences (Cox, 1970). Such trivial basis for division has been mentioned in Buddhist legends: Fishers can be divided into separate castes for using different types of instruments in catching fishes (Fick, 1897). Among the people who make living by making clothing, those who make turbans will have nothing to do with those who make sashes (Bouglé, 1908). According to the 1901 census, there were as many as 2,378 subcastes (Singh, 1976). Out of those, there are about 200 castes that can be grouped in classes recognized by all (Ghurye, 1969).

The question, then, is what determines one caste above or below another. Three main criteria for gauging ranks are: purity of food, fidelity to the traditional occupation, and abstention from forbidden foods (Steele, 1868). And the relative rank of one another has an almost mathematical precision according to the sacred codes (Bouglé, 1908). And the Brahmans are the main gaugers. It is a matter of how the Brahmans make contact with members of other castes such as accepting gifts or a cup of water. If they refuse them with horror, that is a clear sign of the very low status (Bouglé, 1908). However, there can be disputes and even lead to fights. According to Mysore Census in 1901, many members of intermediate castes think or profess to think that their castes are better than their neighbors and should be ranked accordingly (Ghurye, 1969). Among the lower castes, they may set up the "most frivolous distinctions" for despising one another (Nesfield, 1885). Hindu society is "obsessed with the idea of right to be organized hierarchically" (Bouglé, 1908). Some of the basis for the subcaste ranking may be for a trivial reason from the point of view from a foreigner. For instance, there are two Bengal castes who do fishing. The Kaibarta are farmers and fish for themselves and the Namasudralive not only catch fishes but also sell them. The selling the fishes makes the latter inferior to the former (Mukherjee, 1963).

In addition to division of labor and hierarchy, the caste system has a rigid system "separation" in matters of marriage and contact (Dumont, 1966). Marriage outside the caste is forbidden (Bouglé, 1908). Without the centralized body of governing the caste system, it is an assembly of the specific local subcaste that has the judicial capacity to excommunicate its members (Dumont, 1966). The caste council can retry criminal offenses already decided by the courts of the law. The citizen owes moral allegiance to the caste first, rather than to the community as a whole (Ghurye, 1969). The threat of excommunication is a powerful driving force behind the upholding the caste system because the banished ones were treated as contagious and untouchables. Thus, the separation is not enforced by some external authority but it is self-regulated. The moral standards of the various castes are different (Ghurye, 1969) and members of different castes are expected to behave differently and to have different values and ideals up to a point (Béteille, 1969). In effect, "cultural differences" among castes bring out the segmentation (Ghurye, 1969).

The separation, in some instances, may be hardly justifiable. For instance, there may be more than one group of cow herders in a particular area who do not intermarry or eat together (Milner, 1994). From their point of view, however, there is a proper reason for the distinction. One of the main reasons for the endurance of the caste system is a "reciprocal repulsion" (Bouglé, 1908) among the Indian people.

A Critique of Science

The Origin of the Caste System

The origin of the caste system is not easily explained for the lack of historical evidence, especially when it had occurred thousands of years ago. One of the most widely-known explanations of the origin of the caste system is the racial theory in which two different populations—the indigenous people and the invaders from the north—encountered and competed for the control over the Indian continent about 1,500 B.C. The Sanskrit speaking Aryans from the north was able to dominate the Dravidians. According to the racial theory, the victorious invaders sought to preserve the purity of their race and implemented the rigid system of separation and hierarchy. From the modern point of view, the racial theory seems to be quite plausible since we can relate the caste system with the apartheid in South Africa which had recently been abolished. Besides the inter-race conflict, the theory goes that there also was a competition within the Aryans themselves between the warrior class and the priest class in establishing the caste system. The priests (the Brahmans) eventually became dominant over the warriors (the Kshatryas) in the long struggle for power. One of the major problems of the racial theory was there was no archaeological evidence of invasion and conquering of the indigenous people by the northern invaders.

Above all, the racial theory of the origin of the caste system has been all but dismissed by the wake of the recent genetic studies. As DNA evidence can convict or exonerate, it can also help to determine the trueness of the racial theory. Perhaps the most comprehensive study of the genetic diversity of the Indian continent has been conducted by a team of international scientists at the Center for Cellular and Molecular Biology (CCMB) in India, Harvard University, and Massachusetts Institute of Technology (MIT) (Reich et al., 2009). Their study investigated the genetic variations of Indians by using 132 individuals from 25 groups sampling from 15 states and six language families. An additional focus was the comparison of upper and lower castes. The most important finding was that most people in India are essentially a mixture of two ancestral populations: the Ancestral Northern Indians (ANI) and the Ancestral Southern Indians (ASI). Whereas the ANIs are related to the western Eurasians, the ASIs are exclusively related to the indigenous people of the Andaman islands in the Indian Ocean. No one in the study had or even close to having all of one or the other. The degree of ANI/ASI mixture ranges between 39% to 71% in all castes and tribal groups. Thus, there is no systematic distinction among castes which suggests there were no invasion and conquering by one group over another in establishing the caste system. Instead, the study seems to support the view that the caste system grew out of tribal-like organizations during the early development of the Indian society.

Remarkably, Damodar Kosambi, an Indian mathematician, proposed such view in *The Culture and Civilization of Ancient India in Historical Outline* in 1964. Rather than a single point of origin, Kosambi proposes a multiple factors in the tribal society that have contributed to the rise of the caste system. With an adaptation of agricultural settlement, knowledge and ownership of the means of production may have created hierarchy within tribal societies (Kosambi, 1964). Karve (1961) also argues that something close to the caste system existed long before the arrival of the Aryans in India. The once-speculative theory of the tribal origin of the caste system is now bolstered by the recent genetic studies.

The Reference of the Caste System in the Ancient Texts

With the tribal origin of the caste system in mind, let us look at some ancient texts for the formation of the hierarchy. The first known symbolic reference to the emergence the caste system is contained the Purusa Sūkta of the Rgveda, the ancient Hindu hymn, about 5,000 B.C. The four classes of varnas, based on their professions, are compared to the four parts of the Cosmic Being: the Brahmans to mouth, the Kshatriyas to shoulders, the Vaisyas to thighs, and the Sudras to feet. The vocational choice at this time was not based on heredity but was mainly need-based and circumstantial (Narasimhachary, 2002). However, the symbolic reference to human body parts may have come into play that the four castes are in the descending order of importance (Narasimhachary, 2002). About 5,000 years later, Manu the Law Giver offers extensive descriptions of the caste system in twelve chapters. It shows how far the idea of the caste system has evolved since the symbolic reference in the Purusa Sūkta of the Rgveda. In Manu, there are two main thoughts that are crucial in shaping the characteristics of the caste system. First thing that is clearly visible is that Manu's effort to elevate the Brahmans and downgrade other castes, especially the Sudras:

> Let the first part of a Brahman's name denote something auspicious, a Kshatriya's be connected with power, and a Vaishyas with wealth but Shudra's express something contemptible (Chapter 2, verse 31).
>
> Know that a Brahman of ten years and Kshatriya of a hundred years stand to each other in the relation of father and son; but between those two the Brahman is the father (Chapter 2, verse 135).
>
> For he who explains the sacred law (to a Shudra) or dictates to him a penance, will sink together with that (man) into the hell (called) *Asamvrita* (Chapter 4, verse 81).
>
> A once-born man (a Shudra), who insults a twice-born man with gross invective, shall have his tongue cut out; for he is of low origin (Chapter 8, verse 270).

The crucial question is, then, how such an extreme system of stratification ever developed in the Indian culture. In the following section, we will examine three theories.

The Need-Based and the Power-Based Stratifications

For as much as 3 million years, humankind lived in a band of hunter-gatherer. A small group of closely-related roving individuals struggled to survive in the constantly changing environment. Then, about 10,000 years ago, people began to cultivate crops and to domesticate animals. This revolutionary change occurred in separate parts of the world independently. One of the most important outcomes of the transition was that cultivated land was able to sustain much larger population, as much as 60-100 times greater than uncultivated land. The change from the nomadic way of life to sedentary living brought the higher density of people which allowed division of labor. The specialization of specific line of work enabled the burst of technological advances.

A byproduct of division of labor was inequality, that is, some individuals based on their professions began to be valued higher than others. In this sense, inequality has a functional necessity because a society must have the mechanism for motivating some individuals to do tasks that are beneficial to the whole society (Davis & Moore, 1945). This is known as the "functionalist theory" of inequality and it endorses stratification of society so that important positions are filled by the most qualified people (Davis & Moore, 1945).

Now, suppose those individuals who were given higher status for their important contribution to the society amassed a greater wealth and power. Once the higher standing has been achieved, they want to maintain it even if their service is no longer valuable. Furthermore, they want to pass on to their offspring the status and power they held. Through their dominant positions in the society, they try to overpower any challenges. In this scenario, stratification is no longer need-based; instead, it becomes a power-based one. A condition such as this is called the "conflict theory" of stratification (Parsons, 1957). There would be tension between those with power and those without power. Those with less power may be able to challenge and defeat the power by mobilizing other people. If not, those with power may institute a system that legitimizes the inequality, not only for themselves but also for their offspring. If the system of stratification persists for generations, then the inequality, however unfair it may be, can become a norm. In other words, people with lower status may no longer recognize the unjust system as something to overthrow but something to accept as a valid cultural norm. There might even be an elaborate set of rules and etiquettes that state how the people from the two classes should interact. Monarchies and hereditary aristocracies are some of the well-known forms of the established stratification.

The Zero-Sum Impulse and Stratification in the Tribal Society

> Since the general civilization of mankind, I believe there more instance of the abridgment of the freedom of the people by gradual silent encroachments of those in power that by violent and sudden usurpation.
>
> -James Madison

Both functional and conflict features are no doubt major parts of the formation of stratification in the society. I would like to add another significant source in the formation of stratification. It comes from human's impulse to elevate oneself over others. Perhaps my own experience in the Army illustrates the basic urge we all possess. The Airborne training in Ft. Benning, Georgia consisted of three weeks: the Ground Week, the Tower Week, and the Jump Week. The Ground Week training prepared us trainees for a parachute jump and land safely. Parachute jumps from a 250-foot tower culminated the Tower Week. During the Jump Week, the actual jumps from airplanes were made at the drop zone in Alabama. The end of the fifth and the final jump gave us proud and euphoric feelings, and we were now paratroopers. In the bus ride back to the jump school from the drop zone in Alabama, we saw other Infantry soldiers walking on both sides of the road. Someone yelled at them, "Hey, legs." The word "leg" is a derogatory expression by paratroopers to infantry soldiers who have to walk with legs, rather than fly on an airplane and land with parachutes. It was contagious and most of us participated in the taunting of fellow soldiers. In this situation, we were driven by neither need nor power because we probably would never see them again. Rather than effort to achieve some calculated goal, it was driven by an impulse to feel superior and scornful at others for having achieved something that others didn't. This is what I would call the "zero-sum impulse." The phrase "zero-sum" comes out of game theory in the discipline of economics. The pioneers of the game theory, von Neuman and Morgenstern, in the book *Theory of Games and Economic Behavior*, argue that the mathematics used for the physical sciences is inadequate in economics. The crux of their point is that economics is very much like a game where intention and emotion of players are involved. An example of a zero-sum game is poker in which the total sum of winning and losing equals zero. If one player won $10.00, another person or persons must have lost the amount.

Like in a game or economics, we also have an impulse to be zero-sum in valuing and devaluing. Simply put, we have a tendency to value ourselves for having or achieving something that others didn't achieve or possess. The elevation of oneself is accompanied by devaluing others. Suppose one's elevated value is "+5" for, let's say, graduating from the Airborne school. The principle of zero-sum value would predict that the strength of devaluing for non-Airborne members would be "-5." Thus, the sum of the two numbers of assigned values would equal to zero, henceforth, the zero-sum. The zero-sum impulse sometimes can be seen in a football game. A wide receiver makes a touchdown and celebrates his accomplishment by mocking the opponent.

The ecstatic feeling of accomplishing a touchdown came with a compulsion to ridicule his opponent. In the NFL, a referee can penalize the taunting with a fifteen yards penalty.

With the zero-sum impulse in mind, let us now travel back in the time of transition from hunter-gatherer to agricultural society. So-called the "Agricultural Revolution," which is also called the "Neolithic Revolution," is one of the most important events in human history. Domestication of plants and animals independently cropped up numerous places in the world, and the Indus Valley was one of them. The change from the constantly moving hunter-gatherer to the sedentary way of life by farming completely transformed the ways humans live. The efficient way of food production quickly spread and an immediate outcome of the agricultural revolution was the formation of villages and towns. The gathering of a large number of people also brought disparities in wealth and status. Those who settled earlier owned a good parcel of land and became affluent members in the village. The late comers, however, had no choice but to occupy lands that are more difficult and less efficient to farm. The unfortunate group of people who arrived when there was no more arable land had to eke out of living whatever ways they can. One way was to become servants to the affluent people of village in order to avoid going back to the nomadic hunter-gatherer living. The affluent and established families were able to rely on the labor of the hapless people for food production. Thus, the availability of cultivatable land alone would have created inequality among people in villages and towns. This is where the zero-sum impulse could take over and start to shape the society toward the stratified characteristic. The well-to-do group of people would feel themselves superior to the deprived people. Once a group of people attain a higher status, their next impulse is to make them exclusive. They try to restrict others from joining the elite club. They might set up a set of rules that state who might be eligible to enter and who should be rejected. Like fraternity and elite military schools, the more stringent the rules of the entry, the more exclusive the members usually feel. Some status is based on the hereditary; one's birth determines the eligibility of the status. The establishment of a hereditary status system is geared toward giving one's descendants the maintenance of the status quo. The impulse to become superior and maintain superior status over others is one of the main reasons for the development of many hereditary systems in many societies. Once the system of stratification is established by the zero-sum impulse along with need-based and power-based factors, the disparity and injustice can persist hundreds or even thousands of years as we will see in the caste system.

The Obsession with Purity

The British anthropologist Mary Douglas (1966) published *Purity and Danger,* which is devoted to the subject of impurity examining various cultures. For instance, the Lele people of Zaire define things like feces, blood, milk, sexual intercourse, and military groups as polluted. The Old Testament also identifies some of the unclean animals that should not be consumed. She suggests that people give meanings to their reality by classifying between two opposite categories in respect of pollution: what is acceptable and what is unacceptable. Such a symbolic system offers moral order to societies. However, like everything else, moral order based on pollution should be subjected to moderation. The caste system is a good example where it went far beyond the typical need for social order to the point where it can be called "preoccupation" or even "obsession."

The most significant preoccupation of the caste system is impurity and pollution. Dumont (1966) considers that the basis for the caste system is the distinction between pure and impure. Let us look at some of the Manu's verses that show the preoccupation with impurity:

> Garlic, leeks and onions, mushrooms and (all plants), springing from impure (substances), are unfit to be eaten by twice-born man (Chapter 5, verse 5).

> Brahman who is impure must not touch with his hand a cow, a Brahman, or fire; nor, being in good health, let him look at the luminaries in the sky, while he is impure (Chapter 4, verse 142).

> When (a child) dies that has teethed, or that before teething has received (the sacrament of) the tonsure or (of the initiation), all relative (become) impure, and on the birth (of a child) the same (rule) is prescribed (Chapter 5, verse 58).

> Land is purified by (the following) five (modes, viz.) by sweeping, by smearing (it with cowdung), by sprinkling (it with

cows' urine or milk), by scraping, and by cows staying (on it during a day and night) (Chapter 5, verse 124).

How has impurity become such an important issue in the caste system? My focus on the matter of impurity is the role of an individual's obsessive avoidance toward pollution in shaping culture's preoccupation. Someone or a group of people with high status and influence can literally hijack the society toward particular obsession. It is as if society becomes an extension of someone's anxiety and begins to see the world through the deranged person's lens. In the next section, we will attempt to look at the mind of someone with the preoccupation of impurity.

Howard Hughes: Obsessive Compulsive Disorder of the Genius

In order to understand how debilitating the preoccupation with impurity can become, it might be better to examine someone who suffered from obsession of avoiding contact with dirty substances. In this section, we will examine Howard Hughes, American aviator and filmmaker, perhaps one of the most famous persons with the illness known as "obsessive compulsive disorder" (OCD), to better understand the preoccupation of impurity in the caste system.

As in a Hollywood movie *The Aviator*, Howard Hughes was famous for taking bold moves, whether making record-breaking flights, the biggest airplane, or the most expensive movies. His daring feats were only matched by his shrewd business entrepreneurship that brought him an enormous fortune. He was so fearless that even after the numerous crashes he insisted on flying the experimental reconnaissance airplane, XF-11, himself. The crash of XF-11 in 1946 at the age of 41 nearly killed him. He was in the hospital for more than a month for taking care of the burn, broken bones, and facial reconstruction. Even after the near-fatal crash, he continuously flew airplanes, including the world's largest plane, the Spruce Goose.

As he was fearless and daring in flying airplanes and business adventures, he, paradoxically, was plagued by the constant fear of minute germs. It is a trait that his mother had, particularly toward her son. She would thoroughly inspect and clean her son's body over and over every day for fear of germs. She and her son often left Houston to some distant place when there were outbreaks of contagious diseases (Bartlett & Steele, 1979).

A Critique of Science

If Howard Hughes inherited his mother's fear of germs, it increased by manifolds through his life. For instance, Hughes listed nine specific steps for opening the can: 1) preparing the table, 2) procuring of the fruit can, 3) washing of the can, 4) drying the can, 5) processing the hands, 6) opening the can, 7) removing fruit from the can, 8) the fallout rules while around the can, and 9) the conclusion of operation (Bartlett & Steele, 1979). On step number three, he gives a detailed instruction on washing of the can:

> The man in charge then turns the valve in the bathtub on, using his bare hands to do so. He also adjusts the water temperature so that it is not too hot not too cold. He then takes one of the brushes, and, using one of the bars of soap, creates a good lather, and then scrubs the can from a point two inches below the top of the can. He should first soak and remove the label, and then brush the cylindrical part of the can over and over until all particles of dust, pieces of paper label, and, in general, all sources of contamination have been removed. Holding the can in the center all times, he then processes the bottom of the can in the same manner, being very sure that the bristles of the brush have thoroughly cleaned all the small indentations on the perimeter of the bottom of the can. He then rinses the soap from the cylindrical sides and the bottom of the can (Bartlett & Steele, 1979).

His trusted executive, Bill Gay, who put together the inner circle of the Hughes Empire, experienced how paranoid of germs his boss was. In the summer of 1958, Bill Gay's wife had a contagious disease. Hughes ordered Gay to work at home so that he wouldn't pass germs to other staff members then to Hughes. It wasn't until 1973 (fifteen years later) that the quarantine was finally lifted.

After his second nervous breakdown in 1958 (the first one was in 1944), Hughes virtually stopped any contact with the outside world. Once one of the most visible men, often seen on the cover of magazines, he literally turned into a prisoner out of his irrational fear of contamination. He would seclude himself to hotel rooms of which doors and windows were sealed with masking tape to prevent from the flow of germs into his room. The elaborate set of rituals were established and enforced for himself and staff members with preoccupation of preventing germs from reaching him. Those few who had direct contact with him had to go through the "thirty-minute purification ritual" (Drosnin, 1985). They were to "wash four distinct and separate times, using lots of lather each time from individual bars of soap" (Drosnin, 1985). They, then, were to wear white cotton gloves. The final requirement before deliver anything to Hughes was wrapping it with Kleenex tissue. Although Hughes would demand such ridiculous cleanliness to his staffs, paradoxically, he himself almost never bathed, brushed his teeth, or cut his fingernails. He frequently urinated on the floor but he didn't allow janitors clean it. He even stored his urine in capped jars in his garage and in hotel bedroom. It is almost as if whatever comes out of him is sacred and to be preserved. And whatever comes out of others is so disgust and repulsive he irrationally and compulsively went out of way to avoid it. As we shall see later, this kind of paradoxical feature exists in the caste system.

What sets Howard Hughes apart from others with OCD, among other things, was that he was enormously rich and he was able to delegate his compulsions to his employees. His elaborate set of instructions, however bizarre they might have been, was possible to carry out because of his wealth and power. When he died in 1976 aboard an airplane from Acapulco to Houston, he left billions of dollars which became the target of many vultures, some of whom were arch-enemies of Hughes. Another legacy he left was the odd rituals that he wrote down on the yellow note pads became the reminder of the tragic life of the genius who was paralyzed by the demons within himself.

The Caste System: Obsessive Compulsive Disorder of the Society

Let us pose a hypothetical situation. Several thousand years ago, there was a high priest like Howard Hughes who was a genius yet suffers from obsessive compulsive disorder. And like Howard Hughes, his wealth and status allowed him to delegate the impure tasks he was fearful to many servants he had. Unlike Howard Hughes who became the target of ridicule, however, the high priest, by using his intellect and shrewdness, legitimized his compulsions by devising an intricate scheme. The main theme of justification for passing on the impure tasks to others was that some are inherently superior and cannot do the impure works and some others are inferior and are meant to do them. And to ensure his

A Critique of Science

descendants to maintain the higher status, he made the scheme as the hereditary system. Because the servant class was poor and powerless, they grudgingly had to go along to make living. Other classes of people like merchants and craftsman didn't object too much since they could take advantage of the lower class, even though it meant their status was below the priests. As time went on, the hierarchical system based on the preoccupation of purity and impurity progressively became stabilized. The result was that one person's mental illness transformed the entire society into OCD. In this process, there might have been many Hughes through the times, each strengthening from the previous ones.

The notions of purity and pollution affect a wide range of daily activities, especially those having to do with cooking, eating, personal cleanliness, and worship (Milner, 1994). Food contamination is particularly feared so that even someone's mere glance is sufficient to pollute it (Bouglé, 1908). And those who eat food that is forbidden by his caste are "out-cast" and "out-law" (Bouglé, 1908). Something as simple as when to take a shower is geared toward the preoccupation of impurity. For instance, a European traveler recounts that a Brahman preferred to see him before taking his bath because the impurities incurred during the meeting could be cleansed (Bouglé, 1908). Fick (1897) describes that the fear of an impure atmosphere is one of the dominating traits of the Hindu soul. And the dominating theme of the Hindu world is a "force of repulsion" (Bouglé, 1908).

The Hindu preoccupation of impure and pure is embedded into hierarchy and separation. Each caste protects itself from the one below and not at all from the one above (Dumont, 1966). The repulsive force is particularly intense against the Untouchables. Upon sighting two brothers of the Untouchables, the girls flee to wash their eyes (Fick, 1897). People even carry bells which are to announce the presence of the Untouchables (Bouglé, 1908). The Untouchables are usually segregated outside the village, in distinct hamlets and they may not use the same wells as others (Dumont, 1966). Even the shadow of the Untouchables can pollute a member of higher caste (Ghurye, 1969). The situation for the Untouchables had been greatly improved since Mahatma Ghandi has championed for their rights. He introduced the euphemistic term "Harijan" (Children of God Vishnu). And the independent India has declared Untouchability illegal in 1949. Recently in 1997, the first untouchable was elected as the president of India. Yet, the age-old tradition cannot be overturned overnight, particularly in rural areas.

The symptoms of the caste system in India can be considered as the obsessive compulsive disorder at the society level. At the heart of the societal OCD, the repulsion is geared toward the categories of people which permanently set at birth with the preoccupation of impurity and the strict hierarchy. The rigorous inequality in the caste system is accepted as a part of the "Divine Plan" not only by the Brahmans but also all others (Ghurye, 1969). No amount of moral virtue entitles one to move into a higher caste in one's lifetime. Instead, one can be rewarded in the next life if he or she faithfully conforms to his or her role in the caste system (Milner, 1994). The main beneficiaries of the caste system are, of course, Brahmans. They have developed a "science of purity" (Ghurye, 1969) in which their status is permanently supreme. Brahmans are even superior to the king (Dumont, 1966). The caste system as a whole is focused around the prestige accorded to the Brahmans (Hutton, 1946). The Manu states, "From the priority of birth, form superiority of origin, form a more exact knowledge of sacred science, and from a distinction in the sacrificial thread, the Brahman is lord of all classes." The Brahmans, like Howard Hughes, put themselves on the top of the purity ladder: they can only be polluted by others, not the other way around. Oldenberg (1897) have called the Brahmans "the evil geniuses of the Hindu people." I like to add an additional phrase to it: "the evil geniuses with OCD of the Hindu people."

Chapter Three
The Rise and Fall of Communism

> Communism enforces on people a numbing intellectual deadness generated by fear of the state Inquisition.
> -Arthur Versluis

> .....when he wrote at the beginning of the *Communist Manifesto* that communists are "hunted" throughout Europe as dangerous revolutionaries—that Communists themselves in only a handful of decades would become the hunters or inquisitors...
> -Arthur Versluis

Besides the similarity of stratification in the Indian caste system, science also possesses the same basic feature that made communism such detrimental force in the 20th century. Again, I am not suggesting that science is as bad as communism; rather, they both share the underlying cause of intolerance and vilification of certain ideas. The former rebukes capitalism, and the latter rejects metaphysics. By making metaphysical concepts unacceptable, science essentially excluded itself from tenable options, just as communism ruled out capitalistic economic options. In this chapter, we will examine how the idea of communism arose and maintained itself, despite many apparent flaws.

The 20th century saw two major kinds of totalitarianism: (1) fascism and its culmination in the heretofore unknown level of destruction and suffering of World War II and (2) post-World War II communism. After the World War II, the world was engulfed in the cold war, a time of proxy wars between the United States and the Soviet Union. The result of this was decades of bloody conflicts all over the world. Another arena was the battle of ideologies, where each side was seeking to convince the other that its system was superior. The Soviet Union's launch of Sputnik, the first artificial satellite, in 1957 was not just a major technological and military event, but also a demonstration of its superior political and economic systems. International sporting events, particularly the Olympics, became another ideological battleground between the two superpowers. For example, judges of the gymnastic and figure skating events commonly voted according to their ideology if the results were close. But success in space and sports could not make up for the sad realities daily life under communism; the long lines of people waiting to purchase their daily needs were unknown in capitalist societies. The fall of the Berlin Wall in 1989 clearly demonstrated communism's flaws, of which the majority of the world's people, even those living under it, were already aware. Even in Cuba and North Korea, the last remaining Stalinist countries, the enthusiasm for this ideology has largely disappeared.

William F. Buckley, Jr., the conservative founder of *National Review*, called communism "the worst abuse of freedom in human history." Although one may not agree with everything Buckley stands for, it is hard to dispute his statement. In less than one hundred years, more than 100 million people were killed; millions more were imprisoned, tortured, and starved to death because of the regimes' insistence upon adhering to unworkable ideology-based socioeconomic policies. Here is a paradox of communism: The immense human calamity it engendered was caused by the very idea that eliminating the ills of capitalism would bring about utopia. The corrective measures envisioned by Karl Marx and Friedrich Engels created an even greater evil. How could such a flawed and dangerous idea attract so many people, educated and uneducated alike, and be held with such passion? How could the resulting system maintain itself for so long, even when its flaws and failures could no longer be hidden? What can we learn from this tragedy?

Plato and the Reactionary Impulse

> It is impossible to read the history of the petty republics of Greece and Italy without feeling sensations of horror and disgust...If momentary rays of glory break forth from the gloom, while they dazzle us with a transient and fleeting brilliancy, they at the same time admonish us to lament that the vices of government should pervert the direction and tarnish the luster of those bright talents and exalted endowments for which the favored soils that produced them have been so justly celebrated.
> -Alexander Hamilton

Alfred North Whitehead, one of the twentieth century's most influential philosophers, said Plato: "The safest general characterization of the European philosophical tradition is that it consists of a series of footnotes to Plato" (Whitehead, 1960). This might a bit of overstatement, but hardly anyone would deny Plato's timeless influence on philosophy and beyond. That said, let me introduce a side of this man that one might find surprising. In his *The Republic*, he stated:

> The greatest principle of all is that nobody, whether male or female, should be without a leader. Nor should the mind of anybody be habituated to letting him do anything at all on his own initiative; neither out of zeal, nor even playfully. But in war and in the midst of peace—to his leader he shall direct his eye and follow him faithfully. And even in the smallest matter he should stand under leadership. For example, he should get up, or move, or wash, or take his meals...only if he has been told to do so. In a word, he should teach his soul, by long habit, never to dream acting independently, and to become utterly incapable of.

The above paragraph makes one wonder if it was written by the same person who is considered one of the greatest philosophers of all time. It sounds like a totalitarian doctrine that one would expect to see in Nazi Germany, the Soviet Union, and China. Hitler, Stalin, Mao, and many other ruthless dictators have embraced this extreme view of anti-individualism – the result has been the countless horrors and holocausts of the twentieth century.

Perhaps we need to look at what prompted Plato to take such a radical position in politics. His treatise on political philosophy was written during Athens' "Golden Age" (500 to 300 BC), a time of remarkable achievements in architecture, art, and knowledge. Another great legacy of this era was democracy. Unlike the varieties of democracy practiced today, the Athenians did not elect representatives; rather, eligible citizens participated directly in formulating the policies to be implemented. In other words, there were no politicians who campaigned to be elected as the people's representatives in the halls of Congress or Parliament, as is the case today. Another unique feature of Athenian democracy was the selection of public officials by means of a lottery.

But there were negative aspects as well. For example, the Athenians would vote to expel an unpopular fellow citizen even if he or she had not been convicted of any crime. Perhaps the most famous example of this negative side is the trial of Socrates in 399 BC. Put on trial for corrupting the youth, impiety, and for encouraging his students to question well-established beliefs, he was to be judged by a group of public officials chosen by lot. After they found him guilty of the charges, this unrepentant philosopher told them: "The unexamined life is not worth living." This profound statement must have upset them, enough that they sentenced him to death. His followers, including Plato, urged him to flee Athens; most people expected him to do so. Yet he refused to do so, based on his own principles, drank hemlock, and died at the age of seventy.

Plato, who saw his beloved mentor condemned to death by a jury of his fellow citizens, was forced into exile along with other followers of Socrates. A deep thinker in his own right, he must have reflected upon his society; perhaps he even sought an ideal political system that would not be tarnished by the ills of democracy, as it was known in his time. After he returned from an extensive journey, he founded the Athens Academy in 387 BC. While it offered a variety of disciplines like astronomy, biology, ethics, geometry, and rhetoric, it also served as a school for counter-revolutionaries who advocated an alternative to Athenian democracy (Grossman, 1937). In his *The Republic*, his vision of a perfect state ruled by all-wise philosopher-kings, he revealed his loathing for the concept of a democratic system in which an unruly and agitated group of men could execute a perfectly innocent person. His solution was to establish a totalitarian state that kept everyone in his/her place. In other words, an individual should yield his/her own happiness to the happiness of the whole.

In essence, Plato's reactionary political philosophy exemplifies the danger of the human tendency to embrace the opposite extreme. Rather than finding ways to control the potential dangers of having a demagogue in power, he sought to completely dismantle the democratic system. Fortunately, nobody try to implement his totalitarian ideals in Athens. On the other side of the world, however, the Legalist school of thought in China put forth this same extreme political view, and Qin Shi Huangdi, the first emperor in 221 BC implemented it in his brutal campaign to unify his domain. In the 20th century, the Legalist's worldview was carried out by Lenin, Stalin, Mao, and others with cataclysmic consequences to those countries ruled according to communist ideals. If we are to avoid similar disasters in the future, we need to understand the underlying processes that cause such ideas to appear. We should be aware of the possibility that implementing a drastic solution can bring about an even more serious problem.

The Industrial Revolution and Marx's Communism Doctrines

> Communist dogmas have replaced the former religious doctrines, with the same or even greater assurance of certainty and inevitability.
>
> -M. J. Fisher

According to Webster's dictionary, communism is "a theory advocating elimination of private property." Strictly speaking, most early human societies were communist in nature, for small nomadic hunter/gatherer bands roamed the land searching for food and shared among themselves whatever they found. About 10,000 years ago, farming and domestication allowed permanent settlements and private property. One inevitable consequence of private property was inequality among people, for one's prestige and power often depended on one's possessions, particularly land. Eventually, those who had a little began working for those who had a lot. With the subsequent rise of rich landowners, this land-based inequality became institutionalized in a system known as feudalism. As the disparity widened, people began to yearn for a perfect society that would eliminate the misery and suffering brought about by the rich and powerful landlords' exploitation of those below them. Some philosophers advocated the utopian ideal of communal land; some societies actually implemented it, with various results (see Busky, 2002).

The industrial revolution that occurred in the 18th and 19th century Europe created a new form of economic inequality. Beginning in England's textile sector with the advent of the machine run by animals, and subsequently by water and finally steam, the new modes of production gave birth to a class of entrepreneurs known as capitalists or industrialists. These individuals gradually accumulated enormous wealth. In order to stay in business, they had to find ways to become more efficient so that their products could be sold at a rate lower than that of their competitors. Driven by this reality, the entrepreneurs did all they could reduce the cost of production. The influx of inventions during the industrial revolution was largely fueled by this drive for efficiency so that one's rivals could be beaten. However, the blunt of such cost-saving efforts fell on the workers in these industries, who had to work long hours in cramped and squalid conditions just to make enough to survive.

The industrial revolution revealed the obvious pros and cons of capitalism. Those who focused on the pros, like Adam Smith, saw the great benefit of self-interested competition. In his *The Wealth of Nations*, this "father of capitalism" asserted that the free market system could lead national prosperity and public well-being if more goods could be produced at lower prices. He coined the phrase the "invisible hand" to describe how supply and demand based on self-interest would self-regulate the marketplace. This economic philosophy, known as "laissez faire" (Fr. "leave it alone"), states that government should not involve itself industry's affairs.

Those who focused on capitalism's negative sides, among them Karl Marx, condemned the terrible conditions and injustice faced by the workers. Political philosophers who advocated social change pushed the ideas of socialism and communism forward. The one who had the greatest impact was, of course, Marx; *The Communist Manifesto*, written by Marx with aid of Friedrich Engels in 1848, became the most influential writing and inspired both devotion and hatred among its readers (Ebenstein, 1951). It starts by asserting that the existing society is the "history of class struggle" between "oppressor" and "oppressed": of free men and slaves, patrician and plebeian, lord and serf, guild-master and journeyman. After the advent of the industrial revolution, a new form of class struggle arose: the bourgeoisie (capitalists) vs. the proletarians (workers). The *Manifesto* ends with a stunning proposition: "Let the ruling classes tremble at a Communistic revolution. The proletarians have nothing to lose but their chains. They have a world to win. Workingmen of all countries, unite!"

Marx's justification for the violent overthrow of capitalism is its inevitable path: Since the capitalists are forced to exploit the workers as much as possible to survive, the latter lose control over their lives and become foreign to their own world. Marx called this "alienation." The ultimate villain of capitalism, which condemns workers to serve capitalists, lies in the system. Over time, the capitalists who created the system are no longer able to control the forces that it engenders among the workers, just like Dr. Frankenstein's creation of his monster (Hudelson, 1993). Marx's solution is as extreme as his portrayal of capitalism: Kill the monster.

In the *Manifesto*, Marx and Engels call for the complete elimination of the world's basic economic system: "[T]he theory of the Communists may be summed up in the single sentence: Abolition of private property." Marx insisted on this extreme view from 1844 onward, one that abolished buying and selling, as well as trading and bartering (Megill, 2002).

As with most ideologies that promote extreme views, Marx propounded his view of communism with absolute certainty. For his faithful followers, an alternative or even a slight deviation was unacceptable, largely because of the certainty that Marx's vision offered. In fact, a shrewd politician like Vladimir Lenin often attacked his opponents by accusing them of deviation (Fisher, 1952). The *Manifesto*'s promotion of hate for capitalism coupled with extremism, certainty, and intolerance became a ready-made manual for violent uprisings by those who were not contented with their existing circumstances and those who were ambitious for power. Once established, communism's doctrines would become an ideological trap that makes very difficult for nations from taking flexible and corrective actions. As shall see later, China broke out of the trap by the courageous effort of a group of farmers and the reform-minded leader, Deng Xioping in the late 1970s.

The Expediency of the Communist Doctrines for Lenin and Stalin

Marx died in 1883 in England, a poor and obscure figure. And yet his theory would soon become the dominant force behind a revolution in a most unlikely place: Czarist Russia. Unlike most other western European nations that had fully matured as industrial powers, Russia was still in the early stage of industrial development. Thus the villainous capitalists (bourgeoisie) and the heroic factory workers (proletarians) portrayed in his ideology were only small – albeit rapidly growing – forces. Due to the Czarist ruling system, Russia largely lacked the political freedom needed to form a mass labor movement (Perrie, 2000). This is where Lenin came into play. If Marx can be compared to an architect who designed the building, Lenin was the superintendent who carried out the actual construction of the communism edifice.

But he did not just follow Marx's blueprint blindly; rather, he modified and adjusted it to accommodate Russia's various realities. Lenin brought in the revolutionary strategy and tactics needed to mobilize and manage the uprising, rather than waiting for the inevitable conditions that Marx had predicted (Fisher, 1952). Lenin laid out the methodology for achieving the communist revolution in his 1902 pamphlet *What is to be done?* Aware of the drawback of relying on the workers' mass "spontaneity" (Busky, 2002) to achieve a revolution, he set about creating a small group of dedicated full-time revolutionaries who would lead and educate them. This small elite group would concentrate all power in its hands and would rule the majority via a highly centralized hierarchical structure led by a dictator. His disciplined organization was able to outmanoeuvre the opponents. In 1917, as Russia lay prostrate in the wake of World War I, the small group of dedicated Bolsheviks seized power in what has become known as the October Revolution. Kilroy-Silk (1973) makes an ironic point of Lenin's role in this event:

> Lenin, it could be argued, thought of nothing else but revolution and had as an overriding passion the desire for *power* (emphasis added). Marxism then became for him a convenient set of useful slogans with which to galvanize the masses but which were discarded when other, more dramatic, slogans were needed in 1917.

The hate propaganda against capitalism was a convenient tool that a shrewd politician could use to seize and maintain power. In this case, it was Lenin. This was even more the case with his successor Josef Stalin. Leonid Trotsky said Stalin revised Marxism and Leninism not with the theoretician's pen, but with the heel of the Soviet secret police. This makes one wonder what Marx and Engels would think if they were to see how such totalitarian dictators as Lenin, Stalin, Mao, Pol Pot, Kim Il Sung, and Castro used their ideas as instrument to seize power. The real danger lies in absolute power of any kind, be it capitalism or communism (Fisher, 1952). It is particularly poignant that Marx saw himself as a champion of freedom. An even greater irony is that Marx's vision of a world without capitalism did not create a classless society. Instead, the "aristocracy of Communist elite has replaced the former feudal landholders and capitalists" (Fisher, 1952).

Universal Suffrage and Changed Social Conditions

In the year of the *Manifesto*'s publication, France, Switzerland, and other European countries began to extend voting rights to all male adult citizens. Prior to this, only men who owned a certain amount of land or wealth, and who met other certain qualifications, could vote. This historic change allowed poor factory workers and farmers to choose their representatives to Congress or Parliament. As extended voting rights spread over throughout Europe, Marx admitted the possibility of overthrowing the bourgeoisie by peaceful means. However, he saw the use of "force" as the more common road to working-class power (Marx, 1872). This refusal to see the potential implications of extending the vote might have been a result of his intellectual laziness. Laissez faire capitalism, the terrible conditions of which were exposed during the 18th and 19th centuries, would gradually be curtailed through universal suffrage and representative government. Policies designed to protect the workers' well-being, such as the minimum wage law, the minimum age law, the maximum hours law, workmen's compensation, unemployment insurance, and others could dispel the overly pessimistic view of capitalism (Fisher, 1952). Another reason for Marx's pessimistic outlook as regards capitalism was the growth of capitalist monopolies held by a small number of entrepreneurs. This problem was at least partially rectified by formulating and then implementing anti-trust laws. The most famous use of such laws took place in 1911, when the United States Supreme court broke up Rockefeller's Standard Oil.

As the benefits of representative government in the latter part of the 19th and the early 20th century began to offset the capitalists' economic supremacy, the inevitable collapse of capitalism and revolution that Marx had predicted with such certainty obviously had to be reassessed. The radical solution of dismantling the entire capitalistic system may not be necessary to control its abuses. Indeed some Marxists, particularly in western Europe, began to tone down their revolutionary inclinations and put more effort into building a socialist majority in Parliament (Kilroy-Silk, 1973).

When confronted with the discrepancy between reality and what Marx had predicted, Lenin offered his explanation of why the revolution had not occurred. In his *Imperialism, the Highest State of Capitalism* (1916), he shifted the focus to imperialism and the financial institutions. Basing himself on the idea of English economist J. A. Hobson's (1902)

Imperialism, Lenin argued that the final stage of capitalism is imperialism, in which industrialists become subservient to the international banking system. As monopolies intensify through buying and merging, the corporations are eventually consolidated and fall into the hands of a small number of international banks. These international conglomerates, with the help of imperial power, then expand their reaches and compete among themselves to control and exploit the largest number of potential workers in the undeveloped nations.

In other words, the imperial nations become parasites on the bodies of the less powerful nations, as was shown during the colonial era. Just as a herd of wildebeest have to find a new patch of pasture after depleting the local supply of grass, corporations are forced to move into other undeveloped nations after exhausting the resources of their present hosts. Thus, exploitation on a global scale would be the final stage of capitalism. According to Lenin, therefore, capitalism had not yet collapsed because of the availability of exploitable workers in the colonies. In addition, he denounced World War I as a symptom of the final crisis in European capitalism (Perrie, 2000). This meant that the coming revolution would have to be international in scope. Just as Marx did, Lenin also ignored the potential of universal suffrage. Perhaps his desire for power and his preoccupation with revolution made him oblivious to the changing social conditions of capitalism.

The Audacity of the Xiaogang Farmers

Since the domination of Marxism in the Soviet Union and its subsequent spread to other parts of the world, countless events have shown the unfeasibility and impracticality of communist doctrines: millions of people starving to death, waiting in long lines for bread, and fleeing abroad for a better life. These stories were repeated again and again wherever communism became the dominant force. Among them, one story really stands out as a symbol of courageous defiance, of how average people can make a difference. Xiaogang village in Anhuis Province, about a five-hour drive northwest of Shanghai, contains a remarkable group of farmers. One night in November 1978, eighteen heads of households, along with team leader Yan Jinchang, signed a document with their seals and fingerprints. It read: "If any word about this is divulged and the team leader is put in prison, other team members shall share the responsibility to bring up his child till he (or she) is

A Critique of Science

eighteen." In defiance of the very communist ideology that they had lived under for decades, they decided to divide their communal land into plots for each family, who would be allowed to plant whatever they wanted, work whenever they wanted, or however long they wanted. These are basic practices for virtually all farmers around world, except for those living under strict Marxist ideology. In 1958, Mao abolished all of them and unveiled his ambitious economic plan known as the "Great Leap Forward." These farmers of Xiaogang village knew the risk of moving from collective to private ownership of farmland: arrest, long-term imprisonment, hard labor, and even death.

Yan's courageous and seemingly foolhardy plan to challenge sacred communist ideology was hatched in the desperate years of the Great Leap Forward. Ironically, this disastrous plan was launched to engender rapid economic development, the two main thrusts of which were agriculture and steel production. This, unfortunately, turned out to be a lethal combination. All private ownership of farmland was abolished, and by the end of 1958 about 25,000 communes with an average of 5,000 households each were established all over the country. In an effort to increase steel production, several hundreds of thousands of smelters and furnaces were built in the rural areas and the communes, and people living in the communes were to provide the necessary labor. One estimate is that as many as 60 million people were ordered to participate in this effort (Evans, 1993). This meant that the necessary labor required for agriculture was diverted to making steel products that were virtually useless, for the commune workers, just like Mao, had no experience or knowledge of steel technology. As a result, grain production plummeted. By the time this campaign was abandoned in 1961, an estimated 20-41 million had perished (Peng, 1987). The Great Leap Forward turned out to be one of the greatest plunges into hell in recorded history.

Xiaogang was particularly hit hard by this man-made famine; sixty-seven of the 120 villagers died. This needless tragedy caused Yan to doubt the principles of communism and to consider dividing the land into individual households. He had to be cautious and pick the right time to do this, however, because during 1966 Mao launched the Cultural Revolution, a frenzied ideological campaign designed to purge anyone suspected of challenging his authority and his ideology. Mao died in September 9, 1976. Two years later, Yan and the villagers made their move. As much as they tried to keep their daring privatization of farmland secret, it became public knowledge during the following spring. Fortunately, however, the officials who discovered it were sympathetic and allowed the farmers to continue their experiment in privatized farming. The result was a resounding victory for capitalism: they doubled their harvests in the first year. Their daring yet risky move gave the farmers what they had always wanted: enough food for their families. But it also became the impetus to launch a campaign of agricultural reform. The Xiagang farmers' audacity gave the newly elected Deng Xiaoping the place and the ways to start a nationwide agricultural reform. This was not hard for him because, during the Great Leap Forward, he had gone through the same anguish and doubts of communism that the farmers had. As a result, Deng had directly challenged Mao and was purged three times.

Deng Xiaoping's reform of the agricultural sector was a rather unusual event in Chinese history, for the Xiaogang farmers led and Deng's government followed. The bottom-up endeavor resembled the Civil Rights movement in the United States: the bottom-up movement led by Martin Luther King, Jr. and others in the 1960s was met by the top-down initiatives of the Johnson administration. Inspired by the success of the Xiaogang farmers, Deng instituted the "household responsibility system" that gave famers thirty-year leases on their government-owned farmland. As a result, the total grain output increased by 64.6% between 1978 and 2007, despite the migration of large numbers of rural people to major cities.

When Deng Xiaoping died in 1997, Patrick Tyler (1997) of the *New York Times* wrote an unusually long obituary. He quoted J. Stapleton Roy, a former American ambassador to China, that the last fifteen years of Deng Xiaoping's rule were the best years of China's 150 years of modern history. He attributed this phenomenal transformation to Deng. While this may be true, the first shot of Deng's reform was fired by the Xiaogang farmers. Mikhail Gorbachev's glasnost and perestroika may have been launched in 1986, but its seed may have been planted by the farmers of Xiaogang. The Berlin Wall may have fallen in 1989, but the crack may have been started by the reverberations of the Xiaogang farmers' decision to defy communist doctrine.

The Xiaogang farmers did not have college degrees in economics, agriculture, sociology, or psychology, but they knew the basics of human needs and motivation. Placing the individual's fate in the *individual's* hand, rather than in some collective unit, is a powerful engine for efficiency not only for individual, but also for society as a whole. This was what Marx and other political thinkers and politicians who followed him failed to realize. The Xiaogang farmers also understood the need for experimentation, to try both methods (collective farming versus privatized farming) and see which one is better. Witnessing the senseless deaths of their family members and fellow villagers during the Great Leap Forward, their direct experience of life-and-death circumstances gave them conviction that the sacred communist ideology was flawed. Thus they experimented with the capitalistic method and proved that it was better. Although their experiment did not meet the scientific standard, it probably made a greater impact than any other experiment in demonstrating just how flawed communist doctrine really is.

So a poignant question is this: If poor and barely educated farmers could conduct such an experiment and learn from it, why could government leaders and officials throughout the communist countries could not do what the Xiaogang farmer did? Even after millions of deaths, what prevented them from taking corrective action? Why did they refuse to conduct small-scale experiments and compare the results to determine which system worked better? Why was the communist intelligentsia, supposedly the best representatives of the country's educated people and who were entrusted with the people's well-being, unable to help the people? Above all, why could they not see communism for what it was and stop the tragedies that were sure to follow in its wake? The very idea that was supposed to help the working class by getting rid of the ills of the society in fact caused the greatest harm to all people, regardless of their class.

Orogee Dolsenhe

Chapter Four
Hijacking of Minds

> The great problem confronting us today is that we have allowed the means by which we live to outdistance the ends for which we live. We have allowed our civilization to outrun our culture, and so we are in danger now of ending up with guided missiles in the hands of misguided men.
> —Martin Luther King, Jr.

The 20th century has displayed the unparalleled degree of the two paradoxical qualities of individuals. On the positive side, the human intellect has served as a great force in technological progress. On the other hand, the two major threats of totalitarianism—fascism and communism—shows the destructive potential of individuals. The human cataclysms in the 20th century caused by totalitarianisms, with the help of the technological progress, perhaps reveal that we are like teenagers who began to possess "strength without discipline, power without virtue, and knowledge without wisdom" (Dolsenhe, 2005). If we are to grow out of the intellectual adolescence, we have to understand the fundamental flaws responsible for the calamities. Francis Bacon said knowledge is power. It means knowledge is like a tool or a weapon that can be used for good or for bad. Progress of knowledge, therefore, must be accompanied by the growth in wisdom and virtue. We must be able to recognize the possible negative intentions of various leaders who might take advantage of human weaknesses and flaws. In such an event, scientists should not be exempted from ethical judgment. Instead, they must set an example for self-examination in identifying the possible negative side of human capacity. This chapter discusses the human traits that may be most responsible for this destructive potential.

The Two Separate Systems of Decision-Making

In the discussion of the decision-making process of the human mind, there are two basic approaches. One approach uses a computer metaphor in which the mind is compared to a central processing unit (CPU) that carries out instructions of a computer program. The domination of the computer metaphor in modern psychology is largely attributable to the mystique of computers rather than to some empirical evidence, as discussed earlier. As Francis Crick (1994), the co-discoverer of the double helix, declares: "A brain does not look even a little bit like a general-purpose computer."

Another approach to the decision-making process is the use of the homunculi (which literally means "little men" in Latin) metaphor. Remarkably, there are some reasons to believe that humans' decision-making process is closer to homunculi than to computer processing. One of the first homunculi metaphors was offered by Selfridge (1959). Selfridge's "pandemonium model" compares the decision-making process to a shouting contest among basic units, which he calls "demons." Simply put, whoever shouts the loudest, wins. Unlike the frantic setting of Selfridge's model, the homunculi model of Baars (1983, 1988) compares the decision-making process to a more orderly proceeding in a scientific meeting. Each member competes with others to reach the podium. In a similar vein, Loftus & Schooler (1985) used a corporate metaphor in which each person vies for the chance to express his or her view in a meeting.

The question is whether the human decision-making process is more like Selfridge's shouting contest or more like the orderly proceedings of a scientific or corporate meeting? A remarkable aspect of the human mind is that we are equipped with both types of capabilities. Why do we need both? To understand the dual capacity of the human mind, we need to use a different metaphor based on the decision-making process in a ship. A big ship like an ocean liner has a chain of command that is structured to make quick decisions. For instance, if a ship is in a collision course with another ship, a reef, or an iceberg, a speedy decision is paramount. There is no time to have a staff meeting to discuss whether to turn left or right. In the ship's decision-making process, the preset ranking system determines who should have the loudest voice. On the other hand, much of the decisions made on a ship do not require urgent responses. For instance, suppose weather conditions 500 miles ahead are deteriorating. The captain could call for a staff meeting to decide the course of action. He may ask for input from the staff or the National Hurricane Center. Unlike the urgent response to a fast-approaching obstacle, the decision to go left, right, or straight ahead can be discussed at a leisurely pace. They can and should consider all options thoughtfully, by examining all supporting data. If there were insufficient data, the decision might have to be postponed.

A Critique of Science

Like the ship's decision-making process, the human mind has two different systems that are designed for two distinct types of action— one for a fast reaction and another for a leisurely response. The most common problem of human decision-making is that we ignore or forget we have the two systems. Specifically, we tend to make decisions immediately and impulsively, without giving thoughtful consideration even when no urgency is required. Simply put, the fast system takes over the rational system. Thus, to avoid impulsive decisions we regret later, the first thing we have to do is to remind ourselves to slow down. We have got to have discipline to suppress the fast system and use the rational system to thoughtfully examine other options. In the following sections, we will discuss in more detail the fundamental problems, complications, and difficulties of managing the two systems.

The Importance of Emotions

All of us constantly experience a tug-of-war between the two seemingly contradictory parts of our mind, the emotional part and the rational part. For instance, you are angry at someone and you feel like yelling at him or even hitting the person. Another part of your mind, then, tells you that you can't do that because of morality and consequence. It is almost as if the rational part is serving as a brake in controlling the emotional engine. The conflict may show up with the opposite dynamic. For instance, in trying to prepare an exam or write a report, constant urges to watch TV or to play outside might interrupt your concentration. And those who try to lose weight know all too well the powerful drive to eat that interfere with the goal.

No wonder so many philosophers like Plato, Descartes, and Kant have a negative view of emotions. They thought that emotions were a hindrance to the rational part of the mind, like extra baggage that has to be dragged along with it. Plato believed that emotions are irrational urges that need to be controlled by reason. The famous words of Descartes (1637/1960) "I think, therefore I am" put so much emphasis on thinking (or reasoning) that it elevated to the essence of our existence. He didn't say "I am conscious, therefore I am." Kant's influential work, *The Critique of Pure Reason,* exemplifies the significance of "pure reason" without emotional interference. If these philosophers are right, then, there is hardly an adaptive reason for having any emotions at all and Mr. Spock in the TV series *Star Trek* may be an ideal being because he conducts himself in a cool and logical manner without being overwhelmed by human emotions.

On the opposite side of the philosophers with negative views of emotions, Aristotle and Spinoza believed emotion and rationality should be integral parts of a good life. David Hume, an 18th century Scottish philosopher, also took the position that reason alone is "utterly impotent" and that passions excited by morals can produce and prevent actions (Hume, 1984/1739-40). Hume argued that reason is and ought to be the slave of passion. For those who believe in a Mr. Spock-like, super-rational ideal, Hume may sound a bit too strict. However, if Hume's view were differently phrased, it might not only be reasonable but also show the most central but often neglected issue concerning emotion:

> ….emotions matter because if we did not have them nothing else would matter. Creatures without emotion would have no reason for living nor, for that matter, for committing suicide. Emotions are the stuff of life…..Emotions are the most important bond or glue that links us to others (Elster, 1999).

In the latter part of the 20th century, the positive role of emotion in mental function was promoted by several scholars. Herbert A. Simon (1967), a Nobel economist, thought that emotions set priorities among many goals in a given moment. Philosopher Ronald De Sousa (1991) suggests that emotions are inherently rational. A team of emotion researchers, Keith Oatley and Philip N. Johnson-Laird (1987), believe that basic emotions are essential in coordinating among multiple plans and goals under time constraints.

Since there is no real Mr. Spock in the world, how well or how badly an emotionless person would function is largely conjecture. However, a patient named Elliot provided Antonio Damasio, a Portuguese neurosurgeon, with the opportunity to study this question. Elliot was once considered a good husband and father, had a decent job, and was a role model for younger siblings and colleagues. Then, at the onset of severe headaches, his life began to unravel. Once a dependable worker, his work had to be completed or corrected by others. What they found was an orange-sized tumor growing in his brain's frontal lobe. Elliot was then in his thirties and had to undergo the drastic operation to remove the tumor. After the operation, all seemed to be fine. The mental functions of memory, language, and vision were intact. What didn't show up under the normal diagnostic procedure was a radical personality change. Elliot's physicians asked Damasio whether the drastic change was a real disease. Upon examination, Damasio noticed Elliot's flat emotions. When Elliot talked about his own tragedy or discussed potentially embarrassing personal events, he seemed to be detached from his emotions. And emotionally charged pictures, such as people being injured in gory accidents, did not produce emotional agitation. In addition to reduced emotions, Elliot's apparent decision-making failures in his real life were revealed. He couldn't hold jobs and went through several divorces. He was no longer able to make decisions for the betterment of himself and his family (Damasio, 1994). A series of laboratory tests showed that Elliot was unable to choose effectively among a number of options. The tests required a consideration of social conventions and moral values, which were particularly disrupted even though his knowledge of them was intact. Damasio (1994) theorized that the emotional flatness may prevent him from assigning different values to different options. Damasio studied other patients with a similar frontal lobe damage. All of them displayed defects in decision-making accompanied by flat emotions.

Amygdala versus Brodmann Area 10 (BA10)

The numerous negative aspects of emotions we all experience may be, then, side effects we should expect, since most things that were made for a specific function have a set of unwelcome side effects. Take the automobile, for example. Its primary function is to provide us with transportation, but it has a number of associated costs: fuel, insurance, repairs, traffic tickets, accidents, pollution, etc. Similarly, we need to be aware of and prepared to deal with the side effects of emotions.

The most obvious undesirable aspect of emotion is that we often do things out of an emotional outburst that we later regret. These things may be directed at someone else and can take the form of verbal or physical abuse. Relationships may come to an abrupt end or someone may even end up in jail. The problem is that the loss of control happens without person's awareness, as Damasio (2003) points out: "the emotional signal can operate entirely under the radar of consciousness.....The individual may not ever be cognizant of this covert operation. In these conditions we intuit a decision and enact it, speedily and efficiently, without any knowledge of the intermediate steps." Daniel Goleman (1995) aptly refers to this as "emotional hijacking."

The neural mechanism of emotional hijacking was discovered by a neuroscientist Joseph LeDoux at New York University. In LeDoux's experiment, rats were conditioned to fear the sound of a bell accompanied by electric shocks. Once conditioned, rats are fearful of the sounds of a bell even without electric shocks. LeDoux (1994) discovered that even when the rats' acoustic cortex was surgically removed (which means that the rats no longer heard the sound), the fear behavior persisted (LeDoux, 1996). He concluded that there is a neural pathway that bypasses the auditory cortex and goes directly to the amygdala (an almond-shaped structure located deep within the brain).

What is the benefit of a bypassing mechanism that enables the animals to act without even being aware of the sound? It gives animals the capacity to quickly respond in potentially life-threatening situations. If a rat sees a shadow in the dark, it must be able to take evasive action very quickly. The dramas unfolding on the African savannah between predator and prey underscore the importance of quick response. If a zebra glimpses a shadow, it has to quickly determine whether to run or not. The difference between escaping or being trapped by the claws and jaw of a lion may be a matter of fractions of seconds. The capacity to quickly respond is one of the most important features of survival for most animals. And humans also have the basic neural mechanism for quick response, just like other animals.

A downside to having a bypass mechanism is that the amygdala may overreact and take over the slower, thinking part of the brain, even when no urgent reaction is required. One may get fearful or angry and do undesirable or regrettable things without giving the thinking part a chance to analyze the situation. The amygdala's overreaction may result from an emotional memory that may go back to early childhood. LeDoux (1994) believes that he has found the brain system that is responsible for early childhood memories, which were the main target of Sigmund Freud's clinical techniques, or psychoanalysis, which has largely been marginalized for the last few decades. Psychoanalysis emphasizes the memory of early childhood in treating anxiety disorder.

Now, let us discuss the rational part of brain. The field known as comparative anatomy (the study of similarities and differences in the anatomy of organisms) has become an effective tool for those who are interested in finding the source of the cognitive and intellectual capacity of humans. Mammals are considered more intelligent than fishes and reptiles. What mammals have is the cerebral cortex, the outermost layer of the brain, which is absent in fishes and reptiles. Primates particularly have an enlarged cerebral cortex. Humans and great apes (chimpanzees and gorillas) have a greater proportion of frontal cortex to total cortical volume compared to monkeys and other mammals. It was generally assumed that the exceptional cognitive capacities of humans are attributed to a disproportionately large prefrontal cortex. Here is a surprise: The relative size of the human prefrontal cortex is not bigger than that of great apes (Semendferi et al., 2002; Rilling, 2006). Then, what accounts for the difference in intellect between humans and apes? In recent years, the subregion called "Brodmann area 10" (BA 10) in the frontal cortex has received a great deal of interest from many neuroscientists, even though it is the least understood region of the brain. The sudden interest in BA 10 is due to the fact that it is disproportionally larger than those of other great apes (Semendeferi et al., 2001). It is also one of the last to fully mature in humans, developing throughout childhood and adolescence (Dumontheil, Burgess, & Blakemore, 2008). So it is a good bet that BA 10 may be involved in something very important that gives humans exceptional intellect compared to apes. One of the pioneers of studying BA 10 is Jordan Grafman at the National Institute of Health. He and his colleagues discovered that, when subjects engage in task-switching, there was an increased level of blood flow on fMRI (functional magnetic resonance imaging) of BA 10. In plain speaking, BA 10 was involved in task-switching activities. What is important about the task-switching capacity?

In addition to the metaphors of scientific and corporate meetings mentioned earlier, Alan Newell and Herbert A. Simon (1972) introduce a theatre metaphor that addresses the three types of memory. The first type is the focal consciousness that is like the bright spot on the stage. The second type is the vague consciousness that seems to be just beyond the reach, the surrounding fringe area that is dimly lighted. The third type, like the darkened area that the rest of audience sits in, is compared with long-term memory. The significance of their tripartite distinction is that it adds the second type, the vague consciousness that seems to be just beyond reach yet which is effortlessly available if we decide to bring it into our consciousness. Unlike the theatre metaphor, most theories of memory just consist of two types: short-term memory and long-term memory. Short-term memory is what we are presently conscious of; it has been variously referred to as "prehension," "species present," and the presently preferred term, "working memory." Originally it was thought to hold about seven units in short-term memory, and thus the number seven was known as the "magical number seven, plus or minus two" (Miller, 1956). But subsequently, the magic number turned out to be only about four units (Cowan, 2001). Where do the other three units come from? They are not exactly in long-term memory and they are not in short-term memory. They are in what I would call "on-deck memory," to use baseball jargon, the second type of memory. Short-term memory, the first type of memory, is like the batter on the plate; long-term memory, the third type of memory, is in the dugout. But the readily accessible memory, the second type of memory, is the one in the on deck circle who is ready to step into the plate when the batter leaves. On-deck memory is something we all experience when we are engaged in multitasking. While doing one task, we are partially aware of another task we were doing just seconds ago. That partial awareness, even though we may not engage in it at the moment, gives us the capacity to switch tasks at our will.

BA 10 seems to be involved in a task switching process, and it may be the site of on-deck memory. The seeming inflated size of short-term memory is only a minor advantage to having on-deck memory. A crucial function of BA 10's task switching capacity is to allow the integration and comparison of different thoughts. It is quite likely that a human's exceptional power of reasoning may stem from the integrating capacity of BA 10. It gives us the impression that we have the power to choose among options, that we have "free will." Suppose you were feeling one of those amygdala impulses, like unwarranted fear or anger. The integrating capacity of BA 10 would allow you to compare that feeling with an alternative thought that can counteract the potentially destructive impulse. In social settings, it may curve the hostile impulses directed at others. Then, BA 10 may be the seat of the human's exceptional social intelligence and moral faculty. No wonder why Jordan Graham would declare in the interview with *The New York Times* in 2001, "It's what makes us most human." The seats of the seemingly constant battle between emotion and reason may be taking place in BA 10.

BA 10 may be the seat of the human rational capacity that can counter the emotional impulses of amygdala. Goleman (1995) argues that the capacity to control emotions is as important as the intelligence that we are familiar with. He elaborates on the significance of emotional intelligence in endowing us with the ability to "motivate oneself and persist in the face of frustrations; to control impulses and delay gratification; to regulate one's mood and keep distress from swamping the ability to think; to empathize and to hope."

Marx's Personality and the Background of the Development of Communism

> Marx's entire thought structure was largely an emotional reaction and a rationalization and ventilation of personal passions, not an objective analysis of nineteenth-century European capitalist society. Marx dressed this compulsive reaction in rational and even scientific costume because he admired the authority of science, coveted the respect men had for it, and sought to arrogate its certainty for his own writings.
>
> -Eugene H. Methvin

> It is one of history's ripe ironies that Karl Marx was the opposite of what he tried to appear—a dispassionate philosopher rationally blueprinting a "scientific" world transformation. Instead he was a "compulsive revolutionary," condemned by his own frustration and failures to see the world in terms of struggle, bloodshed, and hate.
>
> -Eugene H. Methvin

Nazism and communism share a common feature that makes emotionally hijacked leaders are turned into contagions that infect others. Simply put, emotional hijacking can be turned into what I would call "social hijacking." The difference between Nazism and communism is that, while hatred was directed at Jews in Nazism, communism was aimed against capitalism.

Karl Marx, like many other progressive social thinkers in the 19th century, saw injustice in the capitalist system where workers were "alienated." One of the main reasons that his writings attracted a large number of dedicated followers was that they were saturated with emotionally charged expressions that evoked anger and hatred toward capitalism. His "language of warfare" (Methvin, 1973) became the powerful motivator in arousing people to overthrow the capitalist system. The phrase "the class struggle" was the battle cry for the dissatisfied workers and thinkers in the industrial revolution. Who was Karl Marx? Does his personality have anything to do with communist ideas saturated with emotionally charged rhetoric? In this section, we will examine Marx's personality and his upbringing.

Marx was born 1818 in Trier, Prussia, in a comfortable middle-class home. Both his parents were Jews who converted to Protestantism, and Marx was baptized when he was six years old, yet he lived in a Catholic region. The sense of alienation was further heightened by the fact that he could not obtain a teaching position because of his spirit of dissent and revolt (McLellan, 1973). The various exclusions he faced may have contributed to the critical view he had on society (McLellan, 1973). Marx had three traits that are requisite for many tyrant leaders like Hitler: anger/hate, talent in communication, and a preoccupation with power. Marx's sister Sophie remembers that Karl was "a terrible little tyrant" who turned "his remarkable gift of words into a fearsome weapon" (Methvin, 1973). His father, Heinrich Marx, recognized his son's "splendid natural gifts," but also encountered the stubbornness and harsh defiance that he called "the demon." In college the young Marx rebelled against all that his father stood for (Methvin, 1973).

Against Marx's own wish, Marx's father sent him to the University of Bonn to study law at the age of seventeen. In his first year at the University, Marx was involved in fraternity street fights, fought a duel, spent a day in the campus jail for being a rowdy drunk, and neglected his school work. After one year, his father decided to transfer his son to a more austere school, the University of Berlin (Methvin, 1973). One day, Marx found a place where he could vent his frustration, anger, and ambition when he saw a phrase in a paper: "The poor class, toiling and suffering, transforms itself into a powerful organization, negative and menacing: the proletariat. *Communism looms as a somber and menacing specter*." Ironically, it was written by Lorenz von Stein, the Prussian secret police who was sent to Paris to report on exiled German workers who were attracted to socialism and communism. In essence, the police agent's analysis of the difference between socialism and communism provided Marx with direction (Methvin, 1973):

A Critique of Science

"Socialism wishes to realize its end by the power of the truth, whereas communism wished to do it by the violence of the crowd, by revolution and by crime" (von Stein, 1842). Marx chose the latter, the path of violence and revolution. Meyer (1954) points out Marx's mindset:

> The fiery spirit of revolt that breathes in this passage never abated as long as Marx was alive, and it remained one of the few most essential elements of Marxist ideology, pervading everything the man wrote.

Marx's attraction to revolt and violence after reading von Stein's article might shed some light on his psyche. His anger and hatred toward the establishment might be the source of his preference for communism's aggressive, rapid, confrontational approach over socialism's more subtle and gradual approach to social change. Thus, the choice between the two basic philosophies for changing the unjust situations workers faced might come down his personality. In essence, Marx's impatience led him to choose the path of immediate gratification over socialism's delayed one. Not only did he choose the violent and abrupt method of social change, but he also chose to anticipate an extreme utopia, the total elimination of all the ills of capitalism, even if it meant throwing the baby out with bath water.

An unfortunate historical irony is that Marx was able to pass the extremism that stems from his personality on to others. The extreme solution he envisioned came out of his frustration, anger, and hatred, which he glazed over with eloquent words. His literary skill allowed him to present the very complex problems of economics, history, and sociology with a rather simple set of ideas that average people could understand. In addition, Marx also had a strong attraction to the pursuit of power, as Carl Schurz, a fellow delegate, recalls: "But personal ambition in its most dangerous form has eaten away anything that was good in him. Everything he does is aimed at the acquisition of personal power" (Cited in Methvin, 1973). Schurz also remembers intolerance toward dissent: "Marx would not give even a moment's condescending consideration to any opinion that differed from his own" (Cited in Methvin, 1973).

Orogee Dolsenhe

The Two-Valued Impulse

The late Senator Hayakawa of California, who served from 1977 to 1983, was a renowned semanticist. His book *Language in Action* was selected as Book-of-the-Month in 1941. It was written during the time when the world was in a precarious situation because of the rise of Fascist regimes. Hayakawa's particular concern was the success of Hitler, who was able to persuade millions to share his maniacal and destructive views. In the late edition, Hayakawa cautions against our susceptibility to hate propaganda:

> Hitler is gone, but if the majority of our fellow citizens are more susceptible to the slogans of fear and race hatred than to those of peaceful accommodation and mutual respect, our political liberties remain at the mercy of any eloquent and unscrupulous demagogue (Hayakawa & Hayakawa, 1990).

Hitler and the Nazi propaganda neatly divided almost anything and anyone into two opposite categories (Hayakawa, 1941). The "Jewish-dominated plutocracy" is the category containing hunger, famine, unemployment, crooked capitalism, Germany's defeat in World War I, bad smells, immorality, treachery, selfishness, and any other bad things. On the other hand, any positive features like courage, self-discipline, honor, beauty, health, and joy were with "Aryan." What Hitler approved in art, music, books, people, philosophies, mathematics, physics, dogs, cats, architecture, and cookery was considered "Aryan." And if he disapproved, it was non-Aryan or "Jewish-dominated."

The heart of Hayakawa's discussion concerns the "two-valued orientation," a concept originally proposed by Alfred Korzybski, a Polish-American philosopher. The two-valued orientation is something we constantly use when we contrast the concepts of "right" and "wrong," "good" and "bad." It refers to our tendency to think and express ourselves in terms of two opposites, without considering the in-between positions. It is basically a black/white outlook, without grey. There are some things we can only think of in terms of the two-valued orientation, as in arithmetic when judging whether "$1 + 2 = 3$" is correct or not correct. Another example of the two-valued orientation that we often use is what Aristotle called the "law of identity." The furry animal with four legs walking in front of me is either a dog or not a dog, nothing between.

A Critique of Science

As Korzybski (1933) and Hayakawa (1941) point out, there is no problem using and thinking in terms of the "two-valued logic" in mathematics or in simple identification processes. However, the problem is when we extend the two-valued logic to other areas where we should use the "multi-valued orientation." An example of the multi-valued orientation is judging a person's law-abiding versus law-breaking behavior (Hayakawa, 1941). Both the man who runs through a stop light and a serial murderer have broken the law, but they should not be put in the same category and judged with the same severity or receive the same punishment. As absurd as it may sound, the medieval heresy trials were conducted using the two-valued orientation and the accused was either freed or put to death (Hayakawa, 1941). This type of thinking is known as the "Manichean worldview," named after the Persian prophet Mani. It divides the world into two opposites: the light and the darkness. Either one is with God or with evil.

The main methodology of Marx's persuasion was to divide society between good and evil: the workers (proletariat) and the capitalists (bourgeiose). Such rhetoric that advocates the splitting of the world between the forces of all good versus all evil can become a powerful motivator. It generates an emotional response and people become more susceptible to behaving irrationally and in potentially dangerous ways. It breeds intolerance, closed-mindedness, and inflexibility. Fisher (1951) identifies the source of the passionate attraction of communism:

> One of the reasons for the strength of Communist doctrine lies in its appeal to hatred, one of the most elemental of human emotions. The ideology provides an enemy to hate—the capitalist—and it is hatred which is to inspire the proletariat to arise in revolution to destroy the enemy. Hatred, moreover, justifies the use of violence, terrorism, and ruthless cruelty as methods.

Methvin (1973) makes a similar point:

> Hate is a mighty engine, and the apocalyptic rages of the young Dr. Marx lived on, engulfing succeeding generations of young radicals.

Orogee Dolsenhe

He further states:

> The power and durability of Marx's doctrine is not in its sociological soundness, but in the fact that it rationalizes and justifies hate, and provides visible and accessible hate targets to satisfy one's frustrations with an imperfect and un-ideal world. That is why Marx's ideas appeal so to angry young men; and why so many young Marxists become ex-Marxists as they grow older and understand more (Methvin, 1973).

Another two-valued orientation Marx used was the distinction between now and the future. Marx portrays the present capitalist-dominated time as a hell and the upcoming communism as a heaven on earth. The future, after a revolution, is a utopian society without any of the ills of capitalism. To some, executing hundreds or thousands in achieving the ultimate utopia on earth was worthwhile. The main flaw in Marxism lies in Marx's approach, which I would call the "two-valued impulse" rather than the "two-valued orientation" because we are constantly driven into the black-and-white way of thinking and behaving. Our two-valued impulse is further strengthened by our learning from a very early age. Children learn from fairy tales to distinguish between a "good king" or a "bad king." They grow up watching TV shows that contrast a "good guy" with a "bad guy." Our language itself is permeated with the two-valued inclination. We are told by our leaders that the international world is divided into good and evil. President Ronald Reagan called the Soviet Union the "Evil Empire." And recently President Bush made a speech calling Iran, Iraq, and North Korea, the "Axis of Evil," which a constitutional law expert, Glenn Greenwald (2007), calls "A Tragic Legacy" that destroyed the Bush presidency.

The two-valued impulse is something we all are susceptible to and, as we shall see later, science is not immune to it. Unless we put effort into controlling our two-valued impulse and watch over those who manipulate our innate tendency to divide the world into two opposites, we are likely to repeat the mistakes of the past in religion and politics. So I end this section with Hayakawa's (1941) cautions:

> If we fight, we develop the two-valued orientation; if we develop the two-valued orientation, we begin to want to fight. Under the

A Critique of Science

> influence of the two-valued orientation, we have in place of our normal reactions elaborate sets of signal reactions, lumping together all evils as one Evil, all good things as one Good.
>
> Often when we are hearing or reading impressively worded sermons, speeches, political addresses, essays, or "fine writings," we stop being critical altogether, and simply allow ourselves to feel as excited, sad, joyous, or angry as the author wishes us to feel. Like snakes under the influence of a snake charmer's flute, we swayed away by the musical phrases of the verbal hypnotist.
>
> It is customary for all those whom we call "spread –eagle orators" and "demagogues" to rely upon it [the two valued-orientation] as their principal argumentative technique....In short, the two-valued orientation produces the combative spirit, *but nothing else*.

Then, crucial questions are: Why do we have the two-valued impulse? Is there cure for it?

Group Polarization

> In a polarized environment, individuals may not be able to position themselves on neutral ground even if they try. Each side may perceive a declaration of neutrality or moderation as collaboration with the enemy, leaving moderates exposed to attacks from two directions at once.
> -Timur Kuran

Many proverbs, such as "Two heads are better than one," seem so obvious that one can hardly doubt their veracity. John Heywood, an English writer in the 16th century, is responsible for the exact wording of that proverb; however, the idea behind it has multiple origins. In the Old Testament, it says "Two are better than one, because they have a good return for their work" (Ecclesiastes 5; 9). Aristotle in *Politics* also mentions "Two good men are better than one" (Flexner & Flexner, 1993). In a modern version, John Bordley Rawls (1971), one of the leading political philosophers, gives a more elaborate version of it. He says that the benefits of discussion are to exchange opinion and to combine information with others, which allows us to check our partiality and widens our perspective.

In 1961 a landmark experiment was made by a graduate student named J. A. F. Stoner at the Management School at MIT. For his master's thesis, he chose to compare individual decision making with group decision making. Stoner measured risk-taking behavior, such as choosing a safe or a risky play in a football game and investing money in a low-return, low-risk stock or a high-return, high-risk stock. The subjects in the experiment were asked to write down initial judgments. Afterward, they were put in a group to make a unanimous decision about the same questions. In twelve of the thirteen groups, the group decisions showed greater risk-taking than did individual judgments (Stoner, 1961). He named the phenomenon the "risky shift." However, the shift also occurred toward the cautious side (the "cautious shift") and even when it had nothing to do with risk (e.g., Rabow et.al., 1966; Moscovici & Zavalloni, 1969; Doise, 1969). Moscovici & Zavalloni (1969) renamed it "group polarization."

What Stoner and other researchers found shouldn't be a surprise to anyone who follows the political partisanship in Washington. In an article he wrote when he left the Senate in 1996, former Senator William Cohen calls it the "hollowing out of the middle." Moderates are forced to move left or right or pushed out of politics altogether. Cass R. Sunstein (2009), a law professor at Harvard University, believes group polarization also can explain much extremism around the world, such as fascism, Islamic terrorism, the Rwanda genocide, and the abuse of prisoners at Abu Ghraib. He also points out the polarized differences in the legal system among federal judges and jurors. Group polarization also may be the root cause of the confrontation between capitalism and communism, as we shall discuss later. It has sobering implications. Many institutional forums designed to make better decisions by bringing a group of people together could actually produce unwelcome outcomes, in opposition to what they originally intended. It is almost as if there is an invisible force that puts people under a spell, making them behave in irrational and counter-productive ways when they are in a group setting to exchange information and views. This certainly doesn't mean, however, that we should abandon group discussion altogether. Communication and information exchange are one of the essences of being human. Two heads can be better than one if we can control the potentially polarizing tendency. Therefore, we should find the underlying cause of the hypnotizing spell of polarization and provide a solution—or at least offer a warning—so that people can be better prepared to deal with it.

Among many theories offered to account for group polarization, experimental results more or less eliminated all but two: social comparison and persuasive arguments. The theory of social comparison (also referred as normative processes) posits that group polarization occurs because members try to be favorably perceived. Upon hearing views of others, individuals adjust their position toward the direction of the dominant view. Caving into social pressure may explain people's adjustment of their views, but it does not give a reason for their shifting toward extreme positions. In other words, it does not explain why shifts tend to occur toward an extreme end rather than to the middle (or to a moderate position).

The theory of persuasive arguments (also called information processes) proposes that the group will shift toward the most persuasive position in a collective sense. Since the total amount of information available would be lopsided, the group as a whole would shift toward the initially favored position. Once the group tilts toward one or another position, there will be a shift toward the initial tilt. This means even the minority can offer more persuasive information if members find it to be truthful or if it arouses emotions like fear and anger. Again, the theory of persuasive arguments, while explaining the move to the direction of the initial tilt, doesn't explain the reason for the shift toward one extreme side rather than toward the moderate position. In the following section, I will present a different approach to explaining group polarization.

Visual Neglect

In order to shed some light on the puzzling nature of the tilting and progressive shift toward extremes, I would like to discuss a curious neurological disorder known as "visual neglect." It happens when the right side of brain is damaged, usually by strokes. The patients of visual neglect may behave as if the left side of the visual field is of no great concern or even as if it does not exist: patient may fail to eat food placed on the left side of the dish, may omit to read the left half of each sentence, fail to copy the left side of a drawing, or leave an uncommonly wide margin on the left side of the paper when asked to write (Mesulam, 1985). Yet the patients are usually unaware of the deficit and they often stubbornly deny it (Bartolomeo, 2007). The symptoms of neglect are not confined to the visual sense; rather, they manifest in all sensory modalities, tactile, auditory, olfactory, and gustatory (Robertson & Heutink, 2002). Thus, it can be assumed that neglect is a deficit of the higher level function rather than of the primary sensory or motor functions (Robertson & Heutink, 2002).

One of the most common diagnostic methods is "search/cancellation." In search/cancellation, the patients engage in locating and marking off specific targets, like the letters "A" among a random mixture of letters. It may be done on a sheet of paper by circling or cancelling with a pencil, or on a computer screen by clicking off the target. The patients neglect the left side of the field while not missing any targets on the right side. Paradoxically, if patients are given incentives like one penny for each correct detection, the performance shows significant improvement (Mesulam, 1985). Another interesting fact is that the contrast between the left side and right side is not all-or-none; rather, it has a gradient characteristic. In other words, there is a continually increasing probability of detecting targets further to the right of the visual field (Robertson & Heutink, 2002). The neglect of the left field does not necessarily mean the visual function on the right field is normal. While ignoring the left side of the visual field, some patients will cancel again and again the same items on the right side (Mannan et al., 2005). Thus, the pathological revisiting behavior reveal another side of visual neglect—a preoccupation with the right side of the visual field. Bartolomeo (2007) aptly describes these patients' hyper-attentiveness as their being "magnetically drawn" to the right side.

Another common diagnostic method of visual neglect is line bisection in which patients engage in marking the midpoint of lines. Patients typically mark the midpoint significantly to the right of the actual center. Either patients ignore the majority of the left side of the line or are hyperattentive to the right side of the line (McCourt, Garlinghouse, & Slater, 2000) or both. Interestingly, the rightward deviation may also occur in bisecting number intervals. For instance, when patients are asked to pick the median number between 11 and 19, their answer may be "17" rather than the correct answer "15" (Zorzi et al, 2006).

This brings perhaps the most interesting aspect of visual neglect. Normal individuals also make mistakes in marking the midpoint of the line. This is called "pseudoneglect." Unlike patients with visual neglect who mark the midpoint of lines far to the right of the actual center, pseudoneglect shows a leftward error in a much smaller degree. McCourt (2001) reports that the error is less than five percent in normal right-handed subjects. The leftward bias in normal persons is likely to result in a collision with objects on the right side such as when navigating manual and electric wheelchairs through a narrow doorway (Nicholls et al., 2010). Of course, like patients, normal persons with intact brains are not aware of the leftward bias in their visual systems.

One of the biggest difficulties in treating patients with visual neglect is that they are not aware of their symptoms (Robertson & Heutink, 2002). The only treatment method for visual neglect that went through long-term clinical trials was the "scanning training" originally developed by Weinberg and his colleagues at the Institute of Rehabilitation in New York (Weinberg et al., 1977). The main technique of the scanning training is the use of a vertical line on the left of the text as an anchoring point. Numbers at the beginning of each line serve as an additional cue to help patients start focusing on the left side of the text. The cues were removed after a series of training sessions. The researchers found a significant improvement, particularly in reading and writing tasks.

Intellectual Neglect in Value Judgment

> There is a tendency for issues to turn into dichotomous choices that mask complexity, subtlety, and ambiguity.
> -Timur Kuran

One of the most enduring problems for linguists is the fact that, despite the great difference in environmental conditions, all languages share some universal characteristics. Noam Chomsky, a renowned linguist at MIT, suggests that we have what he calls the "language acquisition device" (LAD) and he proposes that humans are born with the innate faculty for acquiring language. For instance, one of the most dominant characteristics all languages possess is the principle of opposition: good versus bad, hot versus cold, fast versus slow, easy versus difficult, clean versus dirty, young versus old, first versus last, high versus low, light versus dark, black versus white, etc. However, the reality children face everyday hardly fits these neatly dichotomous pairs of opposites. For instance, children experience a large variety of thermal sensations. And as soon as children can read the thermometer, they understand diverse temperature readings. Yet, for some reason, children are predisposed to focus on the two opposing concepts—hot and cold—among diverse temperature settings. Of course, some pairs of opposing concepts, such as male-female, boy-girl, do not have intermediate properties, but they are more the exception than the rule. Most pairs of opposing concepts have in-between properties. Children don't simply learn individual words and concepts; rather they seem to discover the system of knowledge. Children seem to be geared toward making sense out of the chaotic experiences and simplifying knowledge into a form based on the principle of opposition. Children's inclination for making things simpler, interestingly, is similar to the pursuit of simplicity by physicists. Some physicists like Einstein and Feynman see simplicity as something intrinsically beautiful. In fact, Chomsky (1965) equates children's language learning process with the effort scientists' effort to understand the underlying laws of nature.

However, a negative consequence of simplifying experiences into two dichotomously opposing concepts is the tendency to ignore the intermediaries. Like patients with visual neglect, we are "magnetically drawn" to the two extreme ends while neglecting in-betweens. I would call this "intellectual neglect," and it is one of the most prevalent dispositions we have. A by-product of this inclination is to over-emphasize the pair of opposites. Some types of intellectual neglect have more serious consequence than others. Perhaps the most dangerous kind of intellectual neglect is to make a value judgment, as in good versus bad. Putting someone or a group of people into either extreme end of the good/ bad continuum can be very dangerous. As Hayakawa (1941) asserts, one of the dominant characteristics of Nazi propaganda against Jews was categorizing into people into two dichotomous opposites using the two-valued orientation. Because humans already possess a predisposition for simplifying things into a pair of two dichotomous opposites, they can be easily swept away by the extreme attitude of a value judgment based on good versus evil.

At the end of World War II, one form of human cataclysm caused

A Critique of Science

by intellectual neglect replaced another. While the mentality of all good versus all bad was used in the partition of races in World War II's era of Nazism, communism's method of splitting up was based on economic principle. Marx's main approach to solving the unfairness between rich and poor was to demonize capitalism. Just as Hitler's hate propaganda provoked people's hatred of Jews, Marx's rhetoric aroused repulsion toward capitalism. Once the capitalistic system had been demonized, communist countries were forced into the restrictive communist economic position. Even with the apparent massive failure of communism's collectivism-based economic system, communist countries weren't able to make flexible and adaptive decisions. Simply put, they were trapped. The lesson of Marxism is that the economic principle based on a pair of dichotomous opposites can result in disastrous consequences.

With the fall of the Berlin Wall, communism's ideological trap, based on intellectual neglect, came to an abrupt end, with the exception of a few countries like North Korea and Cuba. Yet, the world is experiencing other kinds of intellectual-neglect-caused disasters. The never-ending news of terrorist attacks on innocent people around the world demonstrates the potential danger of extremism. Terrorists believe so fervently in their cause that they are willing to sacrifice their own lives. The main cause of extremism is the two valued-impulse that puts the enemy on one end and themselves at the opposite end. People under the sway of this impulse don't see any compromising or more peaceful options. The September 11 attack on the World Trade Center epitomizes the extreme solution they decided on. Just as in fascism or communism, the very source of terrorism lies in human minds that are "magnetically drawn" to an extreme view while ignoring in-between positions. People may develop the mental habit of extremism in various forms: they may take an extreme view; they may hold this view with extreme certainty; and they may propose an extreme solution. They can turn into fanatics who are difficult to reason with.

Intellectual Neglect in Certainty Judgment

> What gets us into trouble is not what we don't know. It is what know for sure that just ain't so.
>
> -Mark Twain

> I want people to realize you shouldn't think you know it all.
>
> -John Templeton

> Absolute faith corrupts as absolutely as absolute power.
>
> -Eric Hoffer

> The greatest obstacle to discovery is not ignorance – it is the illusion of knowledge.
>
> -Daniel J. Boorstin

> Believe nothing, no matter where you read it, or who said it, no matter if I have said it, unless it agrees with your own reason and your own common sense.
>
> -Buddha

The type of intellectual neglect discussed in the last section is essentially value judgment: how good or bad we feel about a person or a group of people. In this section, we will discuss another form of intellectual neglect. It is associated with a judgment of certainty: how confident we are of our judgment.

We often hear someone say or we find ourselves saying: "I am absolutely positive." "I know it for a fact." "There is no way….." From time to time we hear two persons arguing and both are so sure that they are willing to bet money. For some reason, we are less inclined to use the words like "could be," "may be," "possibly," "likely," "highly likely," and "unlikely." And how often do we find out that, following such extreme certainty, claims turned out to be utterly false. And despite repeated errors, we constantly make statements with absolute confidence. I would argue that the propensity for certainty comes from intellectual neglect, that we are drawn to the extreme position of certainty and neglect intermediate positions.

Notwithstanding our predisposition for unwarranted confidence, meteorologists are unusually advanced in being cautious regarding the "degree of certainty" with which they can make forecasts. Their prediction of chance is put into numerical value. For instance, you hear the weather forecaster say there is a "30% chance of rain." What exactly

A Critique of Science

does that mean? Allan Murphy and his colleagues found that the majority of people (mostly college students in Eugene, Oregon) misunderstood what "30%" means (Murphy et al., 1980). And twenty five years later, Gigerenzer et al. (2005) report that most people in European cities—Amsterdam, Athens, Berlin, and Milan—misunderstood the meaning of a "30% chance of rain." The usual mistakes involve translating this into the statement that it will rain tomorrow "30% of the time" or "in 30% of the area" (Gigerenzer et al., 2005). What meteorologists are trying to convey in stating "30% chance rain" is that, in 3 out of 10 times, there will be at least a trace of rain (Gigerenzer et al., 2005). Dr. Robert Mureau of the Royal Dutch Meteorological Institute was quoted as saying that the general public does not understand meteorologists' probability prediction very well (Gigerenzer et al., 2005). The cause of misunderstanding may be related to the usage of a numerical value in predicting the chance of rain. It is much less likely to misunderstand if numerical values are accompanied by a verbal explanation: 10-30% with "almost none," 30-70% with "possible," and 70-90% with "a fair chance" (KNMI, 2002).

Let me give a historical example of what Mark Twain means by getting into trouble because of erroneous certainty. During the height of the Cold War on May 1, 1960, Eisenhower was informed that the U-2 spy plane that flew deep into the Soviet Union didn't return. Among options from denial to admission, the Eisenhower administration chose an unnecessarily risky action based on the assurance of top CIA officials. Allen Dulles, the CIA director, and Richard Bissell, the deputy director, confidently told Eisenhower: "the Soviet military could not shoot a U-2 down; that even if they did they would never make such an event public; that even if they did, a pilot would never survive, and so the United States could plausibly deny responsibility" (McDermott, 1998). Following their recommendation, the Eisenhower administration issued a report that a weather plane had veered off course. However, the U.S. was embarrassed when Khrushchev released photos of the pilot, Francis Gary Powers, and the pieces of the plane that was shot down over Sverdlovsk, 1,200 miles inside of the Soviet Union. Khrushchev exploited the incident by cancelling the Paris summit conference. This debacle of the Eisenhower administration was largely caused by their absolute confidence in a number of assumptions and so they opted for the cover-up. The real source of embarrassment was not the downing of the U-2 itself, since both superpowers were engaged in spying on each other, but the plot to conceal what actually happened. Khrushchev happily exploited the U-2 incident in propaganda while waging the cold-war. Eisenhower later recalled: "Bissell and Allen, this thing that the plane could *never* (emphasis added) recovered....if he had tried to parachute at 70,000 feet, he'd *never* (emphasis added) survive" (McDermott, 1998).

Let us look at an area with a completely different implication: the criminal justice system where confidence in witnesses has long been a dominant factor in determining a person's guilt or innocence. The Supreme Court in the United States acknowledged in 1972 (Neil vs. Biggers) that the certainty of a witness could be used in evaluating the reliability of perpetrator identification. Other countries like Australia, Canada, and England also weigh a witness's confidence heavily in determining testimonial accuracy (Krug, 2007). Then in the 1990s came the exoneration of wrongfully convicted defendants by newly available forensic DNA testing. By August 1999, sixty-seven wrongfully convicted defendants were vindicated by DNA testing. The majority of these vindications were achieved by the pro bono Innocence Project at the Benjamin Cardoza Law School in New York City. The project was founded by law professor Barry Scheck and defense attorney Peter Neufeld. Mainly using law students to investigate the innocence claims, they exonerated 37 prisoners. As the cases of DNA-supported exoneration began to accumulate, U.S. Attorney General Janet Reno launched a study of the causes of miscarriages of justice. What Wells et al. (1998) found was that a whopping three fourths of wrongful convictions involved mistaken eyewitness identifications. The most disturbing aspect of the finding, perhaps, was that in every case the mistaken witnesses were extremely confident.

In a controlled setting of laboratory experiments that evaluates the relationship between confidence and accuracy, Krug and Weaver (2005) asked participants to mix together eight common cooking ingredients according to a recipe. The subjects were returned for memory tests after five minutes, one week, or two weeks. They were asked to identify products through multiple-choice or to recall by filling in the blank. In addition, participants were asked to choose among five options (0%, 20%, 40%, 60%, 80%,100%) in rating how confident their identifications were. Krug and Weaver (2005) found that the confidence level was unrelated to accuracy. And most studies of confidence and accuracy show the relationship to be relatively weak or nonexistent (Krug & Weaver, 2005).

Then where does the unwarranted certainty come from? In his book *On Being Certain: Believing You Are Right Even When You're Not*, neurologist Robert A. Burton argues that the feeling of certainty comes from mental sensation rather than evidence of fact (Burton, 2008). According to Burton, the mental sensation of being certain comes from the primitive area of brain, and it is independent of the area involved in conscious reasoning. The primitive area of the brain that gives the feeling of certainty is the limbic system and its main function is a quick response.

The Confounding of Value Judgment and Certainty Judgment

As if the complications mentioned in the last two sections are not enough, there is another problem involving value judgment and certainty judgment. It is the mixing the two. Suppose you were a juror deciding a person's innocence or guilt in a horrible crime, let's say murder. You hear testimonies from many witnesses, detectives, and experts. You are quite upset and even angry because of the horrific incident. In your mind, there are two separate judgments. One is a certainty judgment in which you try to determine whether the accused person is guilty or not. Another is a value judgment about how bad the crime is. It is not easy to separate the two judgments taking place simultaneously in your mind. Your certainty judgment can be overshadowed by your value judgment. The confounding of the two types of judgment can also happen when you admire the person you are judging. Consider Ed Wilson's comment in *Charles Rose* that everything Darwin said was right. Everything? As we shall see later, that statement is absolutely wrong. Another example comes from the findings of Peters & Ceci (1982) in which the value judgment about author's names and institutional affiliations trumped the intrinsic merit of articles. Thus, too little respect or too much respect can interfere with optimal judgment: You may have a different standard of the burden of proof as a result.

The Trapping Mechanism and Cultural Hijacking

One reason why public opinion is a source of social pressure is that individual trying to enhance the credibility of their chosen public preferences show approval of people who have made the

> same choice and disapproval of people who have made other choices.
>
> -Timur Kuran

> ....first, the pressure groups will impose substantial penalties on anyone challenging their positions, and second, by definition, a nonactivist with respect to any given issue is someone whose pertinent expressive needs are relatively weak. To keep matters simple, I assume that all nonactivists find it personally advantageous to join one pressure group or the other.
>
> -Timur Kuran

> Who were the prisoners and why couldn't they escape?
>
> -Irving Janis

> ......men are generally more honest in their private than in their public capacity, and will go greater lengths to serve a party, than when their own private interest is alone concerned.
>
> -David Hume

Besides the psychological effects of intellectual neglect, there are other problems with developing extreme positions. One is that those who take extreme end positions with high emotion have a greater motivation. They tend to spend much more energy and time engaging in debate and rallies. Moderates, on the other hand, are more likely to stand on the sideline. Unlike the radicals at both ends who express their views with utmost certainty, moderates are more cautious. Ironically, however, their prudence is a disadvantage in convincing others. Furthermore, they are caught in the middle between the two extremes that both consider moderates their enemies. As time goes on, the moderates in the middle began to defect toward either end. This is known as "group polarization," as mentioned earlier. Here are the paradoxes of group polarization: First is that the people who are most emotional—or emotionally hijacked—are likely to become the leaders and the voices for the group. The second is that the people with greater certainty are likely to overpower people with prudence. The third is that many moderate options are being marginalized or stigmatized so that only the extreme solution becomes acceptable.

A Critique of Science

Once such a norm has been established, it is hard to get out. In essence, it becomes a trap. Here is why. Suppose you were invited to your friend's house for a dinner; would you tell him or his wife what you honestly think if the food is not very good? It is likely that you might even compliment the food you disliked when you leave. We don't want to hurt the hosts' feelings and we don't want to become unpopular. This sort of misrepresentation is a part of social etiquette and it is something we all do. However, the flipside of misrepresentation is that it can have an enormously negative impact in society. Timur Kuran, Professor of Economics and Political Science at Duke University, calls it "preference falsification," and he investigates the negative aspect of misrepresentation in politics in his *Private Truths, Public Lies*. His example of negative consequence of preference falsification is the inflexibility of communism despite the apparent failures. Kuran (1995) reminds us of the points made by many dissident writers who stated that communism would collapse instantly if ever their fellow citizens stopped falsifying their political preferences. For instance, Aleksandr Solzhenitsyn (1974), a Russian writer who won the Nobel Prize in Literature, argues that in the Soviet Union the lie is the vital link holding everything together. However, it is not easy for anyone who lives inside the Iron Curtain to frankly express what they feel. In East Germany, about 300,000 people were informed on or reported to the secret police (Kramer, 1992). And in Czechoslovakia, 20,000 secret police officers and 150,000 informers were on the payroll of the government. An irony was that people actually joined the privileged ruling class by spying on, and destroying, their neighbors, friends, and co-workers in order to earn the right to be left alone (Kuran, 1995). After the fall of the Berlin Wall in 1989, a Soviet citizen was able to admit to having had to wear six faces: "one for my wife; one, less candid, for my children, just in case they blurted out things heard at home; one for close friends; one for acquaintances; one for colleagues at work; and one for public display" (*Economist*, December 23, 1989). Such pretension can also be seen in religion. Czeslaw Milosz, a Polish dissident and a Nobel Laureate, compares the feigning in religion with politics in *Captive Mind*. Milosz (1953) calls an attention to a Persian word *ketman*, which means the enthusiastic faking of piety by heretics in order to avoid censure or punishment. Once such strong social pressure has been established in society, it is very difficult to reverse. The endurance of the Indian caste system illustrates the near irreversibility of established culture. Thus, emotional hijacking turns to social hijacking which then becomes "cultural hijacking."

Individualism and Collectivism

The discussion of communist doctrine requires a basic understanding of the two opposing concepts: individualism and collectivism. "Individualism" is defined in Webster as "a doctrine that the interests of the individual are or ought to be ethically paramount," whereas in Wikipedia "collectivism" is described as "a term used to describe any moral, political, social outlook, that emphasizes the interdependence of every human in some collective group and the priority of group goals over individual goals." For a culture like that of contemporary Americans, individualism is not only a good thing, it is a quintessentially American thing (Oyserman, et al., 2002). On the other side of the Atlantic, soon after the French Revolution, Burke (1790/1973) was worried about the negative aspects of the rising tide of the individual rights movement. He feared that individualism would soon make the community "crumble away" into dust and power. Meanwhile, Max Weber (1930) contrasted Protestantism's individualism that enhanced the development of capitalism with Catholicism's collectivism.

The economic side of the individualism-collectivism debate would provoke one of the most enduring and bloody conflicts in the history of mankind. On the side of the individualistic economics, Adam Smith represents the spokesperson for the capitalistic economy. He believed that self-interest and competition are the driving forces for achieving better conditions, not only for individuals but for the society as a whole. On the opposite side, Marx epitomizes the collective approach to economics. He detested the inhumane conditions workers endured under capitalists who were driven by maximizing profit. Marx's drastic solution was to completely eliminate individualistic capitalism and to build a utopian society of collective ownership. Marx's main role was the promotion of a war-like mentality pitting the economic approaches of individualism against collectivism. He took the two-valued orientation of economics to the extreme end by arousing emotional reactions in the activists.

The issue of individualism-collectivism, however, is a lot more complex than the simple two-valued orientation. Perhaps one of the first

A Critique of Science

studies that showed the multi-valued orientation of individualism-collectivism was put forward by Geert Hofstede, a Dutch cultural anthropologist. He analyzed various cultures based on five dimensions, one of which was their place on the continuum between individualism and collectivism. Hofstede (1980) rated 58 countries based on a scale from 1 to 100, with 100 being the most individualistic. As you might have guessed, the United States ranked the highest with 91 followed by Australia's 90 and the United Kingdom's 89. Guatemala was the lowest with 6 and its nearby countries were not too far behind (e.g., Ecuador, 8; Panama, 11). Asian countries ranged from high (India, 48; Japan, 46) to low (Taiwan, 17; South Korea, 18). And African countries fell between high (Egypt, 38) to low (Sierra Leone, 20; Nigeria, 20). The differences among countries, of course, do not take into account the variations within a country. Hui (1988) found that individuals within a single country may respond with different attitudes of individualism-collectivism, depending on different types of relationships with spouses, parents, neighbors, or coworkers. Thus, individualism and collectivisms attitudes are not mutually exclusive, but context-dependent. In other words, the human mind is not fixed to a specific orientation, but is flexible and adaptive. Triandis (1995) suggests that most people start by being collectivists as a part of their families and gradually learn to detach themselves from the collective. There is also a general trend in human history from collectivistic hunter/gatherer society to individualistic modern society (Triandis, 1995; Yang, 1988). It is ironic that the higher the population density, the greater the preference for individualism over collectivism. People living in a same apartment building for many years may hardly know each other, let alone having the feeling of collectivism.

Let us now get into a discussion of perhaps one of the most divisive and contentious issues in human history, the social and economic policy of individualism-collectivism. People have been pitted against people and country against country in an attempt to force their views on others. The very first thing we have to keep in mind regarding the social and economic policies of a nation are the sheer number of them. For instance, the US Federal Government has fifteen departments: USDA, DOC, DOD, ED, DOE, HHS, DHS, HUD, DOJ, DOL, DOS, Department of Treasury, DOT, and VA. And each department consists of a large number of subunits for a specific domain, each with differing needs and circumstances. What is the best solution for one department may not be the same for another. What is the best method for one subunit of department may not be the same for anther within the department. One policy that may have been the best for a certain time may not be the best in another time. Like people choose a variety of position on a scale from 1 to 100 on the continuum between individualism and collectivism (Hofstede, 1980), the decision-makers in the government at various levels must have flexibility and adaptive capacity to offer the optimum policies depending on circumstances. Government officials shouldn't be forced to make decisions based on various ideological pressures, whether individualism-orientated or collectivism-orientated. Unfortunately, however, the reality is that individualism-collectivism rhetoric dominates the political arena. One may be accused of being socialist for promoting a certain policy. A particular orientation within the individualism-collectivism dimension should not be considered as intrinsically bad.

What Marx did was to take the issue of individualism-collectivism to the most extreme end. Marx's mottos of the "class conflict" and "class war" symbolize his mindset and his confrontational attitude. The enormously complex problem was simplified into the two-valued orientation: individualistic capitalism versus collectivistic communism. The utopian society of communal sharing is to be achieved by the elimination of the evil force of individualistic capitalism. The main problem of Marx's Manichean worldview of economic and social policy is that it loses all flexible and adaptive capacity.

The French Revolution and the Domination of Extremists

> What is objectionable, what is dangerous about extremists is not that they are extreme but that they are intolerant. The evil is not what they say about their cause, but what they say about their opponents.
>
> -Robert F. Kennedy

> .....after each crisis the victors tend to split into a more conservative wing holding power and a more radical one in opposition. Up to a certain stage, each crisis sees the radical opposition triumphant....
>
> -Crane Brinton

A Critique of Science

> Not only is the paranoid position the single most important threat to any existing democratic society but also overcoming the paranoid position is an essential—if not the essential—circumstance for the *establishment* of democracy. Going beyond the paranoid position must be accomplished on both the personal, psychological level and on the social level.
>
> -Eli Sagan

In the late 18th century, two historic "sister revolutions" took place (Dunn, 1999). One was the American Revolution in 1776 and the other was the French Revolution in 1789. Well-known slogans for the American Revolution include the following: "Give me liberty or give me death." "No taxation without representation." "United we stand, divided we fall." The slogans for the French Revolution were "Liberty," "equality," and "fraternity" (brotherhood). The sister revolutions had two phases with completely different icons. The main icon of the first phase of the American Revolution is the Declaration of Independence while that of the second phase is the American Constitution. These two great symbols of the American Revolution can be considered some of the most significant political documents in the history. The French Revolution, on the other hand, consists of the "good phase" and the "bad phase" (Michelet, 1967). The good phase began in 1789 when, with the main principles "justice" and "equality," the French people united and overthrew the tyrannical monarchy. The good phase of the French Revolution went far beyond the American Revolution in introducing a number of pioneering agendas: popular participation, women's participation, the abolition of slavery, and the inauguration of modern socialism (Barry, 1998). The bad phase started in 1793 with domination by the radical faction called the Jacobins. This period, known as the Reign of Terror, is one in which the climate of mistrust, paranoia, and intolerance dominated. In a little more than a year, it is estimated that up to 40,000 people were executed, including Louis XVI and Marie Antoinette, the King and Queen. However, most victims of the guillotine were simple peasants and workers.

The transformation from the good phase to the bad phase is quite perplexing for three reasons. First, the founding members of the American Revolution had carried out rather successfully and instituted one of the greatest political systems in history. The French counterpart could have made a good use of what the American intellects had achieved. Second, the intellects of the French Revolution also knew the writings of Montesquieu, Locke, Rousseau, and Hobbes, which had been the primary sources of inspiration for the American Constitution. Third, Maximilien Robespierre, a young provincial lawyer from Arras, who led the Jacobins on a murderous spree during the Reign of Terror, seems hardly to fit the profile of a ruthless tyrant. He was known as the "incorruptible" because of his honesty and devotion to right and wrong. As a lawyer representing his clients, he often argued for the rights of man and opposed capital punishment.

One of the guiding principles that made Robespierre take the opposite of individual rights was Rousseau's idea of the "general will." Rousseau believed that freedom lies in submission to the general will and that he who does not submit is unfree and must therefore be "forced to be free." Rousseau essentially took the extreme collectivist position that rationalized trampling on individual rights. Yet Rousseau's literary manipulation makes it sounds plausible. Largely influenced by Rousseau, Robespierre opted for extreme collectivism in the paranoia-filled phase of the French Revolution. Like Rousseau, he made a paradoxical statement in advocating extreme collectivism:

> Terror is only justice: prompt, severe and inflexible; it is then an emanation of virtue; it is less a distinct principle than a natural consequence of the general principle of democracy, applied to the most pressing wants of the country.

The fundamental difference between the second phases of the American Revolution and the French Revolution comes down to a single feature: moderate versus extreme. In the American Revolution, the intellectuals debated and compromised between the basic rights of the individual and the authority of the government. No one was prosecuted for presenting an alternative view. On the other hand, the counterparts in the French Revolution continuously took the extreme position of collectivism. They acquired all the characteristics of totalitarian regimes: fanaticism, intolerance, a readiness for violence, hero-worship, and the tremendous conformist pressures that intimidate all moderates and leave the more violent extremists to set the pace (Methvin, 1973). Anyone who dissented was considered an "enemy of the people." The radicals were better organized and more aggressive, whereas the moderates were excluded from the Convention and brought under arrest (Brinton, 1965). In the American Revolution, the cool heads prevailed. In the French Revolution, the hot heads guillotined the cool heads. They declared war on Austria even though they were in no position to fight one. The radicals in the French Revolution even introduced a new calendar, with 1792 being Year I. The Reign of Terror ended with guillotining Robespierre—the very person most responsible for the thousands of guillotined heads. The tragic end of Robespierre may have ended the bad phase of the French Revolution; however, the world would continuously witness calamities in the name of the "will of people." The main driving force behind the catastrophe is intellectual neglect that makes them drawn to extreme positions and ignore or reject moderate positions. As they see the world with extreme outlooks with absolute certitude, they offer extreme solutions.

Obedience to the Authority: Commonness of Evil

> Obedience is as basic an element in the structure of social life as one can point to.....For many people, obedience is a deeply ingrained behavior tendency, indeed a potent impulse overriding training in ethics, sympathy, and moral conduct.
>
> -Stanley Milgram

In1960, a team of Israeli Mossad agents abducted Adolf Eichmann in Argentina and brought him to Jerusalem. Eichmann was known as the "architect of the Holocaust" and was tried on 15 criminal charges including crimes against humanity. He was found guilty and executed by hanging in 1962. The sensational event brought a pair of unanticipated consequences that pointed toward the same direction: averageness of evil. One was brought by Hannah Arendt, a Jewish writer who escaped Germany just before Hitler's rise to power. In her report for the *New Yorker*, she used the famous phrase the "banality of evil" in arguing that the great evils in history including the Holocaust are executed by ordinary people who think their actions are normal. She was struck by Eichmann's average personality, which did not manifest the presumed psychopathic features of Nazi criminals. Eichmann claimed that he was simply following the orders of his superiors. Arendt argued that Eichmann had done the evil act not because of sadistic anti-Semitic ideology but because he failed to think through what he was doing.

Though Arendt's portrayal of Eichmann understandably prompted a firestorm of criticism, another event also induced by Eichmann's trial would corroborate Arendt's position. It came from Stanley Milgram, a social psychologist at Yale, who began an experiment three months after the trial of Eichmann. He designed an ingenious experiment to test how obedient people are to authority. Milgram advertised for subjects in local newspapers who would act as "teachers." Paid $4.00 an hour to conduct a memory experiment, and using an apparatus with dials and buttons labeled "slight shock" and "danger: severe shock," etc., subjects were instructed to deliver electric shocks to a learner if the answers were incorrect. The intensity of the shock would increase each time a wrong answer was given. The learner strapped in to a chair, however, was a confederate actor in the experiment and did not actually receive shocks. Milgram's (1974) interest was how far subjects would go to administer shocks to the victim. A majority of subjects carried on administering shocks even when the learner pounded on the wall: 65 percent of subjects displayed total obedience to the experimenter's commands to apply the electric shock of the maximum 450 volts. In further experiments, as much as 30 percent of subjects would continue to obey the experimenter's commands even when they were told to grasp the victim's hand and force it down to the shock plate. Like Eichmann's claim, the subjects thought that they were doing a good job in following the experimenter under difficult circumstances. One subject's regret after a year captures the essence of the experiment:

> What appalled me was that I could possess this capacity for obedience and compliance to a central idea, i.e., the adherence to this value was at the expense of violation of another value, i.e., don't hurt someone who is helpless and not hurting. As my wife said, "You can call yourself Eichmann," I hope I deal more effectively with any future conflicts of values I encounter (Milgram, 1974).

Bob Altemeyer, a psychologist at the University of Manitoba in Canada who specializes on authoritarian personalities, has this to say:

A Critique of Science

> Sometimes, when I tell people that I study authoritarian personalities, they say things like, "Oh, you mean neo-Nazis and the Klan."…..Most people seem surprised when I say, "No, I study normal folks, not Nazis." Few people, unless they are familiar with the history of fascism, understand that people as ordinary as you and I, and our friends and neighbors, might bring down democracy if the going got tough enough. But *we* are the people who, driven by fear and cuddling in our self-righteousness, could create the wave that would life the monsters among us to power. And once the monsters acquire the powers of the state, their evil will explode (Altemeyer, 1996).

Indeed, too often since the Holocaust evils explode over and over. The genocides in Cambodia, Rwanda, Kosovo, and Sudan make one wonder if the slogan "never again" will ever become the reality.

Hitler's Personality and His Rise to Power

> If it was madness, what made his mind become that way? This question poses a challenge to psychologists and historians alike, who will have to determine the contours of Hitler's personality and sort out the extent to which his behavior was an expression of the idiosyncrasies of his own personality or a reflection of *universal* (emphasis added) aspects of human nature.
> -Edleff H. Schwaab

> …..when people are scared and angry, facts and figures alone are not very convincing.
> -Elliot Aronson

Among the many thing Hitler did, his hatred and murder of Jews particularly stand out and reveal not only his extreme evilness but also his utter irrationality. To give an example, in 1944 the once-mighty German Army was fighting desperate battles against the onslaught of the Russian Army. Needless to say, the German army needed all the help it could get. Yet transporting Hungarian Jews to Auschwitz was given priority over moving war materials and troops. Where does such an illogical and extreme hatred come from? This brings a potentially disturbing question: If Hitler's followers like Eichmann were normal people, then what about Hitler himself. Is there some quality (or qualities) unique to Hitler that made him what he was and caused him to do what he did?

The very first serious attempt to analyze Hitler was initiated by Colonel William J. Donovan, the head of the Office of Strategic Services (OSS), which later became the CIA. Donovan asked Walter C. Langer, a Harvard psychoanalyst, to write a psychological profile of Hitler. The standard practice of psychoanalysis is to trace childhood for any abnormalities, particularly in the relationship with parents. Langer identified the abusive father and the overprotective mother as the main sources of Hitler's sociopathic mind. Langer writes of Hitler's father: "He is brutal, unjust and inconsiderate. He has no respect for anybody or anything. The world is all wrong and an unfit place in which to live. At such times he also plays the part of the bully and whips his wife and children who are unable to defend themselves. Even the dog comes in for his share of his sadistic display." Langer went on to say: "The result is that the whole world appears as extremely dangerous, uncertain and unjust as a place in which to live and the child's impulse is to avoid it as far possible because he feels unable to cope with it." Langer describes the relationship between Hitler and his mother as an "extremely strong attachment" due in part to the fact that she had lost two or three children before Adolph. Langer portrays their unusually strong bond: "Life with his mother during these early years must have been a veritable paradise for Adolph except for the fact that his father would intrude and disrupt the happy relationship." Langer put his finger on the abnormal tripartite relationship as the main cause of Hitler's sociopathology: "little boys frequently fantasize about ways and means of *ridding* (emphasis added) the environment of the intruder."

Since the end of World War II when the immensity of Hitler's madness became obvious to the world, psychologists and historians have attempted to get into Hitler's mind. However, there is hardly any agreement about what motivated him to such monstrosity (Schwaab, 1992). Some psychoanalyst like Alice Miller who specializes in child abuse believes that Hitler's father was a classic child beater and should be held responsible for creating a monster (Miller, 1983). Edleff H. Schwaab (1992), a clinical psychologist who grew up in Nazi Germany, offers a different assessment:

> Contrary to these assumptions, Aloise did not have the reputation of being a mean-tempered brute. The clashes of wills between him and his son were those of two strong-minded and unyielding people. Their conflict may have been selectively violent as Aloise no doubt resorted to giving his son a whack, in line with the prevailing tradition of "Spare the rod and spoil the child." Corporal punishment was a perfectly acceptable means of discipline in Europe at the time.

Fritz Redlich (1999), a Yale psychiatrist, agrees with Schwaab: "One must also consider the reality that, in Germany at the end of the century, physical punishment of children was the rule and not the exception." Hitler's parents were ordinary people who tried to give their children the best care they could (Schwaab, 1992). Historian Bradley F. Smith (1967) adds another view to Hitler's puzzle:

> the young Hitler invites our sympathy, who is a very human little boy and youth, whose chief faults were his laziness and his passion for romantic games. He is someone we all know because we all have felt similar urges and experienced many of the same frustrations.

Robert Waite, a psychohistorian at Brown University who specialize in Hitler and the Nazi movement, came to a frustrating and pessimistic conclusion: "We may doubt that we should ever be able to explain satisfactorily, fully and finally, *why* it was that he did what he did. The question is ultimately unanswerable" (Waite, 1977).

Redlich (1999) identifies five phases of Hitler's anti-Semitism. The first phase is during his childhood in Linz: Hitler had mild to moderate anti-Semitism, like most people in the city. The second phase of his anti-Semitism, in Vienna, was a combination of anger, frustration, and envy, not yet as extreme as that developed after World War I. The third phase grew out of Germany's defeat in World War I; Jews became the main target or scapegoat. One of his principle tutors was Dietrich Eckart, to whom Hitler dedicated the second volume of *Mein Kampf*. Eckart introduced Hitler to Alfred Rosenberg, one of the main architects of Nazi ideologies. Hitler's Jewish conspiracy paranoia may have been the result of Rosenberg's drawing his attention to the *Protocols of the Elders of Zion*, the vicious forgery that argued that the Jews have a diabolical plot to master the world and then destroy it. Hitler completely accepted the authenticity of the *Protocols,* and its contents became an obsession and the main driving force behind his intense hatred of the Jews. The fourth phase was characterized by his anti-Semitic actions after his seizure of power, and the fifth phase was marked by genocide.

During the third phase, Hitler was exposed to anti-Semitic misinformation, which became rampant after the World War I. Like many others, Hitler was eager to blame Jews for the defeat and humiliations suffered during the war. The question is whether Hitler had something that was qualitatively different from others that propelled him to become the monster who brought the fifth phase, the holocaust. Let us look at his personality traits, which some Hitler specialists have addressed: paranoia, hatred, confrontational inclinations, and rigidity (Schwaab, 1992). According to Redlich (1999), paranoid delusions were Hitler's most significant psychological complex. He was particularly paranoid about the Jewish conspiracy aimed at foiling his own design to achieve world dominance for Germany (Schwaab, 1992). Hitler's susceptibility to paranoia was intensified by his high emotionality. Schwaab (1992) says Hitler was "short-tempered and explosive" and this was a "major psychological impairment." His childhood friend August Kubizek describes Hitler as "choking on his catalog of hates" (Methvin, 1973). Waite (1977) points out that the single word "Jew" was enough to set off violent emotional reactions. As far back as 1922, Hitler was suddenly "seized by a sort of paroxysm" in talking about Jews and he promised that when he came to power he would see to it that every Jew in Germany would be killed (Waite, 1977). Hitler's hate also extended to anyone who did not accept his views: "He was consumed by vengeful feelings to punish anyone not willing to support his position. To him, such people were evil, treacherous, and traitorous—deserving of elimination" (Schwaab, 1992). Hitler also was highly confrontational; as mentioned in *Mein Kampf,* "My father did not deviate from his 'never,' and I opposed him ever more vehemently with my 'oh, yes.'" For him, any compromise was an admission of defeat and his relationship to all people was essentially a contest of wills (Schwaab, 1992). Consequently, he could not work in cooperation with anyone (Schwaab, 1992). Hitler's debilitating weakness was the rigidity of his mind: "Hitler developed the habit of screening out whatever did not fit into the framework of his ideas. Feeling compelled to be so selective in absorbing anything he read, he aimed at achieving closure as soon as possible to stop the agony of thought confusion and prevent flooding his mind with complex and unrelated elements of knowledge" (Schwaab, 1992). Rather than seeing his inflexibility as weakness, Hitler was proud of it, calling it a "granite foundation" (Schwaab, 1992).

The source of Hitler's madness and evilness may come down to a combination of mental faculties: paranoia, hatred, a confrontational tendency, and rigidity. If Hitler had a normal level (or even somewhat reduced level) of any one of these four traits, there may not have been the Holocaust. It might be a mistake to look for some crystallized form of evilness within Hitler. Instead, Hitler's faulty decision-making process, which most people experience in a less severe form, might be responsible for the kind of person he was. This position takes the same line as Maslow's (1962) argument about mental patients, that they are not sick, but rather, "cognitively wrong."

Social Hijacking in the Holocaust and Communism

> Some people, that is, never listen to *what* is being said, since they are interested only in what might be called the gentle inward message that the *sound* of words gives them. Just as cats and dogs like to be stroked, so do some human beings like to be verbally stroked at fairly regular intervals; it is a form of rudimentary sensual gratification. Because listeners of this kind are numerous, intellectual shortcomings are rarely barrier to a successful career in public life, on the stage or radio, on the lecture platform, or in the ministry.
>
> -Samuel I. Hayakawa

If Hitler had such potentially debilitating weaknesses, how was he able to become the leader of a nation? Such attributes are almost always detrimental to a person's proper functioning in a workplace or other social setting. Notwithstanding his flaws, Hitler was masterful at manipulating the masses:

Hitler's speeches, his tools for creating power, were impressive masterpieces of polemical rhetoric and political agitation, of emotional appeal and portrayal of strength, of distortion and exaggerations, and of manipulations using dramatic and theatrical performances..... The simplicity and repetitiveness of Hitler's arguments were typical for the demagogic bent of his mind, keenly struggling with good and evil. Life itself was the issue, the worldwide struggle between life-inspiring Germans and life-destroying Jews" (Schwaab, 1992).

Hitler considered the masses inferior, emotional, fickle, forgetful, and in need of guidance like women (Redlich, 1999). He has been called the greatest demagogue in history (Bullock, 1964).

It is tempting to focus on Hitler and blame all the sufferings and deaths of the holocaust on him. Yet, the holocaust consists of two interconnected parts: the contagion and the infection. According to Viereck (1965), the real enigma is not Hitler, but the honest German majority that unleashed the monster. Linholm (1990) believes that an indispensable element of Hitler's rise to power is the Great Depression of 1929. He was able to uplift the spirits of people who were suffering from unemployment and inflation (Schwaab, 1992).

Theodor Heuss, the first President of West Germany after World War II, offered a remarkable insight in comparing Hitler just a year before Hitler became Chancellor of Germany to Marx. He argued that anti-Semitism to Hitler was as important as economics was to Marx. For Marx, history was class conflict; for Hitler, it was racial conflict. Marx considered the bourgeoisie as the enemy of the proletariat; Hitler branded the Jew the enemy of the people. To Marx, the final goal was a capitalist-free society; for Hitler, it was a Jew-free world (Heuss, 1932). The main underlying force behind the shared attributes is the hatred they possessed. In essence, to use Goleman's (1997) expression, they had been "emotionally hijacked." Here is a tragic irony in the domination of communism and Nazism: emotionally hijacked persons use their intellect and communication skills to infect others. Like contagious disease, Marx and Hitler, people who have been emotionally hijacked, could spread their insanities to others. The results were the two greatest cataclysms in the 20th century. I would call it "social hijacking" and it may be potentially the most dangerous and destructive component of the human mind. As much as Marx and Hitler became the slaves of their own extreme emotions, the masses became the extension of the hijacked minds of the two tyrants. The paradox of mankind is, then, that people may follow charismatic leaders who are the least capable of making rational decisions. The greatest danger of mankind is that emotionally hijacked leaders will become contagions and infect the masses. Once social hijacking in completed, the insanities of emotionally hijacked leaders are delegated to the mass. And the hijacked mass passes onto the next generation to become cultural hijacking.

Even without someone like Marx or Hitler, we ourselves constantly deal with paranoia, hatred, confrontation, and rigidity. We often find ourselves engaging in an internal battle between impulses and our rational part. On some occasions, we regret the impulsive choice. At other times, we find ourselves resistant to admitting our own error, to ourselves and to others. We sometimes realize that our conclusion turned out to be utterly wrong and we see the trace of our own paranoia.

A significant part of maturity is the acquiring of the capacity to control emotional hijacking. Thus, what separates most people from Marx and Hitler may come down to a *minute* difference in the capacity to manage urges. The management of our impulses comes from Brodmann area 10, which is enabled by entertaining alternatives in our mind. It gives the capacity to compare and assess more than one possible course of action. It is the seat of our free will. Yet it is an extremely fragile system that can crack under the slight pressure, especially at the hands of skillful communicators who promote hatred. Our challenge is to fortify the fragile system so that we won't be susceptible to becoming an extension of some shady individual.

Orogee Dolsenhe

Chapter Five
The Tripartite Relationship:
Conflict and Cooperation between State, Religion, and Science

Historians often compare different civilizations and attempt to analyze patterns in the rise and fall of nations. In the massive twelve-volume analysis *A Study of History,* British historian Arnold Toynbee presents an overarching perspective on the rise and fall of the world's civilizations. According to Toynbee, the rise of civilizations is the result of a small creative minority overcoming obstacles and responding to challenges. He gives many examples: The Mayans had to prevail over the dense tropical forest. The Chinese had to take control over devastating floods, which repeatedly changed the course of rivers and overran towns and farmland. The Egyptians had to overcome the inhospitable swamps along the Nile River. And the inhabitants of Ceylon had to construct a massive irrigation system because rain fell only on the un-cultivatable side of the island. Toynbee suggests that the decline of civilization is not caused by a loss of control over the environment or by attacks from outside. Instead, it comes from the degeneration of the creative minority and their transformation into the dominant minority. Intoxicated by earlier successes, the creative minority becomes inflexible, prideful, and intolerant. It attempts to hold its power and prestige by force against reason. The now dominant, ruling minority idolizes its achievements and practices, and forces the masses to obey. The downfall of civilization ultimately occurs because the creative capacity of the dominant minority deteriorates and fails to adequately meet new challenges. The question we should be asking is that if the fate of science also fits into Needham's conclusion of the rise and fall of civilization.

Jared Diamond (1997), a geography professor at UCLA, provided a geographer's point of view on the history of civilizations: location, location, location. In his Pulitzer Prize-winning book, *Guns, Germs, and Steel: The Fates of Human Societies*, he argues that the domination of one society over another is caused by geographical and environmental differences rather than cultural or racial ones. The transition from hunter-gatherer to domesticator of animals and cultivator of plants that began about 10,000 years ago is one of the most important events in mankind's history. Those who lived in places where numerous animals and plants could be domesticated would have benefitted not only from increased food production but also from increased technological development such as came with the use of metallurgy, wheels, writing, and the formation of regional groups or states. And those who lived in geographical zones with similar latitudes along the east-west axis (who thus shared similar climates) also gained by easier access to the newly acquired knowledge. They also acquired diseases that originated in livestock, but through the years developed immunity to a wide variety of pathogens. The crux of Diamond's argument is that the domination of the people of the Americas and Africa by the people of the Eurasian continent, particularly by the Europeans, is a consequence of geographical advantage.

Recently, Toby E. Huff (2003), a sociology professor at the University of Massachusetts, Dartmouth, presented *The Rise of Early Modern Science: Islam, China, and the West*. His book attempts to address the enduring question of why "modern" science arose in Europe during the Renaissance and the Enlightenment despite the fact that science was already more advanced in the Islamic world and China in medieval times. In explaining the development of modern science, he highlights the importance of nonscientific cultural domains, such as law, religion, philosophy, theology, etc., and compares nonscientific domains in Europe to nonscientific domains in the Islamic world and China.

A subject that directly relates to the rise and fall of civilizations is the growth and stagnation of knowledge. In certain respects, the rise and fall of civilizations and the growth and stagnation of knowledge are almost synonymous because they typically accompany one another throughout history. Francis Bacon's famous quotation "Knowledge is power" represents the role of knowledge in the rise and fall of civilizations. The knowledge-seeking endeavor has three separate spheres that sometimes cooperate, compete, or dominate one another: science, government, and religion. Depending on the characteristics of their interrelationship, the progress of knowledge is enhanced or stalled. In this chapter, we will discuss the tripartite relationship among three cultures: China, Islam, and Europe.

Did Man Make History or Did History Make Man?

When we see a movie, our main focus is on the actors and

actresses, their actions and emotions on the screen. We are less likely to appreciate the many individuals who play crucial roles behind the screen in creating the magic of the movie experience: the writer, the director, the producer, the set decorator, the costume designer, the musician, the sound effects technician, etc. It is even more difficult to recognize the innovators and risk-takers responsible for much of the progress made by the movie-making industry over the past hundred or so years. Likewise, our understanding of historical events can be dominated by highly conspicuous people at the expense of other anonymous individuals. For instance, our appreciation of the first solo, nonstop transatlantic flight from New York to Paris and the first voyage to the moon are dominated by the most visible personalities: Charles Lindberg and Neil Armstrong. And our grasp of World War II is largely shaped by a few leading players often seen in documentaries: Roosevelt, Churchill, Hitler, Stalin, and Tojo. Historian Thomas Carlyle (1888), using the examples of Shakespeare, Rousseau, Cromwell, Napoleon and others, argues that "[t]he history of the world is but the biography of great men." This view is known as the "great man theory." It obviously overlooks other historical factors, such as economic, social, racial, and cultural ones. Herbert Spencer (1896) argues:

> If it be a fact that the great man may modify his nation in its structure and actions, it is also a fact that there must have been those antecedent modifications constituting national progress before he could be evolved. Before he can re-make his society, his society must make him.

In an accurate understanding of history, the importance of the big picture and underlying features should not be minimized by our tendency to focus on conspicuous and interesting personalities. At the same time, our appreciation of the courage, sacrifice, devotion, and genius of explorers, innovators, and risk-takers should not be curtailed.

Knowledge-Seeking in China

Among many early civilizations in the old world, China possessed two distinctive geographical characteristics that shaped its cultural foundations. First was its remoteness and isolation from other parts of the world (Needham, 1970; Diamond, 1997; McClellan & Dorn, 2006). Second, paradoxically, was the lack of geographical barriers within the country; a wide plain with two large navigable rivers and a relatively smooth coastline made the establishment of a unified government much easier (Diamond, 1997). In comparison, Europe has plentiful natural barriers like mountain ranges, islands, and peninsulas. The unification of China took place in 221 BC under Qin Shi Huangdi, who established the Qin Dynasty. The preceding dynasties (the Zhou Dynasty, 1046-256 BC; the Shang Dynasty, 1600-1046 BC) resembled the feudal system in Europe. Each state enjoyed a great deal of autonomy. A particularly important time for the growth of knowledge in China was the latter part of the Zhou Dynasty, a dynasty consisting of two periods: the first, the Spring and Autumn Period (722-481 BC), and the second, the Warring States Period (475-221 BC). These periods overlap with the period of the One Hundred Schools (551-233 BC), which, as the name suggests, was also a very creative period instrumental in laying the foundations of Chinese thought. As in Italy during the Renaissance, rulers prided themselves on retaining prominent scholars and trying to "outdo one another" in that respect (Lloyd, 2002). During this period, some well-known Chinese philosophies were developed: Confucianism, Mohism, Taoism, Chuang-Tzu's thought, and Legalism. A few centuries later in the Han Dynasty, materialism which paralleled Epicurean physics also appeared. Unfortunately, however, Confucianism, essentially a moral teaching that has little to do with knowledge-seeking, won out over its competitors. The domination of Confucianism would suffocate the progress of knowledge in China for more than a thousand years. How did this happen? To answer this question, let us first briefly examine some basics of Chinese thought.

Confucianism

Confucianism is by far the most influential school of thought in China. Writings on Confucianism are based on the recollections of his disciples; like Buddha and Socrates, Confucius did not leave behind any writings of his own. Unlike Socrates or Buddha, however, Confucius was not the original source of the ideas behind Confucianism. Instead, as he himself said, he was merely a transmitter of general wisdom and traditional values. One of the main reasons that Confucianism became so influential in many east Asian countries—not only in China but also in

Korea, Japan, and Vietnam—was that its emphasis on morality and social relationships demanding obedience and respect for authority. For rulers whose primary concern was social order and cohesion in vast countries containing large numbers of people, Confucianism was an attractive form of thought. Fairbanks & Goldman (2006) explain the reason for the dominance of Confucianism:

> The Confucianists won out over the other schools of Warring States philosophy because they claim to be, and became, indispensable advisers to the emperor. In its broad historical context this meant, as Arthur F. Wright phrased it, that "the literate elite....had entered into an alliance with monarchy. The monarch provided the symbols and the sinews of power: throne, police, army, organs of social control. The literati provided the knowledge of precedent and statecraft that could legitimize power and make the state work.

The imperial court's official adoption of Confucianism greatly strengthened the reverence people already had for Confucius. Criticizing Confucius was almost like attacking one's deity.

Mohism

If Confucius is called the most overrated philosopher in China, the title of most underrated perhaps goes to Mo Tzu, the founder of Mohism. Mo Tzu learned Confucian ideas but rebelled against them, and his thoughts became a rival to Confucianism. He is considered the first true philosopher in China. Although his historical background is quite sketchy and has been the subject of considerable speculation, it is evident that Mo Tzu came from a low social class, unlike most early Chinese philosophers who were high up in the social hierarchy (Graham, 1978). King Hui of Chu refused to meet with him because of his low social status. However, during the Warring States Period, because of his expertise as an engineer and master craftsman, various rulers of small states sought his help in building fortifications, and followers who were mostly craftsman and artisans studied his philosophical and technical teachings. His followers were blue collar workers who were regarded with far less respect than, let us say, Confucians. Those who made a living with their hands were less likely to be educated, less likely to have time to study, and less likely to influence others. Furthermore, once China was unified, the need for fortification was no longer essential. Mohism virtually died out as the once-valued craftsman no longer had an important role to play.

Some of the most unusual features of Mohism are its ethical and political principles. Because of Mo Tzu's lower social background, these were more tuned to the need and welfare of common people. His egalitarian approach, epitomized in such expressions as "everyone is equal before heaven" and "universal love," presented a stark contrast to the Confucian emphasis on a hierarchical view. In addition, he rebelled against the status quo and against traditions that seemed to be wasteful. In particular, he denounced the lavish funerals and lengthy mourning that Confucians promoted. Perhaps the most important contribution is the *Mohist Canon* that spelled out the four branches of knowledge: logic, ethics, mechanics, and the art of disputation. The section on mechanics is quite illuminating because it parallels the development of physics in Europe (Renn & Schemmel, 2003). It presents a concept that is a precursor to Newton's first law of motion, otherwise known as inertia: "The cessation of motion is due to the opposing force ... If there is no opposing force ... the motion will never stop." His unique, hands-on experience as a master craftsman no doubt facilitated the development of such objective and practical knowledge. I am compelled to ask a question: How different would the history of China and East Asia as a whole have been if Mohism had been able to carry on?

Taoism

Taoism's enduring influence is commonly attributed to the fact that Chinese scholars could be Confucian at work but Taoist at home. While Mohism virtually died out after the Warring States Period, Taoism managed to survive by retreating to outside the palace. Lao Tzu is a semi-mystical figure who developed the notion of "tao" which means "path" or "way." Tao is said to be the force that is formless and pervades everything in nature. The way of Tao was to live in accordance with nature. Thus, in contrast to Confucianism, which stressed harmony in human relationships, Taoism stressed the relationship between humans and nature. Taoism may be considered one of the earliest forms of environmentalism. The concept of "chi" meaning "life energy" is another key part of Taoist philosophy. Many other cultures have similar words for similar concepts (for example, "prana" in India, and "entelechy" or "elan vital" in the West). Another well-known concept in Taoism is the yin-yang principle. It stands for two opposing, yet complementary, oscillating elements of nature. The yin-yang principle applied to cosmology explains that Heaven and Earth have been formed by the two fluids, yin and yang. Lieh Tse, a Taoist philosopher elaborates:

> The teacher Lieh Tse said: The sages of old held that the Yang and the Yin govern heaven and earth. Now, from being born out of the formless, form what do heaven and earth take their origin? It is said: There was a great evolution, a great inception, a great beginning, and a great homogeneity. During the great evolution, Vapours were still imperceptible, in the great inception Vapours originate, in the beginning Forms appear and during the great homogeneity Substances are produced. The state when Vapours, Forms, and Substances though existing were still undivided, is called Chaos, which designates the conglomeration and inseparability of things....Evolution is not bound to any forms or limits (cited in *Lunheng* by Wang Chong, 1962).

Chuang-Tzu

Chuang-Tzu is best known for his dream of a butterfly. After waking up from the dream, he questioned whether he was Chung-Tzu who had dreamt he was a butterfly, or a butterfly dreaming he was Chuang-Tzu. This kind of introspection exemplifies Chuang-Tzu's thought, which Burton Watson (1968) sums up with a single word, "freedom." Chuang-Tzu believed that the world did not need to be governed and that social order and harmony arise when things are left alone. Murray Rothbard (1990) calls him the world's first anarchist. Chunag-Tzu also introduced a Chinese view of evolutionary thought: he asserted that life has an innate capacity to transform and adapt to its environment, thus enabling life to change from simple to complex, including into the human. Chuang-Tzu also put forward a Chinese version of skepticism. He cautioned against making hasty generalization and against judging from one's own limited experiences.

Materialism

A Chinese version of materialism also developed in the Han

Dynasty. The most well-known figure of Chinese materialism was Wang Chong (27 A.D.-100 A.D.). As he was born into a poor family and lost his father, he couldn't buy the books he needed to satisfy his hunger for knowledge, so he would stroll in marketplace bookshops and read through the books displayed for sale. His only surviving work is *Lunheng* and his other works (on Macrobitotics, the Censures on Morals, and Government) have all been lost. Wang Chong states in his autobiography that *Lung-heng* has more than a hundred chapters, but only eighty-four chapters survived. He was an uncompromising, nonconforming, original thinker who argued against many of the prevailing assumptions of the time, which was dominated by supernatural and superstitious thinking. For instance, it was widely believed that thunder was a warning given to man from Heaven. Wang Chung set out five proofs to demonstrate that thunder arises from fire or heat rather than from the anger of Heaven. His effort is similar to Thales, an Ionian philosopher, who tried to challenge the prevailing notion that the Greek god Zeus was responsible for thunder and lightning. He repeatedly emphasized that any idea must be backed up by verifiable experimentation. His view of nature and its relationship to man went against the widespread belief in Heaven's will. His view looks a lot like the present scientific approach of unpurposefulness:

> The literati declare that Heaven and Earth produce man on purpose. This assertion is preposterous, for, when Heaven and Earth mix up their fluids, man is born as a matter of course unintentionally (cited in *Lunheng* by Wang Chong, 1962).

He argued against an uncritical acceptance of the words of sages, including that of Confucius, for whom he expressed a great respect. Unlike the lost texts of Mohism, his main work was available, but it had very little influence on Chinese thought.

Legalism

The Warring States Period contrasts with the Spring and the Autumn Period in the way that the larger states forcefully annexed the smaller ones. During this time of upheaval characterized by constant warfare among states, the rulers of the surviving states were amenable to the idea that they should strengthen their control over their people. Those who promoted taking drastic measures to control the masses were known as Legalists, and they served in high positions (such as senior minister or chancellor) in various states. One of the most instrumental Legalist figures was Li Si of Chin, who became a trusted advisor to the thirteen-year-old boy who ascended to the throne after his father's death. While Li Si was a practitioner of Legalism, his rival, Han Fei, was a political philosopher. Han Fei provided an extensive synthesis of Legalist ideas that consists of fifty-five chapters. The fate of Han Fei was like many of Li Si's rivals. Li Si convinced the Chin prince that Han Fei was a threat to the Chin state. Han Fei was arrested and later committed suicide by taking poison sent to him by Li Si. Thus, Han Fei, as brilliant as he might have been in developing totalitarian doctrine, became a victim of the very system he helped to establish.

Because of repeated attempts made to assassinate the Chin prince, the boy-king grew up to be even more receptive to the Legalist's doctrine. A paranoid king with full exercise of brutal Legalist principles, he would unite China and become the first emperor, Qin Shi Huangdi, in 221 BC. His reign also marks the end of the Hundred Schools of Thought. He ordered that most existing books not officially decreed should be burned, that 460 scholars should be buried alive, and that those owning any texts of the Hundred Schools should be subjected to severe punishment. No one will ever know what kinds of treasures were lost forever. Li Si himself became a victim of a false accusation. He was accused of plotting treason and became the victim of the cruelest form of execution devised by Legalists. He reaped the harvest he had sown.

The empire that exercised ruthless control over people lasted only four years after the first emperor's death, largely because the very brutal practices of Legalism initially gave an advantage to other competing states. The Legalist government alienated people who were subjected to forced labor and heavy taxation. The supposedly 10,000 year empire came to an end with revolts all over China. Although the empire died, the totalitarian state of mind it engendered would endure. It merged with Confucianism to form what is known as a Confucian exterior covering a core of Legalism. Professor Zhengyuan Fu at the University of California at Irvine, who was a political prisoner in the 1960s in China, explains its dominance in China:

...the major distinctive characteristic of the Legalist teaching is

> its unabashed insistence on the superordinate status of the ruler and the subordinate position of both his ministers and the ordinary people. The intellectual endeavor of the Legalists is totally directed toward assisting the ruler to maintain his status of domination. This is the major reason that during the subsequent twenty-two centuries of imperial China, most if not all Chinese emperors found Legalist doctrines so appealing….The legacy of the ancient Legalists provided fertile soil for the successful transplantation of Marxism-Leninism into China. It is an inheritance from their ancestors that may well continue to haunt the Chinese people for some time to come (Fu, 1996).

And he ends with this:

> Implementation of the Legalist tenets and the practice of twentieth-century totalitarianism clearly demonstrate the inherent dangers of the well-intentioned social programs championed by vocal intellectual elites…..After all, the road to Hell is paved with good intentions (Fu, 1996).

More than two thousand years later during the Renaissance, a European version of the totalitarian principle appeared in Italy, put forward by Niccolo Machiavelli, a political philosopher. As it was in China, in Italy during a turbulent time smaller Italian city states were threatened by neighbors and foreign powers, and rulers wanted to strengthen their control over their people. Yet Machiavelli's idea of totalitarianism never caught on until the appearance in the 20th century of the fascist regimes in Italy and Germany and the communist regime in the Soviet Union.

The Establishment of the Value System in China: The Palace Examination

The pursuit of diverse philosophical approaches in pre-unified China presents a stark contrast to the constriction imposed on Chinese scholasticism after the unification of China. Once unified, the Chinese intellectual community shifted its focus mainly to Confucian philosophy. To understand this process, we need to understand the establishment of the value system behind it.

One of the predominant factors in determining a society's

direction is the value system it provides. The value system determines who and what kind of knowledge should be promoted and respected. An example of the extent of the impact of the value system on society and culture can be seen in the Imperial China's palace examination system. The palace examination system was instituted in 605 AD by the Emperor of Sui in an attempt to exert power over old aristocratic families; it lasted over thirteen centuries until the end of the Qing Dynasty. Prior to the establishment of the centralized palace examination system, most official appointments were made by recommendations from prominent aristocrats. A major advantage of the new system was that any man regardless of his social status by passing the palace examination was eligible to become a government official. Since the position of a government official was highly prestigious and powerful, passing the examination was extremely competitive: about one in three thousand had the great fortune to pass it (Miyazaki, 1976). The examination tested the mastery of Chinese classics (particularly Confucian classics) and the mastery of classical Chinese calligraphy. An obvious problem with such a lopsided concentration was this: "Science and technology did not figure in the exam system" (McClellan & Dorn, 2006). The value system was geared toward those who could endure what Miyazaki (1976) appropriately calls "examination hell." The educational system that emphasized memorization of classic moral learning (Confucian teaching) and a large number of Chinese characters (more than 400,000) represented "wastage of human effort and talent on a scale vaster than can found in most societies" (Ho, 1967). It was not uncommon for a scholar to have failed a dozen or more times at high-level examinations, which were usually held at three-year intervals (Ho, 1967). Whole lives were wasted in memorizing what we would now consider almost useless knowledge. Such a system would have left almost no room for the reflections on nature seen in mohism, taoism, or other philosophies. In spite of the lopsided value system, China made remarkable technological progress. It is even more remarkable considering the fact that there were very little technological manuscripts available to artisans. The only technological treatise surviving from antiquity is the *K'ao kung chi* or *Artificers' Record*, which deals with the government workshops in the bureaucratic Utopia described by the *Chou li* (Bodde, 1991). Huff (2003) perhaps sums up best the ironic characteristics of the Chinese system: technological progress was made "in spite of, not because of, the official forms of education and examination."

Knowledge-Seeking in the Islamic World

Perhaps no other religion in the world in becoming a major religion achieved such rapid expansion as did Islam. Less than a hundred years after the completion of the *Quran* by the Prophet Mohammad, Islam had expanded over a vast area in three continents, becoming a force that unified diverse races and cultures. The rapid expansion of the Islamic empire was matched only by its swift consolidation of the accessible knowledge of the world. The establishment of the Abbasid Caliphate commenced the Golden Age (also called the Islamic Renaissance) in the middle of the 8th century. The Caliphate moved the Islamic capital from Damascus to Baghdad and provided a remarkable intellectual climate with mottos such as "Seek knowledge, even if it be in China" and "The ink of the scholars is more holy than the blood of martyrs." The establishment of the House of Wisdom, a library and a translation institute, represented the mind-set of the early Islamic Empire. In the early stages, Persian and Indian cultures of knowledge were the main contributors. Of particular significance, the translation of Sanskrit mathematics texts concerning the concept of zero, the decimal number system, and negative numbers played a crucial role not only in Islamic but also in European science. The next stages focused on the translation of Greek knowledge, preserving the works of Plato, Aristotle, and others. Muslim and non-Muslim scholars from all over the world were welcomed. The fact that "Islam made science international" (Jones & Wilson, 1995) may be one of the most important contributions of Islamic science. An era of consolidation and translation was followed by an era of superb creative thinking that brought remarkable advancements to various areas: medicine, mathematics, agriculture, chemistry, optics, and architecture. The tremendous accumulation of knowledge is evident in the sheer numbers and sizes of libraries throughout the empire. For instance, the House of Wisdom at Cairo in the 10th century contained more than two million books (McClellan & Dorn, 2006). However, the high level of productivity produced by Islamic science in the Golden Age was followed by a decline in the 12th and 13th centuries. How did such a magnificent system of knowledge begin to decline? A number of explanations have been proposed: foreign invasions (Mongol and crusades), conflicts between Sunni and Shia Muslims, and economic decline. A popular theory of the decline cites the domination of religious conservatism (e.g., Sabra, 1987; Huff, 2003).

Multicultural Contribution of the Birth of Modern Science

> We ought to be grateful to the progenitors of those who have imparted to us a measure of truth, just as we are to the latter, in so far as they have been the causes of their being, and consequently of our discovery of the truth.
> -Aristotle

> We owe great thanks to those who have imparted to us even a small measure of truth, let alone those who have taught us more, since they have given us a share in the fruits of their reflection and simplified the complex questions bearing on the nature of reality. If they had not provided us with those premises that pave the way to truth, we would have been unable, despite our assiduous lifelong investigations, to find those true primary principles from which the conclusions of our obscure inquiries have resulted, and which have taken generation upon generation to come to light heretofore.
> -Al-Kindi

Most discussions of the history of science start with ancient Greece and thinkers like Socrates, Plato, and Aristotle, and then jump to Renaissance scientists like Galileo and Newton, as if nothing really important happened between the two ages separated by more than a millennium. Only some authors give credit to Islamic scholars for preserving Greek thought. In the discussion of the birth of modern science in Europe, contributions from cultures outside Europe have largely been ignored or marginalized. Hall (1962) makes this point about the Eurocentric focus in the history of science:

> Europe took nothing from the East without which modern science could not have been created; on the other hand, what is borrowed was valuable only because it was incorporated in the

European intellectual tradition. And this, of course, was founded in Greece.

The shift away from a Eurocentric version of the birth of modern science took off with Joseph Needham's *Science and Civilization in China*, followed by Seyyed Hossein Nasr's *Science and Civilization in Islam*. The next text to follow was *A Concise History of Science in India* by Bose, Sen, & Subbarayappa. Recently, the multicultural roots of modern science have been reemphasized by Goonatilake (1998) and Bala (2006). Bala (2006) gives credit to French philosopher Pierre Duhem (1996) for presenting a thoughtful analysis of the origins of modern science. In his massive ten volume published between 1913-1959 *The System of World: A History of Cosmological Doctrines from Plato to Copernicus*, Duhem challenges the general assumption that the birth of modern science occurred with Copernicus, Kepler, and Galileo. Before Duhem's work, the general view supported the so-called "War of the Ancients and the Moderns" in which modern science was believed to have been achieved by winning out over the Hellenistic tradition. Greek thought was viewed as a hindrance rather than as a precursor. Duhem's view of the prominent figures (Copernicus, Kepler, and Galileo) was that "we are watching the triumph, prepared long in advance...." Bala (2006) suggests that Duhem opened the door to examining how far away and how long ago the preparations began; he initiated the way for studies of the multicultural contributions to the birth of modern science by Needham, Nasr, and Bose, Sen, & Subbarayappa.

The Neglect of Experimentation in the Ancient World

Experimentation is considered a cornerstone of modern science. Rather than simply making guesses or following authority, science engages in a process of verifying proposed ideas through rigorous testing. The ancient Greeks, who dominated the intellectual scene of the Axial Age in Europe, disliked experiments. They considered experiment not only irrelevant but also a hindrance to the process of beautiful, pure deduction. They were content with accepting seemingly obvious natural facts as starting points for their reasoning. Plato believed that the physical world is an imperfect and impure copy of the perfect, non-material, and abstract world of "Forms." The difference between the two worlds can be illustrated with the idea of a circle. We have an idea of a perfect circle in our mind. A circle drawn on a sheet of paper, however, is far from perfect. The Greeks preferred the more perfect process of mental deduction over experiment.

There was also another reason for neglecting experimentation. Throughout history, there have been two classes of people. One class included the vast majority of people who made their living with their hands: farmers, artisans, craftsmen, masons, butchers, etc. They were manual workers who had to work long hours just to survive; for these people it was difficult to find time to study or money to buy books. They were mostly illiterate and, consequently, they had far less influence in ancient society. The other class was comprised of a small group of privileged people who didn't have to use their hands to make their living: politicians, administrators, teachers, astrologers, etc. These "white collar" workers generally were richer, better educated, and had a higher standing in society. This small group of people enjoyed higher status and usually looked down upon manual workers, the zero-sum impulse. In ancient Greece, the laboring class was referred to as *banausoi* meaning "cramped in body." In some Greek communities, *banausoi* did not even have full citizenship rights (Bodde, 1991). In democratic Athens in the 5th century B.C., one popular way to attack a political opponent was to accuse him of being the son of *banausoi* (Bodde, 1991). In such a cultural climate, having to dirty one's hands to build an instrument to conduct an experiment was not a preferred option.

In early China when engineering expertise was required to fortify castles, Mohists who were mostly craftsmen and artisans enjoyed a brief period of respect. But when the unification of China made castle fortification unnecessary, this rare respect enjoyed by blue-collar workers virtually disappeared. The social status of manual workers in general had been no better than that in Greece. Mencius considered it the natural duty of the manual worker to feed the brain worker (Bodde, 1991). The most despised blue-collar workers were the butchers and skinners. Influenced by Buddhism and respect for ancestors, the aversion to cutting bodies may have contributed to the peculiar way Chinese medicine put an emphasis on herbal medicine and acupuncture. For practitioners of Chinese medicine, the idea of operating on a patient was as repugnant as the idea of butchers cutting up a slaughtered animal. Even to this date in developed countries like Japan, the stigmatized status of the skinner still persists.

Generally speaking throughout the world there was a disconnection between the white-collar scholar and the blue-collar artisan. The knowledge obtained by the artisan's direct and practical experience was not made use of by scholars. Blue-collar workers without the time and resource weren't able to carry out intellectual pursuits. Furthermore, the illiteracy of artisans prevented them from recording and passing on the knowledge they obtained. The shipbuilders of the Ming dynasty who manufactured an amazing fleet of huge ships, for instance, were illiterate and unable to record their skill (Needham, 1971, cited by Bodde, 1991). The value systems of both Greece and China had a low regard for the artisan class, yet they evolved into completely different forms. The contrast between the different paths taken by the development of knowledge in China and Greece was noted by the Islamic scholar al-Jahiz as early as the 9th century:

> The curious thing is that the Greeks are interested in theory but do not bother about practice, whereas the Chinese are very interested in practice and do not bother much about the theory (quoted in Needham, 1970).

Due to the emphasis on consolidating the knowledge of different cultures, Islamic scholars were in a position to compare the different characteristics of Greece and China and to recognize the lack of balance in each culture. And perhaps because of such realizations, Islamic scholars would overcome the age-old pattern of lopsided knowledge derived from the zero-sum impulse and combine theorizing with experimentation. We will examine how they did that in the next section.

The Scientific Methodology of Islamic Science

A remarkable occurrence in the development of Islamic science was the overcoming of the obstacle to combining the two modes of knowledge-seeking that had been separated by the two classes of people, blue collar and white collar workers. Three Islamic scientists played critical roles in advancing this synthesis: Ibn al-Haytham, al-Biruni, and Ibn Sina. All three scientists possessed artisan skills and were able to produce experimental apparatus to test their ideas. This also happened in Europe in the Renaissance. Galileo and Newton were skilled technicians who were able to construct the necessary equipment on their own.

Ibn al-Haytham, whose Latinized name was *Alhazen*, was born in Basra in present day Iraq in 965 AD. As a devout Muslim, he was troubled by the disagreement between the Sunnis and Shiites over the Prophet's rightful successor. After spending many years studying the doctrines of different sects, he came to the disappointing conclusion that doctrinal disagreements derived from sociological factors rather than from anything having to do with the truth. After this setback, however, he found a new interest in philosophy when he read Aristotle's works. He realized that there were areas of knowledge in which it was possible to determine the truth. His most important work was *The Book of Optics*, which was translated into Latin in the late 12th or early 13th century by an unidentified translator. This work would have a profound influence on European science. In it, he consolidated into a single framework the different approaches of the Greeks (Lindberg, 1976): the emphasis on the physical nature of light proposed by Aristotle and Epicurus, Euclid's mathematical theory, and Galen's physiological structure of the eye. He developed a synthesis of mathematical, physical, and physiological models of perception and was able to account for a wide variety of optical phenomena.

Perhaps Alhazen's most important contribution was the manufacturing of curved lenses and mirrors to conduct investigations. His construction of lenses and mirrors in his lab for the sake of testing ideas is an important advancement. It broke down the customary barrier to obtaining knowledge that separated the blue collar from the white collar worker. Doing artisan work to verify ideas was no longer something to be disdained. Rather, it would become a mark of pride and identify the intellectual elite. The change of value may have a far greater significance than the specific knowledge of optics Alhazen put forward. Once the barrier between the two classes of people had been broken down, others like al-Biruni and Avicenna would follow in putting emphasis on experimentation.

Perhaps due to the frustration of studying the doctrines of different sects, Alhazen also adopted the strict approach that we should only study what can be investigated directly through experimentation and mathematics and avoid what's beyond our experience. In essence, he was not only boosting the value of experimentation but also devaluing the phenomena beyond direct experience, the zero-sum impulse. The rejection of what is beyond reach would be referred as "positivism" and would haunt the world of modern science. Steffens (2007) calls Alhazen "the first scientist." He might also be called the first positivist.

The Establishment of Madrasas and the Shift of the Value System in Islamic Science

In the early years of Islam, the motto "The ink of the scholars is more holy than the blood of martyrs" exemplifies the high value Islamic society put on the scholastic domain. In such a climate, scholars enjoyed intellectual freedom and were even able to express skepticism about religion. For example, al-Rawandi in the 9th century rejected the grand supernatural themes of revelation and miracle (Fakhry, 2004). However, there was a gradual degeneration of intellectual freedom in the Islamic scholastic culture. Let us examine the underlying causes of the decline.

Both Christianity and Islam achieved a high degree of uniformity through completely different processes. In Christianity, there was a centralized body, a governing organization that exerted an absolute power over a large territory that consisted of a diverse group of people. The papacy had institutional dominance and could force people into complete obedience to the major tenets of Christian orthodoxy. Any dissenting views were not tolerated, and the papacy had the power to punish those who did not conform. In Islamic civilization, however, the lack of a centralized religious institution like the papacy in Western Europe made it difficult to achieve a uniform doctrine and religious law. Yet, uniformity was achieved by religious scholars. The process was far more complicated than the Christian process of uniformity because it was enmeshed with political power, religious authority, patronage, and competition among elites, as well with as individual epistemological commitments (Dhanani, 2002). Consequently, it was hard to achieve and took a considerable amount of time.

In achieving uniformity, perhaps the most crucial factor was the establishment of madrasa. The word *madrasa* simply means school in Arabic and it is a precursor to the university. The Fatimid Shia dynasty founded Al-Azhar, one of the first madrasas in Cairo in 975 AD. The main motivation for the creation of an institution higher learning was that education was central to the Islamic virtue, and madrasas quickly spread to other parts of the empire. Madrasas were typically a part of a large complex, usually including a mosque, a library, a teaching hospital, a Sufi convent, a hostel for travelers, and the tomb of the founder (Arjomand, 1999). One of the most important factors in determining the future course of scholars' intellectual direction was how the educational system was financed. To understand the system of financing the madrasas, the Islamic legal system called "wacq" needs to be understood. The origin of wacq can be traced back to the Prophet Mohammed who directed a caliph, Farouk Omar ibn al-Khattab, to make his property inalienable so that the income could forever be distributed as charity (Schoenblum, 1999). The noble effort of the Prophet established the system of the philanthropic foundation in Islamic society. The hospital is one of the most valuable products of the wacq system. By the 11th century, almost every Islamic city had a hospital. The funding by the wacq trust was the main reason why Islam had the most advanced medical technology and services.

The wacq trust typically consisted of a founder (waqif), a trustee (mutawillis), a judge (qadi), and beneficiaries. The Islamic law of waqf had three basic principles that protected the endowment entity: irrevocable, perpetual, and inalienable. On the surface, the wacq is similar to modern trust law. However, the wacq was different from the English trust law in two respects, which would play a critical role in shaping society and the madrasas. First, the wacq derives from religious authority and furthers what are perceived to be religious, pious, and charitable goals (Schoenblum, 1999). Second, for more than a thousand years the law was inflexible (Schoenblum, 1999). In contrast, the English trust is a vaguely defined legal construct that has altered with changing circumstances in a manner that has addressed the individual's and the society's wealth management demands (Schoenblum, 1999). Kuran (2010) argues that, while the wacq provided essential services, its inflexibility made an impediment to the growth of corporations.

The intellectual path of the Islamic Empire would be determined by the marriage of the two innovative systems: madrasa, emphasizing higher learning, and wacq, providing the philanthropic foundation. The credit or blame for combining the two systems that provided financing for the educational complex beginning in 1040 belongs to the great vizier, Nizam al-Murk (Arjomand, 1999). Madrasas were mostly founded by the members of the ruling dynasty who appointed the first professors (Arjomand, 1999). The founder could set the regulations for endowments such as teacher salaries, stipends for students, expenses for maintenance, paper and ink, bread and water, and even sleeping mats (Leiser, 1986). Once instituted, the regulations were not to be changed. The endowed property could not be sold, mortgaged, or leased under Islamic law (Cattan, 1955). The waqif could dictate who could teach and what subjects could be taught (Leiser, 1986). Philosophy often became a victim of the founder's directions, with some founders specifically prohibiting the study of philosophy (Arjomand, 1999). Madrasas did not have any admission requirements, prescribed courses of study, or examinations before the late 19th century (Winter, 1992). This meant that there was a tendency to favor religious commitment over competency. Because of the founders, most endowments were religious endowments, and a religious agenda inevitably dominated the intellectual climate of madrasas. Although the early period had provided a secular and tolerant intellectual climate, scholars increasingly faced charges from religious orthodoxy that their studies, particularly the study of philosophy, logic, and ancient sciences, might be disrespectful of Islam; studies like arithmetic, geometry, and astronomy, on the other hand, useful for determining the time and direction of prayer, had a religious utility and were received positively. Some scholars resorted to concealing their philosophical studies under the guise of some discipline that had better standing (Goldziher, 1981). Gradually and surely, "all forms of knowledge became subordinate" to religious doctrines (Sabra, 1987). Huff (2003) sums up: "Islamicizing of the ancient science had the consequence of setting limits on intellectual innovation, in effect halting the free range of the imagination" and "the utility and usefulness of knowledge is narrowly construed to mean knowledge useful in a strictly religious context."

In the process of Islamicizing madrasas, the most fundamental and significant factor in shaping the intellectual community was the shift in the value system. The once high value put on philosophers' intellect was redirected toward theologians' faith. As time went on, the grip of religious conservatism gradually tightened around the intellectual freedom of scholars. Ibn al-Salah, a prominent theologian in the 13th century, denounced philosophy: "The basis of foolishness and degeneration, a topic of confusion and misguidance which is motivated by perversion and blasphemy" (al-Fasi, 1990). Just two centuries after Alhazen's death, his monumental works were condemned as *fatwa*, a legal opinion on Islamic law by Ibn al-Salah, a prominent theologian in the 13th century. This may explain why most of his works were lost. Out of more than 200 works, only a quarter of them survived.

It is truly ironic that Islamic institutions of higher learning may have actually played a dominant role in the demise of Islamic science rather than helped to promote it. The history of the Islamic Empire and even the world might have turned out very different if the two noble concepts (madrasa and waqf) had not been consolidated or if the waqf system had allowed the flexibility seen in English trust law.

The Conflict between Aristotelian Philosophy and Islamic Thought

The expansion of Islam into a large area meant that Islamic scholars encountered diverse sorts of knowledge, cultures, and religious traditions. From the onset of the Islamic empire, there was a division between "foreign sciences" and "Islamic sciences" that persisted throughout Islamic intellectual history (Saliba, 2007). Among "foreign" sources, Greek philosophy in particular provided intellectual stimulation yet conflicted with Islamic thinkers' own system of thought. The challenge faced by Islamic thinkers in dealing with Greek philosophy was especially great in the subject of God. This was because for anyone educated in monotheistic religions (Judaism, Christianity, and Islam), Aristotle's description of God "would have been strange, perhaps repugnant" (Grant, 2002). Aristotle's notion of divinity as "Unmoved Mover" has a deistic or impersonal characteristic that would have sounded disrespectful to those who considered the divinity an object of worship. Therefore, it was inevitable that the introduction of Aristotelian philosophy would evoke heated debates among Islamic scholars on the subject of divinity. At any rate, as we will discuss later, the very same source of conflict would also play out all over again in the Western intellectual community a few centuries later. Despite the similarity, however, the development of Islamic and Christian intellectual communities would have completely different outcomes.

The foremost contentious issue was what God can do. The monotheistic view was that God had omnipotent power; Aristotelian philosophy put limitations on God's power. The central issue concerning God's role was whether the world is eternal or created. In Aristotelian philosophy, time is eternal and God could not have created at a particular time. Another critical issue was the structure of the world that has causes and effects. If the world operates under the rule of cause and effect, then, the world would be rational that more or less leaves out the intervention of the arbitrary will of God. In the Aristotelian cosmos based on the principles of cause and effect, there is no room for miracles. The conflict between the two Persian scholars, Ibn Sina and al-Ghazali, represents the fundamental problem facing the three monotheistic cultures: Islam, Jewish, and Christian. Let us take closer look at two scholars who never directly met: the greatest Islamic philosopher, Ibn Sina (981-1037), and the greatest Islamic theologian, al-Ghazali (1058-1111).

Ibn Sina

Ibn Sina is better known by his Latinized name "Avicenna." He has been said to be "among the greatest in the history of philosophy" (Maurer, 1962). Avicenna was a rare genius with diverse interests in many subjects. He became a physician at age sixteen, and expanded his interest to philosophy, astronomy, mathematics, logic, chemistry, geology, paleontology, physics, poetry, and psychology. In particular, Avicenna's crucial role in medicine deserves recognition. His massive medical encyclopedia, *The Canons of Medicine,* which contained more than a million words, dominated the curriculum of European medical schools from the 12th to the 17th century and laid the foundation of modern medical science.

Avicenna professed to have read Aristotle's *Metaphysics* forty times without understanding it. Considering his reputation as one of the greatest philosophers, perhaps his confession can be understood as reflecting his experience of a long period of conflict and then reconciliation between the Islamic thought he was raised with and the Aristotelian view he faced later. Many of us, brought up with different types of religions, can probably relate to Avicenna's sense of internal discord because of the two different worldviews—religion and science. What emerged from his long period of consideration was a mixture of Islamic ideas, Aristotelian philosophy, and his own thought. His independence would earn him a branding as an infidel philosopher from al-Ghazali, a scholar to be discussed in the next section.

His endorsement of one transcendent God was consistent with Islamic thought. Attributing God to the existence of goodness could be another way in which Avicenna's thoughts accord with Islamic belief (Arieti & Wilson, 2003). Following the initial agreement with monotheistic religion, he divides humanity into two groups of people: philosophers and the common people. Philosophers can use the Aristotelian approach in attempting to understand the nature of God using reason. For the common people, however, the nature of God may be too difficult to comprehend. The role of prophets is to communicate with the common people in language they can understand, emphasizing morality and social harmony (Arieti & Wilson, 2003).

Al-Ghazali

For those with devout faith, Avicenna's ideas seemed to challenge the sacredness and omnipotent power of God. One such pious person was al-Ghazali who was raised and deeply educated in the Islamic faith. Born two decades after Avicenna's death, al-Ghazali studied jurisprudence and Islamic theology and did not consider himself a philosopher, nor liked to be considered such (Campanini, 1996). One of the most influential Islamic theologians, his reputation is largely built on confrontations and attacks against philosophers. Al-Ghazali has sometimes been acclaimed in both the East and the West as the greatest Muslim after Muhammad (Watt, 1963). His fiercest attacks were directed against philosophers and waged in defense of orthodoxy. His best known work of theological objection is *The Incoherence of the Philosophers*. His main objection to Avicenna and other philosophers who viewed the rational world of cause and effect was this: While philosophers may continually talk about God, they give God no role to perform or a "pseudo-role" with respect to the world (Leaman, 1999). Limiting God's scope of action was what really irritated al-Ghazali (Leaman, 1999).

One of the most astonishing arguments that al-Ghazali made against philosophy is that the rational search for truth is dangerous because it can lead man to think that he is sufficient unto himself (Peters, 1968). Emphasizing the *Quran* and the prophets, he classifies knowledge into two major categories: the sacred and the profane (Dhanani, 2002). Sacred sciences are acquired through prophets. Profane sciences are subcategorized into praiseworthy, tolerated, or blameworthy. Praiseworthy sciences are ones like medicine with clear utility. He considers mathematics a tolerated science as long as it doesn't lead to the blameworthy sciences, whereas metaphysics is blameworthy.

There is another reason that al-Ghazali's idea would soon dominate the Islamic intellectual community. Like the Confucian scholars in China, he allied himself with the reigning political authority, grand-vizier Nizam al-Mulk, which earned him a prestigious position at a Madrasa in Baghdad. He not only professed a sincere loyalty but stigmatized any revolt, even against an oppressive and evil monarch (Campanini, 1996). A large number of students gathered around him at the Madrasa and he became the most influential theologian. Yet he also experienced a crisis of spiritual doubt at the height of his fame. He realized that the high religious standards he preached were not compatible with serving political authority. His concerns were serious enough for him to abandon his prestigious position, and he went into seclusion for several years. Upon his return, he taught at private and state-sponsored schools until his death.

Al-Ghazali is commonly credited with having destroyed philosophy in the Islamic world (Campanini, 1996). Specifically, al-Ghazali's denunciation of philosophy in *The Incoherence of the Philosophers* silenced philosophy in the Islamic world (Arieti & Wilson, 2003). Dhanani (2002), however, points out that the relationship between science and religion was quite complex. He argues that the thesis that al-Ghazali's critique was responsible for the decline of science is a portrayal by historians using the lens of the seventeenth-century scientific revolution in Europe.

The main cause of the decline may lie in the value system endorsed by Islamic intellectuals rather than any in theologians. Ironically, the decline may have started with a good intention. Education was central to the Islamic idea of the common good, which motivated the action of the patrician houses and the policy of the patrimonial state (Arjomand, 1999). The members of the ruling dynasty took pride in the establishment of institutions of higher learning called "madrasa." Unfortunately, however, the value system of madrasa was mainly dictated by the wishes of the founder. Unlike scholars in European universities, Islamic scholars had very little control over the direction of the value system.

The Rise of Universities and the Changing of the Value System

The scientific revolution in Europe that began in the 14th century is commonly associated with the names of great men like Copernicus, Galileo, Newton, etc. To reiterate, our tendency to focus on these famous men may cause us to overlook underlying factors. The rise of European science in the 14th century is a part of the cultural movement known as the Renaissance, which began in Italy and spread to other parts of Europe. The birth of this remarkable intellectual revolution poses a set of puzzling questions: What are the factors that prompted the development of European science? Why didn't this revolution happen before the 14th century? And why didn't it happen elsewhere? Why did it begin Italy? How were the circumstances there different or similar to the circumstances affecting Islamic science? What were the fundamental differences between Islamic science and European science such that they went in completely opposite directions in their knowledge-seeking endeavors? The "opposite directions" I am referring to is this: The high level of productivity that Islamic science reached in the Golden Age was followed by a decline in the 12th and 13th centuries. European science, on the other hand, rose out of the Dark Age and began to grow in the Renaissance and further prospered in the Enlightenment. It is curious that the opposite course occurred despite some common characteristics: First, both societies were dominated by monotheistic religions. Second, the intellectuals raised in these monotheistic cultures had to clash with Greek thought, particularly, the Aristotelian notions of God and nature. Third, both Islam and Christian societies put a great importance on education and created high-level learning centers, madrasas in Islam and universities in Europe.

Christianity's Transformation from the Persecuted to the Persecutor

In the first century at the time of the birth of Christianity, Judaism was by no means monolithic but consisted of numerous sects (Hinson, 1996): the Sadducee (the priestly aristocracy), the Pharisees (the party most zealous for keeping the Law), the Essenes (who emphasized messianic hope), the Zealots (the extreme Jewish nationalists), and the Herodians (who were supported by the royal family). Christianity differed from the other sects in the way that it looked backward from the vantage point of the death and resurrection of Jesus of Nazareth (Hinson, 1996). An even more crucial difference was that Christianity incorporated Gentiles through Paul, who is known as the "apostle to the Gentiles." Initially, Christianity was mainly popular with the lower classes. However, as it began to draw people from the middle and upper classes of the Roman Empire in about 175-180 AD,

Christianity acquired a completely different status (Hinson, 1996). The rapid growth of Christianity throughout the large area of the Mediterranean inevitably provoked violent responses from officials and common people alike in the Roman Empire because Christianity demanded allegiance to a law and throne other than those of Rome. The first major persecution of the Christians by the Romans began with Nero after the great fire broke out in Rome. Although Nero's persecution might be most well-known, the largest scale of persecutions occurred during Diocletian's reign in 303-311 AD. It is estimated that over 20,000 died during the Great Persecution under Diocletian and his successor, Galerius.

With the inauguration of Constantine, Christianity's fate completely turned around after centuries of persecution. Constantine's issuing of the Edict of Milan in 313 marked the beginning of a new era for Christianity. Constantine's edict stressed religious tolerance of all kinds. The tolerance of religion, however, brought unanticipated consequences: fierce rivalries among sects. Bishops attacked each other and populations rose against each other for theological differences. Constantine could not believe such "idle and trivial" speculation could cause so much unrest, and he worried that hostilities would threaten his dream of political unity and stability (Freeman, 2004). He complained: "You [the bishops] do nothing but that which encourages discord and hatred and, to speak frankly, which leads to the destruction of the human race" (Freeman, 2004). One of the main reasons for Christians' different views was that most communities were remote from each other and the varied cultural and religious traditions shaped local theologies (Freeman, 2004). Faced with potentially dangerous divisiveness, Constantine began to actively intervene in religious conflicts to bring stability. In the process of bringing stability, however, Constantine's initially tolerant approach to religion became very strict and repressive. Under Constantine's rule, once persecuted Christianity would become the persecutor.

The East Roman Empire and the West Roman Empire

After winning over his rival and becoming emperor in 284 AD, Diocletian decided the empire was too big to be ruled by one person. Its enormous size posed difficulties in communication and transportation. His novel solution was to divide the empire into two parts. He kept the wealthier part in the east for himself and gave the other part in the west to his general, Maximilan. By the time of Constantine, the vital center of the Roman Empire had already shifted to the east. Constantine had established the capital of the Empire in the east and in so doing further diminished the political importance of Rome. Even without the shift, easterners already enjoyed an intellectual advantage over westerners. While Latin was the dominant language in the west, easterners used the Greek language and upheld the spirit of Hellenism more faithfully than their western counterparts. Like the ancient Greeks, the average easterner had a better education and delighted in speculative thought (Hinson, 1996). Indeed, for many centuries the Latin western world lagged behind the Hellenistic east in virtually every measure: cultural, economic, political, and military (Woodhead, 2004).

Constantine's abandonment of Rome, like his predecessor's, created a power vacuum that would be filled by Rome's bishop, the Pope. There were two other factors that strengthened the Pope's power in the west. First was Rome's authority through its double apostolic succession from Peter and Paul (Manschreck, 1985). In settling disputes, the Pope, as successor of Peter, would have the final say (Freeman, 2004). Second was the further weakening of the state by barbarian invasions. The fractured west was held together by the Pope and the church, which served as a link between diverse people (Hinson, 1996). Consequently, in the west, Pope Gelasius I (492-496) was powerful and bold enough to claim that, in civil affairs, clergy are to submit to the emperor, but in ecclesiastical affairs, the emperor to submit to the Pope (Manschreck, 1985). This sort of thing was unimaginable in the east. This division of power between church and state would serve as a form of check and balance (or separation of power) in the west.

In the East, the relations between church and state were completely different. In the wake of Constantine's involvement in religious controversies, the pattern of the subordination of ecclesiastical affairs to imperial control had been laid down. The consequence was a church made into a virtual department of the state apparatus (Manschreck, 1985). Unlike the counterpart in the west, there was no clear distinction between religion and politics (Freeman, 2004). The emperor controlled elections of the patriarch and conciliar decisions of the church (Manschreck, 1985).

Orogee Dolsenhe

The Regression of the Greek Knowledge and the Dark Age

While the status of the church in relation to the state differed drastically in the two empires, both east and west shared a common contempt for secular scholastic activities. This was a complete turnaround of the Greek emphasis on curiosity and rational approaches. Greek philosophy had become the regular target of some Christian writers long before the Roman Empire embraced Christianity. Perhaps the most influential such figure was Tertullian. Like al-Ghazali in the Islamic empire, Tertullian was most responsible for destroying philosophy in Europe. As the first early Christian writer to write in Latin, he condemned heresy as a consequence of the influence of philosophy: the Gnostics derived from/were influenced by the Platonists and Marcion's doctrine of God derived from/were influenced by the Stoics (Gaukroger, 2006). Ironically, however, like many early Christians, Tertullian had been converted from Neoplatonism (Gaukroger, 2006). Essentially, converts abandoned a philosophical system of inquiry in favor of a theological one preoccupied with the perils of worldliness (Gaukroger, 2006). In the treatise *On the Soul*, he denounced philosophers as "patriarchs of the heretics" (Hinson, 1996). He said "What does Athens have to do with Jerusalem" and he called Aristotle "Wretched."

Tertullian's intolerance also extended to other Christian sects and possessed some unfortunate elements. First, in *To Scapula* 2, he considers religious liberty a right: "a fundamental human right, a privilege of nature, that every person should worship according to his or her own convictions. One person's religion neither helps nor harms another religion. It is not in the nature of religion to coerce religion, which must be adopted freely and not by force" (cited by Hinson, 1996). Second, his works were influenced by the Stoics. Third, Tertullian's anti-heretical attitude largely derived from the Roman legal system, which had been used to prosecute Christians (Versluis, 2006). While Tertullian decried the unjust conduct of Romans against Christians, he used the same type of intolerance to attack fellow Christians with differing views. Fourth, the word "heresy" comes from *hairein*, a Greek word meaning "to choose." A "heretic," then, is one who chooses, who therefore exemplifies freedom of individual thought, and by implication, who does not accept at least some of the doctrines of the corporate Church (Versluis, 2006). Yet through the works of Tertullian and others, the innocuous word became associated with demonic influences or with the devil (Versluis, 2006). The demonization of heresy led to the hideous treatment of heretics for centuries.

As Christians' power increased with imperial assistance, Christians' tolerance of Gentiles non-Christians decreased (Hinson, 1996). St. John Chrysostom, archbishop of Constantinople, wrote that Christians should "restrain our own reasoning, and empty our mind of secular learning, in order to provide a mind swept clear for the reception of divine words" (Lim, 1995). In 529, Justinian closed the legendary Athens Academy which was founded by Plato. Additionally, he offered professors the options of baptism or exile. Seven chose exile. Greek philosophy was suppressed; it did not simply fade away (Freeman, 2004). One of Augustine most famous Christian writings advocates "nothing short of a total translation of all philosophy into Christian terms" and attempts to show that Christianity is able to answer all the questions of classical metaphysics (Gaukroger, 2006). In *Confessions*, Augustine describes curiosity as "lust of the eyes" and groups it together with "lust of flesh." He even saw the "disease of curiosity" as no different from perverse entertainments, such as those featuring freaks or mutilated bodies. In a similar tone, Emperor Leo I advised bishops to avoid difficult questions beyond human grasp: "Clever theologians soon become heretics" (Freeman, 2004). Pope Gregory the Great also warned those with rational minds against seeking causes and effects in the natural world (Freeman, 2004). Explaining earthquakes by natural phenomena rather than attributing them to God's command was considered a heretical approach (MacMullen, 1997). By the mid 5th century, it was no longer possible to have an open-ended discussion as to the nature of god, and Aristotle vanished from the western world with the exception of two works of logic (Freeman, 2004). With the intellectual atmosphere dominated by religious dogma that emphasized following authority and rejecting inquiry, it was inevitable that the Dark Ages would descend on Europe for centuries.

Yet there was a sudden intellectual awakening in the so-called Renaissance (Knowles, 1962). How did Western Europe break free of a seemingly impregnable intellectual trap? There must have been something very crucial that enabled it to escape the straightjacket.

Orogee Dolsenhe

The Establishment of Universities and the Reconfiguration of the Value System

Unlike the Byzantine or Eastern Roman Empire's maintenance of sovereignty and order over a vast territory and a large number of subjects for many centuries, the Western Roman Empire's unity and infrastructure dissolved due to invasions. Its fall in the 5th century marks the beginning of the period known as the Middle Ages, which lasted nearly a thousand years until the onset of the Renaissance. Although for many historians this long period of religious domination, called an "age of faith," was "static and unprogressive," Haskins (1927) argues that the institutional and systemic innovation of the last two centuries of the Middle Ages played a crucial role in bringing about an intellectual revolution. He appropriately calls it the "Renaissance of the 12th century." What were the institutional and systemic innovations? At their heart was the distinctive legal principle, by and large a result of the increased power of the church, that put restraints on secular authorities. By emphasizing legal supremacy, the church declared itself free from the influence of emperors, kings, and princes. In the process, it created its own canonical law which, in turn, stimulated the development of a parallel but secular legal system. Perhaps the most important component of the legal system that developed in the 11th century was the notion of treating a collective entity or group of people as an artificial personality called a "corporation." In essence, by declaring itself autonomous from secular authorities, the church jump-started the same for secular organizations such as guilds, associations of merchants, trade unions, etc.

Many autonomous collective entities that emerged as a result of the legal revolution would bring drastic transformations in intellectual, social, political, and economical realms in Europe. But the formation of universities would have the most profound impact upon the pursuit of knowledge. A remarkable thing about the rise of European universities is that they "just grew" slowly and silently without a founder or fixed date (Haskins, 1927). Historians generally give credit to the University of Bologna for being the first European university at the end of the 11th century. At Bologna, a northern Italian city, a group of zealous students not only from Italy but also from beyond the Alps gathered around famous teachers to learn grammar, rhetoric, and logic. In particular, the teaching of Roman law (the Justinian code) that had been rediscovered in a library in Italy was one of most important attractions.

A completely different set of circumstances from those involved in the rise of Islamic learning institutions, madrasas, precipitated the rise of European universities. Unlike madrasas, which were established by religious endowments, a collective body of students trying to protect themselves against local establishments created universities in Europe. Madrasas had the support of religious endowments; students were not burdened with paying tuition and room and board. In Europe, however, university students had to come up with their own money. Consequently, they became more assertive in their demand to get their money's worth. As their informal schools gradually grew, students had to unite for mutual protection, particularly against the rising prices of rooms and other necessities. By collectively threatening departure, students were able to obtain favorable deals from townspeople. Emboldened by their success, they also demanded guarantees from professors for what they were paid. One such demand was for professional qualification and certificates to teach, the earliest forms of academic degree.

The establishment of teaching certificates by the demand of the student union illustrates the crucial role students played in the development of the academic system in Europe. Thus, the gild of students became the nucleus and the driving engine of the university and established the unique characteristics that the present academic system takes for granted. Like many other bottom-up movements, such as labor unions, consumer protection and civil rights movements, the founding of universities came from average people who asserted their rights. Unlike madrasas, which were established by a top-down process, European universities and their basic characteristics were formed by input from students. They would become the main force behind breaking out of the religious straightjacket of the Dark Ages and the driving engine for the Renaissance and the Enlightenment.

As the size of universities grew, they no longer had to rely on a few famous teachers to attract students. The city of Paris was first to become a "city of teachers" and, by the 13th century, the University of Paris had grown to include up to 6,000 students, becoming the "mother of universities as well as the mother of sciences" and the model for other universities (Haskin, 1927). The University of Paris had grown around Notre Dame Cathedral as a corporation in the latter part of the 12th century. Universities provided not only teaching by masters but also "neutral space" (Huff, 2003) for open discussions among scholars. The gathering of a large number of intellectuals in an environment that fostered stimulation and competition no doubt played a central role in the explosion of knowledge following the rise of universities. The large size of the university and its large "academic market" also enabled the development of "internal differentiation" and even allowed the maintenance of a "marginal intellectual field" (Ben-David, 1971). The masters of specialized fields garnered prestige and respect in addition to financial security. The University of Paris achieved the critical mass possessing multiple features important to the progress of knowledge: stimulation, competition, cooperation, propagation, preservation, and consolidation.

The unique legal system in Europe played a critical role in protecting the autonomous status of the corporate entity of the university. As a consequence of the reform of the legal system during the 12th century, universities were one of many emergent autonomous collective entities. Because of the university's autonomous status, scholars in universities enjoyed far greater intellectual freedom than scholars in madrasas. One of the most important outcomes of the autonomous university was the rearrangement of the value system in knowledge-seeking. Unlike the imperial dictation in China and the religious intrusion in Islamic madrasas, the autonomous status of Europe's universities gave scholars an opportunity to set the value system on their own.

One of the most important features of the rearrangement of the value system involved "redefining the goals of enquiry" (Gaukroger, 2006). The reconfiguring of the value system was recognized by Gibbon in the 18th century. In *Essai sur l'étude de la literature*, Gibbon (1761) describes the gradual shift over the previous one hundred years from a focus upon the study of ancient literature like philosophy, history, poetry, and oratory to physics and mathematics. An even more crucial shift was the flipping of the value system between theology and science. During the Middle Ages, theology was the queen of the sciences and science was viewed as a "handmaiden to theology"; the role of science was to provide an aid to the interpretation of Holy Scripture rather than be studied for its own sake (Grant, 1986). This is similar to the notion of "science in the service of Islam" (Grant, 1986). As we shall discuss later, the flipping of the value system between religion and science would have a profound impact on the intellectual community in Europe.

There are two different approaches to establishing the value system: Externally-driven and internally-driven approaches. In the externally-driven approach to the value system, someone else from outside determines what is to be valued and what not. The Palace Examination of China is a good example. Imperial officials were the ones who choose what items should be on the test, while scholars themselves had no say in the matter. Preoccupied with harmony, stability, and status quo, Confucian moral teaching became the primary subject of the palace examination. The imperial reward system forced scholars to focus on studying the highly conservative moral learning that has almost nothing to do with understanding nature.

In the Islamic empire, the external source of establishing the value system was religion. The selection of students and teachers at madrasas was mainly driven by the inflexible wacq system that privileged religious conviction at the expense of intellectual freedom. Unlike China's imperial examination system, the innovative systems of madrasas and wacq had not started out with the goal of infringing upon the value system of scholars; rather, they were initiated with a noble emphasis on education and charity. Unfortunately, however, the marriage of the two systems created an external control over scholars' value system.

What happened in European universities was a self-determining of the value system. Universities were able to set up their own curricula and set selection standards for students and teachers. And the legal system of the corporation protected university scholars from religious infringement, although it took some time for universities to gain complete independence. Self-determination of the value system is one of the most important features of European universities. The explorers themselves are the best persons to judge what to investigate. With a self-determining value system, scholars in European universities were able to reconfigure the value system.

The Aristotelian Doctrine and Reconciliation

Despite the university's legal status as an autonomous collective entity, however, the realization of full intellectual freedom did not come immediately. It often had to face censorship from the Catholic Church. One of the earliest well-documented cases of conflict after the establishment of the university between university scholars and religious authorities concerned the appropriateness of Aristotle's work, soon after his works in natural philosophy had become available in Latin from translations of Islamic manuscripts. The availability of a large body of Greek knowledge (particularly Aristotelian philosophy) translated into Latin from Arabic reignited the clash between religion and Greek thought as in the Islamic intellectual community. Virtually every person in Europe in the Middle Ages had been raised in a deeply monotheistic setting. Exposure to Greek thought later in one's life would no doubt precipitate intellectual conflict. The first exposure to Greek thoughts must have felt like a cultural shock.

Greek philosophy had virtually vanished during the Dark Ages in Europe as religious fundamentalism became deeply entrenched in society. Had it not been for Islamic science, Greek knowledge might have been lost forever. As the institutional innovation of the university provided a more tolerant intellectual atmosphere, there was a renewed interest in learning forgotten Greek philosophy. Inevitably, however, as in Islamic science, Aristotelian philosophy would become a lightening rod in the clash between religion and philosophy. As mentioned earlier, at the heart of the problem was the monotheistic notion shared by both Islam and Christianity of an omnipotent God that demanded sacredness and reverence. Christian theologians had an aversion to the Aristotelian approach, which put a limitation on God's capacity and denied divine providence. Like Islamic scholars, Western intellectuals faced two major contradictions in the juxtaposition of a monotheistic conception of God and Aristotelian philosophy: the fundamental nature of time and the role of God in a world governed by cause-and-effect.

In facing the conflict between religion and Aristotle's philosophy, one may choose between one of four different outcomes: 1) steadfastly holding onto the religious learning of the early years of one's life, 2) choosing to embrace Aristotelian philosophy, 3) reconciling the two, or 4) maintaining the two views side by side by setting aside the inconsistencies. In Europe, unlike in the Islamic community, the effort to reconcile the two (the third option) began as soon as the works of Aristotle and his commentaries became available (Lindberg, 2007). Although several scholars attempted this reconciliation, Thomas Aquinas is the most well known. He essentially explained how he resolved the conflict between his faith and Aristotelian philosophy in language that would make sense to others. For instance, he argued that God is free to create the world He wants, but He must follow the established order except for miracles and the free actions of humans.

In 1210, the provincial synod of Sens in France decreed that the books of Aristotle on natural philosophy and all commentaries were not to be read at Paris, either in public or in secret, under penalty of excommunication (Grant, 1996). In 1270, the bishop of Paris repeated the formal condemnation of Aristotle's works and any ideas derived from it. The teachings of Thomas Aquinas at the University of Paris were among those works targeted by the bishop. The condemnation was echoed in England by the bishop of Oxford a few months later. The main point of contention was whether human reason is adequate to understand the will of God. Unlike the orthodox Augustinian theology which put faith above reason, Aquinas believed in a close connection between faith and reason. Taking an approach similar to that of Aristotle, he argued that studying nature could reveal the truth about God. He found a common ground that both science and religion could share. Aquinas' effort to bridge the gap between Aristotelian philosophy and Christianity may have played a crucial role in allowing the coexistence of science and religion in Europe. Fifty years after his death, Pope John XXII declared Aquinas a saint of the Catholic Church and Aquinas's once-vilified idea would become a canon of the Catholic scholasticism. Thus, unlike the situation in the Islamic intellectual community, the European scholars faced involved a compromise between the two contrary approaches. While Aquinas wanted to find common ground between Aristotelian philosophy and Christian faith, a few centuries later Galileo dared to place experimental results before religious authority particularly when it came to the orbit of celestial bodies. Galileo (1989) argued by quoting Cardinal Baronius: "the intention of the Holy Ghost is to teach us how one goes to heaven, not how heaven goes." The argument separates two different value systems, one for theology and another for science. The famous Galileo affair was viewed as science's attempt to proclaim its own autonomous domain. In the following section, we will examine the Galileo affair more closely.

Examining the Conflict between Science and Religion in the Galileo Affair

It has been my experience that folks who have no vices have

very few virtues.

-Abraham Lincoln

Human salvation lies in the hands of the creatively maladjusted.
-Martin Luther King, Jr.

The standard account of Galileo's trial depicting science as a victim of religion's ignorance and cruelty is one of the most famous cases in the conflict between science and religion. Galileo symbolizes martyrdom to science. With the phrase "refusal to look through Galileo's telescope," Cardinal Bellarmine represents religion's irrationality and ignorance. A close look at the famous episode, however, reveals that the caricature is quite exaggerated and erroneous. As we will discuss, this famous conflict is much less about the clash between science and religion than about the clash between two egocentric personalities.

Galileo studied medicine at his father's wish but could not obtain a scholarship to continue. He later turned to mathematics. During his student days, Galileo had already earned the nickname the "wrangler" for his combative style (Schmutzer & Schlitz, 1983). While teaching at the University of Pisa, his rebellious spirit began to cause him trouble. He refused to wear a black robe, the customary uniform for a professor at that time, and was repeatedly fined by university officials. In 1592, he became the chair of the Mathematics Department at the University of Padua. Three years later, Galileo developed one of his first major ideas—an explanation of the tides—which he got the insight for while traveling on a barge. He reasoned that tides are caused by the earth's dual motion (daily rotation and annual revolution). According to his model, there ought to be one high tide at noon. This is in direct contradiction with the fact that there are two tides each day and these shift around the clock. He bitterly quarreled with Kepler, who had the correct idea that tides are caused by the gravitational pull of the moon. Galileo rejected Kepler's idea, calling it astrological superstition. It is ironic that from his mistaken explanation of the tides he developed an idea supporting the Copernican system. We will talk further about his erroneous theory of the tides later.

Galileo had two faculties that were rather unusual among professors at that time: engineering aptitude and entrepreneurship. The combination of the two was a great asset and allowed him to construct such things as a military compass (a precursor of the slide ruler), a

A Critique of Science

hydrostatic balance, and a thermometer. The same talents would serve him superbly in making a telescope. The invention of the telescope by two spectacle makers took place in the Netherlands in 1608. Galileo greatly improved the telescope's magnification power. He then took advantage of the new technology in various ways, selling it to brokers in Venice who could then sell it to people who wanted to see approaching ships from further away. The most important application of the telescope, of course, was to observing the sky. It would become known as the instrument that launched the scientific revolution.

Galileo's publication of *Sidereus Nuncius* (*The Starry Messenger* in its English translation), the account of his astronomical observations in 1609, catapulted him into instant celebrity. Galileo's discoveries of the moon's rough surface, the sunspots, and the phases of Venus would contradict Aristotle's philosophy and centuries-old beliefs that all heavenly bodies were perfect spheres that moved around the earth. Galileo's discoveries that the moons of Jupiter and Venus had a full range of phases disproved the Ptolemaic system based on celestial bodies revolving around the earth. Contrary to popular belief, however, his observations did not confirm the Copernican system. In order to correctly understand the circumstances surrounding the Galileo affair, we must be aware of the three main theories of the solar system at that time. Besides the well-known Ptolemaic system (the geocenturic system) and the Copernican system (the heliocenturic system), there was the Tychonic system (the geoheliocenturic model), which was proposed by Tycho Brahe, the Danish astronomer. It puts the earth as the center of the universe, with the sun and the moon revolving around it. And other planets (Mercury, Venus, Mars, Jupiter, and Saturn) orbit around the sun. The Tychonic system was popular because it could accommodate the mathematically advantageous Copernican system and the discovery of the phases of Venus without altering the earth-centered view of the universe. Although Galileo made disparaging remarks about anyone who opposed the Copernican system, *the view that* Galileo endorsed the Copernican system actually comes from his erroneous theory of the tides. Not only did Galileo reject Kepler's correct idea that the moon causes the tides but he also erroneously believed that the earth's motion was responsible for the ebb and flow of the sea. *In the Dialogue*, his erroneous theory of the tides was the "secret weapon" and the "conclusive physical proof" of the motion of the earth (Koestler, 1959).

The main challenge to Galileo's discoveries, however, came at first from Aristotelian natural philosophers in universities rather than from clerics. Once vilified, Aristotelian philosophy had become entrenched dogma by Galileo's time, and it suppressed the new discoveries. Besides the Aristotelian dogma, the hierarchy of the academic system didn't help Galileo's idea. As the chair of mathematics at the University of Padua, Galileo was paid 180 florins a year, later raised to 520. In comparison, the leading natural philosophers were paid 2,000. The salary disparity gives some indication of the low status of mathematicians in universities at that time. Philosophers had vested interests in the Aristotelian dogma and didn't take too kindly challenges from a person they considered a lowly mathematics professor. The Aristotelian philosophers, not just critical of Galileo, formed a group called the "Pigeon League" named after the chief campaigner, Lodovico delle Colombe. One of the main reasons for the conflict becoming so intense was Galileo's egocentric and caustic personality. Galileo had been known for making a "laughing stock out his opponent," which made him many life-long enemies (Koestler, 1959). The most regrettable action of the league was the utilization of Scripture to denounce the Copernican system that Galileo embraced. For instance, Colombe used a verse in Psalm 105:5: "He set the earth on its foundations; it can never be moved." Rather than using logical argument, the members of the Pigeon League tried to get Galileo in trouble by invoking theological authority. The league's chanting of heresy to muster support from the clerics didn't go unnoticed. One person in particular stood out among them: Thommaso Caccini, a Dominican monk, who had previously been censured in Bologna as a "rabble-rouser" (Koestler, 1959). When Caccini joined the Pigeon League, the controversy took a new turn. Caccini was the Joe McCarthy of the 17th century and his fanaticism and ambition would serve as a catalyst of Galileo's trial.

Galileo formally responded to the accusation that the Copernican system contradicted Scripture in his *Letter to Castelli* (Castelli was his former student and an ardent supporter of Galileo). Galileo thought that this would silence all objections to Copernicus, but unfortunately it brought about the exact opposite result. The letter did not address astronomical discussions of the Copernican system. Instead, the basic theme of the letter was this: "Scripture cannot err, but its interpretation can." Galileo behaved as if the Copernican system was already proven and shifted the 'burden of proof' to the other side (Koestler, 1959). Demanding corroborating evidence from the academic and clerical authorities, particularly at a time of religious upheaval, underscores Galileo's audacity and naiveté. When Nicolo Lorini, another famous member of the Pigeon League, read a copy of the letter, he saw it as an opportunity to condemn Galileo. Lorini made a copy of the letter with two deliberate errors: the original text's "taken in the strict literal meaning, look as if they differed from the truth," was changed to "which are false in the literal meaning" and "overshadows" was changed to "pervert." This shows how far the members of the Pigeon League were willing to go to destroy Galileo. The partially fabricated letter was forwarded to the Consultor of the Holy Office. The condemnations of Caccini and Lorini prompted what historians call the "first trial." However, the proceedings never went beyond gathering evidence and Galileo was never formally charged, nor even informed (Shea & Artigas, 2003). The case was dismissed.

However, the controversy did not go away for two reasons: The members of the Pigeon League, particularly Caccini and Lorini, persistently put pressure on the Holy Office to denunciate Galileo. Another more important reason concerned Galileo's uncompromising stance; he essentially demanded either acceptance or condemnation of the Copernican system. One of the main reasons for Galileo's zeal in pushing for the Copernican system was that he had heard rumors that he had been forced to recant it. He felt that his name and pride were on the line. Caught between Galileo and his critics, those high in the church hierarchy showed much restraint. Cardinal Bellarmine, for instance, the most respected theologian with the official function of "Master of Controversial Questions," was in constant touch with the leading astronomers like Clavius and Grienberger, who were among the first to be convinced by Galileo's discoveries. Bellarmine did not jump in with a hasty conclusion but had a wait-and-see approach to the controversy. So long as heliocentrism was treated as a working hypothesis and had no conclusive evidence to support it, he had no problem. Then there was Paolo Antonio Foscarini, a Camelite Father who was willing to stick his neck out for Galileo. He published a book titled *Opinion of the Pythagoreans and Copernicus Regarding the Motion of the Earth* that argued compatibility of the Copernican system with Scripture. Galileo also had friends, Cardinal Piero Dini and Monsignor Giovanni Ciampoli who warned him to stay out of an entanglement with Scripture. Several cardinals from the Holy Office had tried to persuade him to be quiet and not to irritate the issue (Drake, 1957).

Yet Galileo would wage an aggressive campaign to promote the Copernican system with the argument that it didn't contradict Scripture. Keep in mind that this was a time of religious turmoil, when religious traditions were being challenged by Lutheranism and Calvinism. In such a combustible atmosphere, Dominicans expressed outright hostility and alarm that Galileo's idea was even more dangerous than Calvinism (Rowland, 2003). As time went on, even those Jesuits who had lauded Galileo a few years before were told by their superiors to avoid supporting any position that might weaken Aristotelian dogma (Rowland, 2003). Galileo resorted to sending a letter to his former friends and allies in the Church hierarchy to plead his position because they were reluctant to see him (Rowland, 2003).

The final straw that broke Galileo's back came ironically when his friend and sympathizer Cardinal Maffeo Barberini became the Pope Urban VIII, in 1623. Galileo had six long talks with the Pope. This gives some indication of the respect the new Pope had for the troublesome scientist. Galileo was given permission to write about the Copernican theory so long as it was treated as a hypothesis. After years of labor often interrupted by illness, the *manuscript* was completed in January 1630. Galileo met with the Pope again and received assurance that he had no objection to the discussion of the Copernican system as long as it was treated as a hypothesis. The Pope, however, thought that the original title, *Dialogue on the Flux and Reflux of the Tides,* gave too much emphasis to physical proof and suggested that it should be called *Dialogue on the Two Chief World Systems. However,* the Pope was too busy to read the *Dialogue* and it was sent to the Inquisition for approval. The person who was in charge of reviewing Galileo's work was Father Niccolo Riccardi, the Chief Censor and Licensor, who would play a critical role in Galileo's trouble, not by malice, but by his bungled handling of the situation. Riccardi's lack of knowledge on the subject and his indecisiveness made him vulnerable to pressure from Galileo and his friends. With minor corrections, the *Dialogue* was published in February 1632.

An even greater blunder was Galileo's typical manner of mocking his opponents. Although Galileo may not have intended, the

A Critique of Science

Pope thought his view of cosmology was portrayed as that of a simpleton by the Aristotelian philosopher Simplicio. Offending an enormously powerful Pope whose ego matched the size of his own ego was the fatal error that sealed Galileo's fate. The Urban VIII is described as "cynical, vainglorious, and lusting for secular power" (Koestler, 1959). He is the first Pope to allow a monument to him to be erected in his lifetime. And he is famous for saying that he "knew better than all the Cardinals put together." It was a complete turnaround from the earlier cozy relationship with full mutual adulation. Once an admirer of Galileo, the Urban VIII turned into a prosecutor. The Pope was so angry that he treated the insults/the mockery as personal. The significance of the conflict between two colossal egos in this battle has been reduced by science's idealistic understanding to a conflict between science and religion. Koestler (1959) sums up Galileo's trial:

> [T]he conflict between Church and Galileo (or Copernicus) was not inevitable; that it was not in the nature of a fatal collision between two opposite philosophies of existence, which was bound to occur sooner or later, but rather a clash of individual temperaments aggravated by unlucky coincidences....the idea that Galileo's trial was a kind of Greek tragedy, a showdown between "blind faith" and "enlightened reason", to be naively erroneous.

The Pope also had another problem to worry about: maintaining an uneasy neutrality between two major contenders of the Thirty Years War (Shea & Artigas, 2003). A particular concern was the criticism by Spanish cardinals for being too tolerant of heretics.

For being uncompromising and caustic, Galileo had to pay with his freedom. He was confined to his house for the rest of his life. And anyone who owned a copy of the *Dialogue* was to hand it over to the local inquisitors. However, the decree was rarely followed by people who owned the book. Instead, the book became more popular. The price of the book was originally half a scudo but rose to six by the summer of 1633 (Shea & Artigas, 2003). The translated copies were rapidly sold out in Frankfurt and Paris. Rather than being treated as a heretic, Galileo became popular all over Europe. He was visited by prominent scholars like Thomas Hobbes and John Milton (Rowland, 2003).

The Galileo affair also had an effect on shaping the relationship between science and religion that still endures. While Galileo was elevated to a mythic stature, the Church became the symbol of an anti-intellectual and anti-science entity. Ben-David (1971) argues that the Galileo affair has been exploited for purposes of Protestant propaganda against the Catholic Church. The Protestant exploitation was readily picked up by the scientific community in validating the church's religious cruelty and irrationality. As time went on, science rallied around the Galileo affair against all religions, including Protestantism. It is a historical irony that the Protestants became a victim of their own creation; the scientific community later would use the ideological version of the Galileo affair that the Protestants helped to generate to discredit religion as a whole.

Spinoza's God: The View of the Boldest Scholar in 17th Century Europe

Einstein was once asked by Rabbi Herbert Goldstein: "Do you believe in God?" Einstein answered: "I believe in Spinoza's God" (Einstein, 1934). Who is Spinoza and what exactly is Spinoza's God? Einstein was referring to Baruch Spinoza, a 17th century Jewish Dutch philosopher who held a radical view on nature and God. It was radical enough to be excommunicated by the Jewish community in Amsterdam and denunciated as a "menace to all piety and morals." The punishment for advocating it was the harshest imposed by his community with no possibility of reconciliation or redemption (Goldstein, 2006). But Spinoza's work also earned him respect from a segment of academia and he was offered a chair in philosophy at the University of Heidelberg, which he courteously declined. He opted to live as a lens grinder, which likely resulted in a lung illness caused by breathing in glass dust. He died prematurely at the age of 44.

Spinoza's God is interchangeable with Nature. In this interpretation, God is not personal and is not concerned with the fate and happiness of man. The world works in accordance with necessity; therefore, it is determined. Human actions are determined by prior causes which have no room for freedom and choice. Spinoza considers the notion of free will an illusion arising from ignorance.

Besides his revolutionary conception of God, Spinoza held another radical position. He dared to examine the Bible in the manner in which any other book was studied (Strauss, 1965). It may not sound that extraordinary now, but in 17th century Europe it was almost unthinkable. In a society dominated by a monotheistic religion in which God was the object of worship, the very idea that someone might examine the Holy Book as an investigating journalist was repulsive to say the least to most people. Spinoza's historical criticism of the scripture, such as the authenticity of Moses as the author of Torah, attempted to reject its claim of divinity. He argued that religion has been created by self-serving clergy who take advantage of people's fear and hope. His conception of God and Nature contradicted the Biblical miracles that most people embraced at the time.

Since the establishment of universities, the value system at the root of the conflict between religion and science had gone through a great deal of transformation. Aquinas, Galileo, and Spinoza perhaps influenced the direction and degree of the shift taking place in European science. Aquinas wanted to find common ground between Aristotelian philosophy and religion, whereas Galileo wanted to place the results of experiment before religious authority. Spinoza, on the other hand, rejected a personal God concerned with human fate and happiness. Furthermore, he dared to examine the authenticity of Christian Scripture. As we shall see later, the value system at the root of the conflict between religion and science would evolve further as time went on.

Orogee Dolsenhe

Chapter Six
The Flipping of the Value System and the Inception of the Science's Caste System in the 18th and 19th Century

Nearly all men can stand the test of adversity, but if you really want to test a man's character, give him power.

-Abraham Lincoln

Status groups, which are always to some degree concerned about their boundaries, and which typically use physical distance as a metaphor for social distance, often draw on notion of pollution. Where these ideas of pollution become highly elaborated, they significantly reinforce the already strong tendency toward restriction and regulation of cross-group associations and social mobility......in India, status boundaries are unusually significant, and much of the preoccupation with pollution is related to this concern.

-Murray Milner, Jr.

When science won its battle with the Church for the freedom to entertain its own hypothesis, it in turn became the principal repository of the idea that particular forms of knowledge could either be absolute truths or at least could approach absolute truths. Such a belief in the ultimate power of scientific knowledge evoked strong feelings of comforting security in many people, almost comparable with the feelings experienced by those who have an absolute faith in the truths of religion. Naturally there was an extreme reluctance to question the very foundations upon which the whole basis of this sort of truth rested.

-David Bohm & David Peat

The fate of an idea or person is often tied to the overarching scheme that is simultaneously taking place. The intrinsic merit of idea may have little to do with the recognition and value it may receive. Some of the great ideas may be buried for several decades or even centuries before appreciation. Conversely, some unworthy ideas may become entrenched and drive out alternative approaches as seen in the domination of behaviorism. As mentioned earlier, the domination of behaviorism is just the tip of the iceberg. In this chapter, we will examine the bigger submerged part of the iceberg with emphasis on the historical circumstances.

In the 18th and 19th century, a number of interrelated thoughts were taken up by the intellectual community which became the main characteristics of modern science. They would become the dogmatic scheme that would overpower any idea that may contradict them regardless of their intrinsic merits. The most crucial part of the newly emerging system is the overcoming of religious dogma that has been entrenched for more than millennium. The triumph over the age-old value system based on monotheism was an extremely difficult one considering the entity, that is, the object of reverence and worship by large number of people. Anyone who explicitly criticized apostle or denied God would no doubt invite animosity and hostility. Yet, the most unlikely event took place in the Enlightenment with an enormous consequence. Paradoxically, however, the new system of thoughts would become as dogmatic and intolerant as the previous one. Furthermore, the new set of ideologies was accompanied by the formation of a quasi-caste system that would have a long-lasting impact on the knowledge-seeking community. As we shall see later, the change was mainly through the flipping of value systems that had little to do with some new understandings or discoveries.

Naturalism, Atheism, and Anti-religion

What's the difference between naturalism and atheism? According to dictionary.com, naturalism is defined as "the view of the world that takes account only of natural elements and forces, excluding the supernatural or spiritual whereas atheism is "the doctrine or belief that there is no god." Although their meanings are almost the same, atheism has an undertone of "anti-religion." Let us take a brief look at the history of naturalism/atheism.

In Greek mythology, Zeus was the ruler of the sky and he created thunder and lightning whereas Poseidon who was trying to expand his power from sea to land was the force behind earthquakes. Thales, a Greek philosopher, offered an alternative to the mythical gods. He suggested that the earth floats on water and earthquakes occur when the earth is rocked by waves like a boat on waves. This type of approach regarding the world as the product of physical forces is a naturalistic explanation, opposed to a supernatural one. During the time of Socrates, the conflict between natural philosophy and religion became open and explicit (Thrower, 1971). He was accused of corrupting the young and neglecting the gods by Miletus. However, the real reason was his close association with Critias and Alcibiades who happens to be on the wrong side in the political turmoil of the time (Davies, 2003).

Once the Roman Empire adopted Christianity as the state religion, the fate of naturalistic philosophy became intolerable as other pagan religions. With the closing of the Athens Academy in 529 AD by Justinian, naturalistic philosophy would nearly become extinct in Europe. In the Medieval Age and many centuries afterward, any explicit expression of doubt on the existence and sacredness of God was almost unheard of. It wasn't until the 14th and 15th century that the open expression of naturalism would re-emerge (Thrower, 1971) with a different expression, "atheism." The term "atheism" (*athéisme*) was coined to use as an insult directed at those do not believe in God in France in the 16th century. Pierre Bayle, a French scholar in the late 17th century, advocated a view that atheistic society could achieve secular morality as much as Christian society does. In the latter part of the Enlightenment, atheism would begin not only to establish but attack religion among the intellectuals particularly in France. Bear in mind that openly attacking religious sanctity was quite dangerous thing as Thomas Aikenhead, an eighteen-year-old Scottish student in Edinburgh, unfortunately had to find out in 1697. He was hanged for saying that religion was "ill-invented nonsense." In Paris in the 18th century, three scholars would play decisive roles in establishing atheism in Europe and put religion on the defense: Jean Meslier (1664-1729), Denis Diderot (1713-1789) and D'Holbach (1723-1789). Their atheism stance was explicitly anti-religious.

The first person to play a pivotal role in altering the intellectual climate regarding divinity was Jean Meslier. Jean Meslier lived a double-life. One life was a Catholic priest of a small country parish who was exemplary in its austerities and virtues (Buckley, 1987). He gave away every year whatever remained of his salary to the poor of his parishes. He once refused to pray for the lord of his village for having ill-treated

Orogee Dolsenhe

some peasants. He performed his duty as a priest for forty years. Nobody knew he had another life until his death at the age of 55 in 1729 when he left three copies of 366 pages-long hand-written manuscripts titled *My Testament* that vehemently attacks religion. He gave the reason he lived a double-life:

> I was nevertheless compelled to teach you your religion and to tell you about it at least sometimes, to carry out that false duty that I had committed myself to do as the vicar of your parish, and then I had the displeasure of finding myself annoyingly obliged to act and speak totally against my own feelings, to entertain you with foolish nonsense and vain superstition that I hated, condemned and disliked in my heart. I however protest that I never did it without great pain and extreme repugnance; this is also why I hated that much the vain functions of my ministry, particularly all those idolatrous and superstitious celebrations of masses, and those vain and ridiculous administerings of sacraments, that I had to carry out (Meslier, 1729/2009).

He was the first avowed anti-religious person who has written a coherent work on the subject. He considers all the religions of the world to be: "only human inventions, and that everything they teach us or make us believe are only errors, illusions, lies, and impostures invented by scoffers, swindlers, and hypocrites to deceive men, or by shrewd and crafty politicians to hold men in check...." The tone of his writing is full of embitterment and anger against establishments: religion, monarch, and aristocrat. Consider Meslier's famous quotation has often been mistakenly assumed to be authored by Denis Diderot: "Men will never be free until the last king is strangled with the entrails of the last priest." He also believed private possession of wealth was the source of evil and advocated communalism. Michel Onfray (2006) gives Meslier multiple titles: atheist, deChristianizer, anarchist, communist, communalist, materialist, internationalist, revolutionary, leftist, and philosopher. Voltaire, one of the most influential philosophers of the Enlightenment, tells D'Alembert:

> The Testament of Meslier ought to be in the pocket of all honest men; a good priest, full of candor, who asked God's pardon for

deceiving himself, must enlighten those who deceive themselves (From *Superstition in All Ages*)

Voltaire added his own twist to Meslier's contention. Meslier did not ask God's pardon for he did not even believe in God (Buckely, 1987). Meslier's vindictive and angry tone was a real game-changer in the relationship between science and religion. *Testament* became a part of the clandestine literature circulating in France and it was condemned by the Paris Parliament in 1775. It was only in the middle of the 19th century that it was published openly. To sum up, Meslier's *Testament* was on the attacking mode against religion which might have quickened breaking away from religion's influence on science.

Denis Diderot was another key player in promoting atheism. He was born in a devout religious family and he was going to become a Jesuit. That commitment lasted only a half year. After a brief interest in law, he finally settled into becoming a writer. His first original work *Pensée philosophiques* in 1746 was anti-Christian which resulted in a condemnation by the *Parlement* and it was publicly burnt in Paris (Israel, 2001). The next important work that got Diderot in even bigger trouble with the Church was *Lettre sur les aveugles* (Letter on the Blind) in 1749. It introduced the notion of self-generation, variation, and natural evolution of species without Creation or supernatural intervention (Israel, 2001). Although the work was anonymously published, the authority swiftly identified the author because he was under police surveillance since 1747. He was arrested and was imprisoned in the dungeons at Vincennes for many months. He was released after signing a letter never to write anything against religion again (Israel, 2001). Buckley (1987) considers him the "first of the atheists..... as an initial and premier advocate and influence." He gives the mechanistic philosophy more elevated capacity: "Now matter is the creative source of all change" (Buckley, 1987). What kind of role did Meslier have on Diderot's thought? Buckley (1987) points out that Diderot read the entire text of Meslier's book and transposed a citation from it into his poem without mentioning the name of the atheistic priest. Just as Galileo's book attracted a greater interest after it became banned, the police's attempt to search for its author meant that Diderot's book made the list of hot books. Like many atheistic philosophers during the Enlightenment and afterward, his book had two sides. On the one side, he argues that the world can be explained without supernatural agency and he challenges the proofs of God offered by Descartes, Malebranche, and Newton. On the other side, he is highly critical of religion for hindering the progress of knowledge:

> How many discoveries in the great science of natural philosophy has mankind progressively made, which the ignorant prejudices of our forefathers on their first announcement considered as impious, as displeasing to the Divinity, as heretical profanations, which could only be expiated by the sacrifice of the enquiring individuals; to whose labour their posterity owes such an infinity of gratitude? Even in modern days we have seen a SOCRATES destroyed, a GALLILEO condemned, whilst multitudes of other benefactors to mankind have been held in contempt by their uninformed cotemporaries, for those very researches into nature which the present generation hold in the highest veneration. Whenever ignorant priests are permitted to guide the opinions of nations, science can make but a very slender progress: natural discoveries will be always held inimical to the interest of bigotted superstitious men.

Paul Henry d'Holbach, a French-German author and a close friend of Diderot, is best known for *The System of Nature*. D'Holbach argued that there is no necessity of supernatural powers to account for the formation of things. Furthermore, he attempts to explain the emergence of the idea of divine within human consciousness (Buckley, 1987). D'Holbach considered religion as obstacle for the progress of society. D'Holbach was much more successful in evading the authority than Diderot was. *The System of Nature* was published in a pseudonym Jean-Baptiste de Mirabaud, the secretary of the French Academy of Science who had died ten years ago. Alarmed by its anti-religious agenda, the French Catholic put pressure on the monarchy to suppress the circulation of the book. However, the French police's attempt to establish the real author was in vain. In 1770, seven books were ordered publicly torn to pieces and burned by the official executioner (Buckley, 1987). In spite of the authority's frantic effort to stop it, *the System of Nature* became so popular that it had four printings in the first year. *The System of Nature* led the antireligious "invasion" (Buckley, 1987).

A Critique of Science

In the proliferation of atheistic idea, the informal gatherings at salons and cafés played crucial roles. There were more than 380 gatherings in the 1720s in Paris (Terrall, 2002). Intellectuals pride themselves on drinking coffee in contrast with the irrationality of drunkard (Terrall, 2002). The lively discussions among intellectuals in a relaxed atmosphere without the pressure of censorship that is normally faced in publication helped to set a new course in the relationship between science and religion. Buckley (1987) elaborates:

> Around this major assault swirled the general and undisciplined currents of discussion and debate, of circuitous satire and comfortably fashionable opinions, the talk in the café, the gossip in the salon, and the whispers at the court—never committed to paper or print, but forming a climate of opinion in which presuppositions could crystallize and exist without searching internal criticism. Such criticism always came from the outside, dowdy and irrelevant, lacking crispness, smartness, and even good taste.

Bear in mind that this is the time without TV, radio, or professional sports that the present people are occupied with. The discussion of atheism might have been as much sensational as the trial of O.J. Simpson. It was during this time David Hume, a renowned Scottish philosopher, was introduced to atheism by Baron Paul d'Holbach's. Hume professed that he never met an atheist at a table, to be told that he was in the company of seventeen (Thrower, 1971). The persuasive arguments raised by a small number of atheistic intellectuals in Paris, however, would increasingly become the "mark of an elite" throughout the next century (Buckley, 1987). What was changing was the value system of supernatural agency, not the knowledge of it.

The rise of atheism in Paris is not just offering a differing view regarding God. It is the explicit and hard-nosed criticism that would set the tone of anti-religious ideology in academia. We can only guess the contents of gossip in cafés and salons in the 18th century but consider the published words of Denis Diderot (1777): "The most dangerous *madmen* (emphasis added) are those created by religion, and ... people whose aim is to disrupt society always know how to make good use of them on occasion." The atheistic rhetoric was "irrevocably hostile" to supernatural revelation and confessional beliefs (Buckley, 1987).

It was during this time that the value system of religion and atheism beginning to flip at least in the intellectual community. Once enjoyed revered status, religion was viewed as anti-intellectual and hindrance to the progress of knowledge. Whereas, once atheism has been associated with derogatory connotation, it increasingly becomes a sign of intellectual elite. Denial of god, among the intellectuals gathered in salons and café in Paris, was something to be proud. Before the flipping, there was too much reverence regarding divinity that hampered the progress of knowledge. After the flipping, the impediment was the peer-pressure to deny the existence of supernatural agency. It is a sad irony that science really has not had a chance to examine divinity without the suffocating pressure of the value system. The intellectual era may be divided into two separate periods: BF (before the flipping) and AF (after the flipping).

The completely different reactions from Spinoza's ideas— the excommunication from the religious community and the invitation as a chair of philosophy at the University of Heidelberg—illustrates the divergence of the academia from the society. The establishment of autonomous universities enabled to reset the value system. The new value system, however, had two sides: the promotion of science and the demotion of religion. The assault on religion in the Enlightenment was done in large part by emphasizing the contrast between reason and faith. While religion has been tagged with the words like "ignorance," "superstition," and "irrational," science was equated with "reason" and "rational." The Galileo episode energized the disparity between science and religion. By the end of 18th century, especially in France, there was a growing aversion among natural philosophers to talk of divine intervention. And medical elites in Britain viewed extreme religious enthusiasm as a symptom of madness (Brooke, 1991). The notion of divine intervention had become aversive among natural philosophers. And science was becoming more respected not only because of its results but as a way of thinking which can offer the correction of past error (Brooke, 1991). The latter part of the 18th century may be quite significant because the "popularization of science and rationalist movement threw Christianity on the defense" (Brooke, 1991). Meanwhile, scientific knowledge was prized as "one mark of gentleman" (Brooke, 1991).

A Critique of Science

In the Enlightenment, there was a basic change in the relationship between science and religion. Science began to take an aggressive mode against religion. One factor that played a significant role was a general dissatisfaction with established churches for lending support to tyrannical monarchs among other things (Brooke, 1991). The challenge to the church, thus, came from two different directions: scientific and nonscientific. The discontent toward religion in the Enlightenment was a significant factor in challenging the authority and prestige of Christianity. Brooke (1991) points out that the "grievances expressed by critics of Christianity often had nothing to do with science." In essence, they were mixing up the two separate issues: the religious role in hindering the progress of knowledge and the possibility of supernatural existence. The former should not be the basis for the dismissal of the latter. Then, the supposedly rational science was gradually taking an irrational stance. Much like religion's faith in God's existence, science rejected the possibility of supernatural existence with certainty dominated by the anti-religious theme. Naturalism has transformed into anti-religious dogma.

To sum up, the ideology of God hampered in two opposite ways: too much respect and too little respect. Dissenting views were not allowed in the name of God or in the name of science and "ungodly" was replaced with "unscientific" as a way of intolerance and discrimination.

The Shift from Supernatural to Natural Approach in Astronomy

Nothing is so firmly believed as that which we least know.
-Michel de Montaigne

One of the principal shifts in the intellectual community during the Enlightenment was the popularization of the mechanistic philosophy which is directly related to naturalism. The mechanistic philosophy in conjunction with naturalism would play a dominant role in virtually all fields of science in the future. The notion of the machinery universe is one of the many metaphors borrowed from human inventions to describe nature. During the Enlightenment, the most complex gadgetry was the spring-driven clock and Newton equating the universe with a giant clock. The planetary motions of the solar system were like the gears ticking along. The laws of physics govern the motions of planets as well as the clock. And just as the clock was built by the master-builder, the universe was built by God. Newton expressed in *Opticks*: "the order of nature becomes needful, in the course of time, of a renewal by God because as a result of the inertia of matter its irregularities increase." For many philosophers like Kant and Leibniz, however, the idea that God has to resort to continuously correcting the irregularities was repulsive because it implied an imperfection of God. The omnipotent God would not need to intervene in the perfect universe. Once the universe is created, it ran all by itself. Whether one believes the notion of active God or absentee God after the initial creation, they all accepted the divine's role in the universe. When Newton discovered the fundamental laws of physics, he saw them as proofs of God's sovereignty over nature. For Newton, the descriptions of gravity and motion with mathematical precision were the findings of what God had implemented. Newton saw that the clockwork universe requires the clockmaker. In addition, he argued the continuous intervention of supernatural agency is necessary to keep the clockwork universe working. Other 17th century scholars like Marin Mersenne, French theologian, philosopher, and mathematician, saw physical laws as expressions of divine will and the ultimate source of order. Likewise, Robert Boyle, British chemist and physicist, thought the experimental approach many of which he pioneered would demonstrate the Book of Nature that was written by God. Another English philosopher, Ralph Cudworth, argued that mechanistic philosophy demonstrates the necessity of spiritual agents.

  A great irony, however, was that the very reasons that Newton believed to be evidence of divinity were used to promote the atheistic mechanism. The mathematical equations used to promote supernatural agency was utilized to argue for naturalistic explanation that the source of movement was inherent in matter. By looking at the exactly same universe, Newton saw intelligence behind it whereas Laplace saw a naturalistic process that runs by itself. The religious approach to nature was slowly being displaced from the center to its periphery (Gay, 1977). And the mechanistic philosophy became one of the leading ideas to confront the long-held theistic approach to nature. The Greek philosophers like Leucippus and Democritus believed that the world is the result of chance collisions of atoms which have no room for the divinity. During the Enlightenment, the mechanistic philosophy, despite the lack of consensus, was viewed as a new theory of matter, causality, and method in which the ultimate components of things were particles (Brooke, 1991). What was different from the early Greek atomists was that the new particle physics had a much higher degree of prediction which was enabled by mathematics and experimentation.

At any rate, Newton's discoveries were instrumental in setting off a new course in academia. The notion that matter inherently possesses the capacity to move would displace the role of supernatural agency. This is ironic because Newton was religiously devout and he thought his discoveries were the validation God's presence in nature. He argued, without God's intervention, "the Bodies of the Earth, Planets, Comets, Sun, and all things in them would grow cold and freeze, and become inactive Masses" (quoted in Deason, 1986). Descartes also played a similar role in establishing the notion that the source of movement of matter is autonomous within itself. As Buckley (1987) points out, Descartes' world was "as godless as his physics was autonomous—not because god was not necessary for either, but because the world gave no convincing evidence of his presence or even his existence." The mechanistic philosophy was not initially intended to be godless but gradually gave power to atheistic philosophy.

Another crucial part of the shift of cosmology was the fact that the planets in the solar system orbit around the sun in almost the same plane and the same direction. For Newton and others, such an exquisite system was a manifestation of God's design. It was a French mathematician and astronomer Pierre-Simon Laplace who took out the need of a supernatural agent for the planetary motions of the solar system with the nebular hypothesis. It was originally proposed by Emanuel Swedenborg in 1734 and developed further by Kant in 1755 by extending to the entire universe. The crux of the nebular hypothesis is that the formation of the solar system was created by the gaseous clouds (nebula) which slowly rotated and collapsed into the flattened shape and some of them collapsed by the gravitational force and eventually formed the planets. Laplace in 1796 added mathematical dimension to the nebular hypothesis. A greater role of Laplace, however, was the displacing of the divine role in the formation of the solar system with a famous episode involving Napoleon. Napoleon asked Laplace why there is no mention of God in his work the *System of the World*. Laplace bluntly answered: "Sire, I had no need of that hypothesis." Napoleon later commented: "Ah, it is a fine hypothesis; it explains many things."

Orogee Dolsenhe

Positivism and the Beginning of the Caste System in the 19th Century

>...as science has acquired more secular power, it has tended toward the self-perpetuation of existing regimes, as dominant research programmes are pursued by default, a situation that the sociologist Robert Merton has dignified as the 'principle of cumulative advantage'. Kuhn and Merton, both Harvard products of the Conant years, were kindred spirits. They saw science as mainly about those few who rise above the rest and constitute themselves as a self-perpetuating community. 'Self-stratifying' more than 'self-organizing' describes the situation....
>
> -Steve Fuller

> There is no such thing as philosophy-free science; there is only science whose philosophical baggage is taken on board without examination.
>
> -Daniel Dennett

    One of the main criteria for determining the qualification of science has been positivism. Perhaps Auguste Comte, a French pioneer of sociology, may best represent the main drive that caused the split between science and the rest of subjects in emphasizing the superiority of positive knowledge contrast with speculative ones. In *The Course in Positive Philosophy*, he distinguished three stages that are mutually exclusive in quest for the truth: the theological, the metaphysical, and the positive phases. The theological stage refers to the period in which explanations are offered by animistic or supernatural notions. The metaphysical stage offers explanations of events by abstractions that replaced supernatural agency. The positive stage is the final and highest period which does not go beyond observable and measureable facts. As the name suggests, he argued that knowledge should be confined to being only positively given avoiding all speculation. Comte's declaration was based on the zero-sum impulse and it had a huge repercussion and was the signal for the inception of the caste system in knowledge-seeking.

    Comte may have been the first to use the term "positive" in emphasizing observable certainty in pursuit of knowledge but the movement has been gradually building up for some time. As discussed earlier, Alhazen in the 10th century who had experienced the religious

A Critique of Science

dispute between the Sunnis and Shiites over the Prophet's rightful successor deeply affected him. His disappointing conclusion, after many years of studying religious doctrines, was that the disagreement has more to do with the sociological than the truth. Consequently, he turned toward the methodological investigation that combined experiments and mathematics which are the hallmarks of science. And his book *The Book of Optics* perhaps may represent the beginning of so called "scientific investigation." At the same time, Alhazen took a rigorous position that anything beyond what we can observe should be excluded from studying. In this sense, he was the first positivist as much as he was the first scientist.

Perhaps Isaac Newton may have contributed more to jump starting the movement of positivism than anyone before Comte in Europe. In fact, Newton has been called the first great positivist (Brewster, 1855). The brilliance of Newton also had a flip side as most geniuses do. Besides being deeply introverted, he had overwhelming insecurity and a violent temper which was a mixture prone to intolerance of any kind of criticism. Some of his ideas were held off for publication for fear of criticism. Nothing angered him more than some of his doctrines being called a "hypothesis" (Burtt, 1924). An excerpt from Newton's masterpiece *Principia* reveals an attack mode in dealing with a speculative idea that beyond direct observation:

> Whatever is not deduced from the phenomena is to be called and hypothesis; an hypotheses, whether metaphysical or physical, whether of occult qualities or mechanical, have no place in experimental philosophy.

Thus, long before Comte, Newton's insecurity of criticism drove him to declare such strong stances on rejecting speculative hypothesis from experimental philosophy (later to be called science). The gradual and steady march toward positivistic ideology would result profound impacts in many areas of the intellectual community.

It might be startling to hear that what happens in the 19th century and onward is quite similar to the beginning of the caste system in India. Both were driven by the zero-sum impulse. And just as people in India who has been raised in the caste culture, most people—whether they are scientists, philosophers, and laypersons—are not aware of the unjust stratification among knowledge-seekers. And just as India's caste system, it may take outsiders to notice the systemic flaws. As we will discuss further, the caste-mentality in knowledge-seeking has created the culture that has become a profound hindrance to progress of knowledge in numerous disciplines in a number of ways. Fischer (2008) argues that "social stratification" interferes with development of research.

The Indian caste system is largely based on the value put on the types of work people do. Likewise, the caste system in knowledge-seeking is based on the rearrangement of the value system on the types of subject scholars engage in. The preoccupation of the Indian caste system was "purity" while science's caste system was the positive knowledge. That meant metaphysics would be excluded from science no matter how useful or necessary they may be. Burtt (1924), however, argues: "...there is no escape from metaphysics, that is, from the final implications of any proposition or set of propositions. The only way to avoid becoming a metaphysician is to say nothing." Ben-David (1971) argues: "the idea leading to basic scientific change are often derived from general, nonscientific metaphysical speculation. Thus the transitions from Aristotelian to Newtonian physics and from classical Newtonian to modern physics are not explicable by the immanent logic of scientific thought and empirical verification." Smocovitis (1996) calls for the return to the "metaphysical origins." Redner (1987) believes that positivism has proved itself "ineffective and useless" for working scientists. Redner (1986) further argues:

> Physics is the science of the natural world; metaphysics is not the science of a world beyond, a supernatural world, but rather the science of the most fundamental principles that underlie the natural world.

As the positivistic ideology began to dominate, something odd happened. Perhaps it should have been expected if we understand the zero-sum impulse. Science broke away from philosophy. However, both terms "science" and "philosophy" have long been used without distinction until the partition (Ross, 1990). Malik (2002) echoes:

> For many people, science deals with 'hard facts', while philosophy revels in speculation. In the seventeenth century no

such distinction existed. Science and philosophy formed a common endeavor and created a common body of knowledge...in Descartes' day, what we now call science was labeled 'natural philosophy' – the philosophy of nature.

Van den Daele (1977) also agrees that, prior to the partition in the 19th century, what we now call science used to be called "natural philosophy," "experimental philosophy," and "mechanical philosophy". In fact, physics is still considered "natural philosophy" in Scottish universities (Ross, 1990). A doctor of science has an acronym "Ph.D." (*philosophiae doctor* in Latin) which is a vestige of the bygone era that philosophy was not considered inferior to science. As the disciplines began to leave philosophy to join the upper caste, they left issues they cannot resolve for philosophy to deal (Balashov & Rosenberg, 2002). Another interesting consequence of the establishment of science club was the formation of the pecking order. The elevated status of physics particularly in the 19th century became the envies of other disciplines. Chemists like Boyle, Lavoisier, and Mendeleev with stress on laboratory experiment demonstrated their worthiness of joining the breakaway science club. The established disciplines sometimes responded with condescending remarks to other less developed disciplines. For instance, physicist Ernest Rutherford, called biology not much better than "postage stamp collecting." At best it was viewed as a second-class, "provincial" science by physicists (Mayr, 1988). Remarkably, some biologist would openly express deference to physics as James D. Watson, the co-discoverer of double helix: "There is only one science, physics: everything else is social work" (quoted by Rose, 1997).

The Study of Fossils and the Seed of Doubt

People all over the world have been puzzled by the prodigious findings of marine fossils on top of mountains. They would wonder how such a thing was possible. Ancient Greek scholars left good records about their puzzlement and theories. Xanthus of Lydia, for instance, suggested that such mountains must have been under the sea at one time and that the land and sea areas were constantly undergoing a change of positions. Others thought that the planet's waters were diminishing. As Christianity emerged as a dominant thought, such findings were held to prove Noah's Flood. This was first stated by Tertullian. The Flood theory of fossil origins went on to become the prevailing view and fit in well with the main theme of Biblical creation. In 1650, Irish archbishop James Ussher actually proclaimed that creation had occurred less than 6,000 years ago.

During the Enlightenment, the newly emerging fields of geology and paleontology would play critical roles in shaping a new way of thinking, one that directly challenged the Bible's authority. In 1669, Steno argued that the Flood had been so brief and so violent that it could not possibly be responsible for the regular arrangement of the rocks beds. Rather, he suggested that the geological strata were slowly deposited by calm waters (Gregory, 2007). In 1720, René Réaumur submitted a report to the Paris Academy of Sciences proposing that a brief Noachian flood could not account for the thick sedimentary layers. He suggested that the region was once covered by the sea (Gregory, 2007). D'Holbach also pointed out that some types of fossils were constantly found together and thus could not have been thrown together by chance during the Flood. Orthodox theologians attempted to explain away such inconsistencies. For instance, Elias Camerarius in 1712 suggested that God had placed various fossils while creating the world (Haber, 1959). Voltaire even suggested that they were dropped by pilgrims and Crusaders who habitually wore shells in their hats (Haber, 1959). One interesting finding was that the fossils of primitive organisms appear in the lower strata while those of higher organisms are found in the upper strata. Charles Lyell's discussion of the fossil records of different layers in his *Principles of Geology* played a crucial role in challenging the biblical account of life and humanity. Although he himself was religious and struggled with the implications of these fossil findings, his book helped promote the idea of evolution. Notwithstanding his own religious faith, he would become one of the prominent scientists to help Darwin's idea of evolution enter mainstream science.

In the long and enduring conflict between naturalism and supernaturalism, the issue of evolution is the ultimate arena in which people make their cases for life arising by itself or due to some supernatural agency. This is where science and religion often collide, where physics and biology meet, where we try to find out where we come from and who we are. For over 150 years, Darwinism has been the foundation of biology which may also be a significant obstacle to scientific progress. In fact, it may turn out to be one of the greatest, if not *the* greatest, embarrassments of science. In the next three chapters, I will discuss how Darwin's notion of natural selection became the foundation of biology, a position that it continues to hold despite its problems.

Orogee Dolsenhe

Chapter Seven
The First Phase of the Evolutionary Absurdity:
The Survival of the Slipperiest

> As Darwin himself remarked, he was a master "wiggler." Any scientist who is incapable of wiggling a bit will never succeed in science, but there are also limits to the wiggling. If it becomes too pervasive, the scientist ceases to be a "scientist." There more ways than one for science to degenerate into nonsense.
>
> -David Hull

> Darwin was not a thinker and he did not originate the ideas he used. He vacillated, added, retracted, and confused his own traces. As soon as he crossed the dividing line between the realm of events and the realm of reality he came 'metaphysical' in the bad sense.....
>
> -Jacques Barzun

> Thus, it may come as a shock to discover that many—if not most—of the books about Darwin engage in outrageous errors of both omission and commission.
>
> -Arnold C. Brackman

> Darwin was a muddled thinker, and that his main work is crammed with misunderstandings, contradictions.....as a thinker he was confused and inconsistent....
>
> -Søren Løvtrup

> .....the way Darwin handled his facts; when they did not fit his views he ignored them or tried to explain them away.
>
> -Søren Løvtrup

The three concepts of "evolution," "Darwinism," and "natural selection" are used almost interchangeably in modern biology. Their virtually identical meaning indicates the acceptance of Darwin's principal role in explaining evolution through his theory of natural selection. However, such textbook accounts have long been controversial. The proponents and opponents of Darwin's idea have clashed for the last 150 years. Darwinism actually lost its dominant position in biology beginning with the 1880s and into the first couple of decades of the 20th century. Alternative views eclipsed Darwinism until the advent of "modern evolutionary synthesis" or "neo-Darwinism." Darwinism's dominance in the 19th century and neo-Darwinism's dominance in the 20th century, however, are strikingly different in one crucial way. The degree of intolerance in the latter is far greater than in the former, as can be seen in the fact that biologists now pay very little, if any, attention to alternative views of evolution.

Even with neo-Darwinism's near total supremacy, there is a continuous stream of challenges in a variety of issues. The gap between proponents and opponents often are so wide that one cannot help but to compare it with doctrinal disagreements among religious sects. Consider a comment made in *Darwin's Dangerous Idea* by Daniel Dennett (1995), a well-known philosopher at Tufts University: "If I were to give an award for the single best idea anyone has ever had, I'd give it to Darwin, ahead of Newton and Einstein and everyone else." Richard Dawkins, another ultra-Darwinist, says that Darwinism is "the most powerful idea" ever occurred to human mind in his video "The Genius of Charles Darwin." Compare their adoration of Darwin with the completely different view of C. D. Darlington (1959), who used the word "myth" to describe the erroneous portrayal of Darwin. Søren Løvtrup, a Swedish embryologist at the University of Umea, not only echoed this word in his *Darwinism: The Refutation of a Myth*, but went one step further: he believes that the Darwinian myth will be ranked the "greatest deceit in the history of science" (Løvtrup, 1987). Which side is right? Is Darwin's idea really the greatest ever, or is it just a myth? Unfortunately, Dennett ignores Løvtrup's argument. Disagreement over Darwin and his idea is nothing new; it has been ongoing since he published *On the Origin of Species* in 1859. Why is this so? Why are both sides looking at exactly the same book and reaching completely different conclusions?

Henslow's Role in Darwin's Success

Perhaps no one played a greater role in making Darwin a great naturalist than his mentor and friend John Stevens Henslow. Their life-long friendship began at Cambridge University, when Darwin enrolled in a botany class taught by him. They became so close that Darwin picked up a nickname: "the man who walks with Henslow." This class was quite popular, because Henslow emphasized direct experience through fieldtrips. His distinctive feature was "collating," a botanical procedure that compared specimens. These collated sheets usually carried two or three plants; however, there could be as many as thirty-two (Kohn et al., 2005). Rather than just identifying plants, Henslow's technique proved to be quite effective in studying continuous variation within a single species. This interest in finding the limits between which any species of plants might vary would lay the foundation for the framework of Darwin's theory (Kohn et. al., 2005). After the class ended, Henslow advised Darwin to study geology and arranged for him to accompany Adam Sedgwick on a geological trip to North Wales. Thus, Darwin learned the field techniques of reading rock formations from one of the top geologists at that time. When Henslow received a letter from Captain Fitzroy looking for a gentleman companion on a two-year trip to survey South America aboard the HMS Beagle, he urged his favorite pupil to go.

Contrary to the account presented in many textbooks, Darwin initially signed on not as a naturalist but as Captain Fitzroy's socially acceptable companion. The captain required such a companion because it was thought that such prolonged social isolation might take its toll on his mental health: his family had a history of suicide. When Robert McCormick, the ship's surgeon and designated naturalist who was assigned to collect specimens, left the ship in Rio de Janeiro just after a few months after the voyage began because Darwin had downgraded his role, Darwin stepped in as the de facto and unpaid naturalist (McComas, 1997). Charles Darwin, a 22-year old amateur naturalist, did far better than simply replace him. He was a "born naturalist," as he called himself (Colp, 1980). Darwin's patient and keen observation was matched by his zeal for investigating nature. The five-year voyage gave him ample opportunities to collect various animals and fossils that were unknown before. Darwin and Henslow were in constant contact, and most of Darwin's collections were sent to Henslow, who published the analysis of the specimen with excerpts from Darwin's letters. Thanks in large part to Henslow, Darwin was already considered a reputable naturalist even before he returned home.

Orogee Dolsenhe

The Tautological Issue of Natural Selection

> It is a well-established fact that language is not only our servant, when we wish to express—or even to conceal—our thoughts, but that it may also be our master, overpowering us by means of the notions attached to the current words.
>
> -Wilhelm Johannsen

> I suppose natural selection was a bad term; but to change it now, I think, would make confusion worse confounded, nor can I think of a better.
>
> -Charles Darwin

Perhaps one of the most serious problems facing the concept of natural selection is its tautological implication. "Tautology," derived from the Greek *tauta*, means "the same." A tautological statement is necessarily true and, therefore, cannot be wrong. For instance, "parents have children" or a "triangle has three sides" is a needless repetition of an idea and cannot be denied. Waddington (1960) argues that natural selection is a tautological statement:

> Natural selection, which was at first considered as though it were a hypothesis that was in need of experimental or observational confirmation, turns out on closer inspection to be a tautology, a statement of inevitable although previously unrecognized relation. It states that the fittest individuals in a population (defined as those which leave the most offspring) will leave the most offspring. Once the statement is made, its truth is apparent…

Stanley (1979) agrees:

> I tend to agree with those who have viewed natural selection as a tautology rather than a theory. It is essentially a description of what happened, with only weak powers of prediction, in that the kinds of individuals that are favored can often be recognized only in retrospect. The circularity in no way impugns the heuristic value of natural selection as a generation-by-generation

description of evolutionary change.

The phrase "the survival of the fittest" was coined by Herbert Spender after he read *Origin* to draw a parallel between his economic theories and Darwin theory of evolution. Since then, it has become more than a synonym of natural selection. Darwin (1876) himself recognized in the later edition: "I have often called this principle....by the term natural selection. But the expression often used by Mr. Herbert Spencer, of the Survival of the Fittest, is more accurate, and is sometimes equally convenient." It is a clearer and more vivid expression of the concept of natural selection. Consequently, the tautological implication is even clearer in Spencer's phrase. For instance, Popper (1972) makes the following point:

> Yet there does not seem to be much difference, if any, between the assertion "those that survive are the fittest" and the tautology "those that survive are those that survive." For we have, I am afraid, no other criterion of fitness than actual survival, so that we can conclude from the fact that some organisms have survived that there were the fittest, or those best adopted....

Margulis & Sagan (2002) elaborate the basic problem of the concept of natural selection:

> What then does the selecting in natural selection? Just as many modern evolutionists permit themselves an unscientific vagueness about the role of natural selection in evolution, they also remain vague about the identity of the natural selector. It is all too easy to wave one's arms and say "the environment selects, the fittest survive." What does "fit" really mean? What parts of the environment select? How far does the environment extend? Questions like these tend to be answered only in generalizations or in an ad hoc manner, case by case. A staunch resistance to any systematic effort to identify the agent or agents of natural selection take place.

Despite the obvious flaw, Darwin's supporters see it quite differently even though they recognize the tautology of the concept. For instance,

Haldane (1935) put it this way:

> The phrase, 'the survival of the fittest,' is something of a tautology...There is no harm in stating the same truth in different ways.

Even a more startling comment comes from Dobzhansky (1968):

> Genotypes whose carriers differ in Darwinian fitness is a given environment are transmitted from generation to generation at different rates. The last statement is frankly tautological, yet it is *illuminating* (emphasis added).

Macbeth (1971), a retired lawyer, seems better at judging the tautological problem of natural selection than dogma-prone scientists:

> Someone asked how we determine who are the fittest. The answer came back that we determine this by the test of survival; there is no other criterion. But this means that a species survives because it is the fittest and is the fittest because it survives, which is circular reasoning and equivalent to saying that whatever is, is fit. The gist is that some survive and some die, but we knew this at the onset. Nothing has been explained.

Another outsider, a journalist named Bethell (1976), put it bluntly:

> What he really discovered was nothing more than the Victorian propensity to believe in progress...'the survival of the fittest' is not a testable theory, but a tautology. Which one survives? The fittest. Who are they? Those that survive.

The tautological statement may be useful when used as a definition (Riddiford & Penny, 1984). If it is used as a causal statement, however, it becomes problematic. I am not aware of any other scientific theory with such a tautological connotation. The tautological issue exposes the inherent flaw of natural selection: it is an evolutionary statement rather than the mechanism of evolution. Richard Owen raised this argument in 1860:

The sign of such intellectual power we look for in clearness of expression, and in the absence of all ambiguous or unmeaning terms. Now, the present work is occupied by arguments, belief, and speculations on the origin of species, in which, as it seems to us, the fundamental mistake is committed, of confounding the questions, of species being the result of a secondary cause or law, and of the nature of that creative law (Owen, 1860).

Even some of Darwin's supporters were aware of the problem and informed him in more tactful way (Løvtrup, 1987).

"A Special Difficulty" and Darwin's Delay
Although Darwin may not have noticed his circular logic, he did see another major problem. In fact, he initially considered that this particular flaw might be "fatal to my whole theory": the apparent altruism found in social insects like honey bees, termites, and ants. The worker members of social insects are sterile, work diligently, and sacrifice their lives for the colony's sake. Such extreme altruism seems to contradict the very basic idea of natural selection: the ruthless struggle for one's survival and maximization of one's offspring. The trouble with the rise of an altruistic caste of social insects is this: Such self-sacrificing individuals, being inherently less fit to survive than selfish individuals, would, according to Darwin, have been eliminated from the population. He felt it was "the greatest *special* difficulty I have met with," and it drove him "half mad" (Browne, 1995).

This brings up another question: Why did he take so long to publish *Origin*? During his five-year voyage on the Beagle, Darwin was a creationist, just as his mentor Henslow and most of his contemporaries were. His serious inquiry into the transmutation of species began six months after his return to England. He completed, but did not publish, an essay on the transmutation of species in 1844. Why? There are several theories (Richards, 1983): He needed more time to collect supporting evidence. He required the services of several correspondents and associates, among them Lyell, Hooker, and Huxley. Other projects keep him from working on *Origin*. Some have suggested that the negative reception of Chambers' *Vestiges of the Natural History of Creation* caused his hesitation. In the public domain, however, Chambers' book was quite popular – 26,700 copies had been sold by 1860, 10,000 more than the *Origin* were sold during Darwin's lifetime.

I would like to address Richard's (1983) contention that Darwin delayed publication mainly because he had to find a way to deal with the altruistic instinct of social insects. His manuscript shows that he was aware of this problem in 1843. Only in late December of 1857 did he present his solution: "kin selection" or "group selection." He wrote that the difficulty "disappears when it is remembered that selection may be applied *to the family*, as well as the individual, and may thus gain the desired end." In other words, selection can work at the group level.

This notion has long been controversial. G. C. Williams (1966), J. Maynard Smith (1964), and other evolutionists have argued that kin selection is an inherently weak evolutionary force, for whether the advantage of group selection can override the effects of individual selection within group or not has never been proved. Consider the oft-used example of the evolution in alarm calls among vervet monkeys, who use different sounds to warn their fellows about the presence of a snake, leopard, or eagle. From the Darwinian point of view, a monkey's best survival strategy would be to quickly run away and hide as soon as it sights a predator. However, it sounds the alarm before it runs away. Although such an altruistic action may help others, it increases the risk to the monkey by both delaying its escape and attracting the predator's attention. Thus the altruistic monkeys would be less fit and the selfish ones would have a better chance of producing offspring. The concept of kin selection holds that a group of monkeys with an altruistic member would have an advantage over another group composed only of selfish ones. The problem here is that the former group would continuously lose its altruistic member. Dawkins (1976) has called this "subversion from within."

This notion took a new turn when William David Hamilton, a British evolutionary biologist, posited "Hamilton's rule." Its simple mathematic equation makes it different from other models of kin selection. Perhaps his own words can best illustrate what makes his model special: "University people sometimes don't react well to common sense, and in any case most of them listen to it harder if you first intimidate them with equations." And intimidate them he did. Dugatkin (2006), for instance, believed that Hamilton's rule solved the puzzle of kin selection. The notion of kin selection, which started out with

Darwin's muddled thinking, was capped with mathematical intimidation.

The Confusion between Artificial Selection and Natural Selection

Darwin drew his concept of natural selection from the artificial selection that can be observed among domesticated animals and plants. In fact, after a brief historical sketch and introduction, the first chapter of *Origin* is a "variation under domestication." In the case of artificial selection, the breeder is the active agent who is doing the selection. Thus, the breeder's conscious effort is the essential part of the selective process. But according to Darwin, the active agent in the natural world is the "selective process"! Unlike Darwin, Alfred Wallace, the co-discoverer of natural selection, thought that the two types of selective processes were completely different. A year before the publication of *Origin*, he wrote:

> ...no inferences as to varieties in a state of nature can be deduced from the observation of those occurring among domestic animals. The two are so much opposed to each other in every circumstance of their existence, that what applies to the one is almost sure not to apply to the other (Wallace, 1958).

The Ambiguity of the Teleological Implication of the Concept of Natural Selection

Darwin's *Origin* can be discussed in terms of theory and evidence. Its theoretical part can be subdivided into (1) the idea of species transmutation and (2) the explanation of transmutation by natural selection. What Darwin provided, perhaps better than anyone else, was evidence that supported the idea of species transmutation. His main contribution was to challenge biblical creationism by providing empirical evidence that species could evolve. But when it came to explaining the mechanism for this evolution, namely, natural selection, problems arose. As simple as the concept may be, it can lead to a great deal of confusion. For instance, opponents of evolution often dismiss evolution altogether due to the difficulties inherent in natural selection, whereas proponents often accept natural selection based upon empirical evidence that only supports the idea of transmutation.

With that in mind, I will now discuss the concept of natural selection. The first problem encountered is the phrase's very ambiguity. It sounds like it is implying an anthropomorphic or personified agent behind the selective force. Artificial selection features an agent (breeder) who intentionally chooses which animals should be bred. Darwin compared the variations he observed in domesticated animals to explain the selective process in nature.

Other material in his book also engender confusion. For example, while he often invokes "chance" or "accident" in discussing how new variations arise from time to time among the organisms of a species, in the last section of *Origin* he says:

> There is grandeur in this view of life, with its several powers, having been originally breathed into a few forms or into one; and that, whilst this planet has gone cycling on according to the fixed law of gravity, for so simple a beginning endless form most beautiful and most wonderful have been, and are being, evolved.

The phrase "breathed into" may be interpreted as some sort of purpose or goal in nature, and therefore stands in contradiction to the "chance" and "accident." This became a source of disagreement even among Darwin's supporters. While Huxley and Hackel praised Darwin for rejecting teleology, American botanist Asa Gray praised him for placing teleology on a sounder foundation. His son Francis Darwin, in his *Autobiography*, went so far as to say: "One of the greatest services rendered by my father to the study of Natural History is the revival of Teleology." German anatomist A. Kölliker, however, criticized Darwin for championing teleological thinking (Kölliker, 1864). Although it is now generally agreed that Darwin offered a non-teleological explanation of evolution, a few scientists still see him as a teleologist (e.g., Lennox, 1993). This reminds me of the different reactions to the 1960 Nixon-Kennedy debate: Kennedy's supporters thought he had won, and Nixon's supporters thought he had won. Whenever there is ambiguity, people tend to interpret the event or idea according to their preference. Thus, what did Darwin himself actually believe? A conversation that he had with the Duke of Argyll in 1882, the last year of Darwin's life, may reveal his own ambiguity on the nature:

> ...in the course of that conversation I said to Mr. Darwin, with reference to some of his own remarkable works on the *Fertilization of Orchids*, and upon the *Earthworms*, and various

> other observations he made of the wonderful contrivances for certain purposes in nature – I said it was impossible to look at these without seeing that they were the effect and the expression of mind. I shall never forget Mr. Darwin's answer. He looked at me very hard and said, 'Well, that often comes over me with overwhelming force; but at other times,' and he shook his head vaguely, adding, 'it seems to go away' (The Duke of Argyll, Good Words, April 1885, Cited by F. Darwin, ed., *Autobiography*).

Perhaps the ambiguity of teleology in *Origin* comes from Darwin's own ambivalence.

From "My Theory" to "the Theory"

> ...although deeply interested in his own priority, he never realized that his own ideas were second hand. He thought he had worked them out himself, even when he had only sorted them out. Moreover his ideas were less clearly sorted out and less clearly expressed and, worst of all, less strictly and less openly held and maintained than the ideas of those who first thought of them.
>
> -Cyril Dean Darlington

The first edition of *Origin* states "my theory" 45 times; in later editions, however, this is gradually replaced with "the theory." By the sixth edition, "my theory" no longer appears. Since "my" was replace by "the" without any explanation, no one noticed and no one cared (Darlington, 1959). Samuel Butler (1887), a former Darwin enthusiast turned critic, pointed out this change two years after Darwin died. Darlington (1959) argues that this reconstruction of a proposition, which is unique in the history of literature, reveals Darwin's character. So why did he stealthily delete "my"? It is likely that he realized that he could not really claim priority for his idea of evolutionary mechanism offered in *Origin*. There are two reasons for this change: his predecessors and what Darlington (1959) aptly called the "retreat from natural selection."

The first person to point out Darwin's overlooking of forerunners was Patrick Matthew, a little-known Scotch botanical writer. Soon after the publication of *Origin* in 1860, Matthew claimed in *The Gardener's Chronicle* that he had posited the idea of evolution 29 years earlier in *On Naval Timber and Arboriculture*. Darwin was forced to admit Matthew's priority over Wallace and himself in a letter to Wallace. Wallace also agrees with Matthew's claim of priority in a letter to Samuel Butler: "To my mind your quotations from Mr. Patrick Matthew are the most remarkable things in your whole book, because he appears to have completely anticipated the main ideas of the *Origin of Species*..." (Wallace, 1905).

Matthew might have been justified in claiming priority in the case of natural selection; however, he also failed to acknowledge his predecessor, William Lawrence (Darlington, 1959). Four decades before the publication of Darwin's *Origin*, Lawrence published his *Natural History of Man*. Although the intellectual climate of the time was moving rapidly toward secularism, it was still hazardous time to put forward the idea of evolution in the early 19th century. Lawrence especially put himself at risk because he proposed evolution based on man and indicated all of the implications for the society of his day (Darlington, 1959). Lord Chancellor Eldon rejected his request for copyright because it contradicted the Bible (Darlington, 1959). Unwilling to put his professional career as a surgeon in possible jeopardy, he chose to suppress his own book. This might have been a wise move, for later on he would become the President of the Royal College of Surgeons and Sergeant-Surgeon to Queen Victoria. Darlington (1959) contends that Darwin could have written a better book if he had read Lawrence's. This brings up a "what if" question: Had Lawrence chosen the path of continuous study and promotion of evolution, would he have avoided his current status of seemingly near oblivion?

The "Historical Sketch" found in the *Origin*'s fourth edition acknowledges William Charles Wells as the originator of the concept of natural selection:

> In 1813, Dr. W. C. Wells read before the Royal Society 'An Account of a White Female, part of whose skin resembled that of a Negro;' but his paper was not published until his famous 'Two Essays upon Dew and Single Vision' appeared in 1818. In this paper he distinctly recognizes the principle of natural selection, and this is the first recognition which has been indicated; but he

applies it only to the races of man, and to certain character alone.

An examination of Darwin's note reveals an interesting point: Darwin had read Lawrence, but not Wells (Vorzimmer, 1977), and yet mentions Wells as the originator of this concept. One reason for this might be that Wells applied natural selection to man, and so Darwin might have seen a chance to claim his priority in applying this concept to all life.

Edward Blyth, a chemist turned zoologist who spent much of his time in India as a curator of the museum in Bengal, discussed natural selection in three papers published in 1835 and 1837 in *The Magazine of Natural History*. Eiseley (1959) argues that Blyth discussed the main ideas of the struggle for existence, variation, natural selection, and sexual selection in these papers. Eiseley (1959) contends that Blyth was Darwin's main (but unquoted) source of inspiration for natural selection based upon the matching of a number of rare words, similarities of phrasing, and the use of similar examples. Darlington (1959) adds that Darwin's inspiration from Malthus was to "avoid mentioning his significant inspiration which was Blyth." Unlike Matthew, however, Blyth did not claim priority and remained a friend of Darwin. At any rate, no one can disagree with Eiseley's contention that Blyth should be given more credit for his contribution to evolutionary biology.

Moreover, Darwin did not put much effort into examining existing concepts of evolution. Butler calls the "Historical Sketch" in *Origin* "meager and slovenly" (cited by Løvtrup, 1987). Even a reference to his own grandfather, Erasmus Darwin, was treated with a mere footnote. Irvine (1955) has this to say:

> Darwin's "Historical Sketch" was a jewel of studied or unstudied detraction, damning many – among them his own grandfather – in little space and small type. His ambiguities, intended or unintended, were not simply baffling, but interested, deceptive, and insidious, enabling him to insinuate….

Butler (1924) was perhaps the first one to bring this to the public's attention:

> Mr. Darwin has a habit, borrowed, perhaps, *mutatis mutandis*, from the framers of our collects, of every now and then adding

the words "through natural selection," as though this squared everything, and descent with modification thus became his theory at once. This is not the case. Buffon, Erasmus Darwin, and Lamarck believed in natural selection to the full as much as any follower of Mr. Charles Darwin can do. They did not use the actual words, but the idea underlying them is the essence of their system.

Another reason for changing "my theory" to "the theory" was that Darwin's emphasis was gradually shifting from natural selection to Lamarck's "use and disuse," which is also known as the "inheritance of acquired characters." Darlington (1959) calls this the "retreat from natural selection." A widely used example of this principle is the giraffe's long neck and legs, which is held to be the result of its ancestors' constant efforts to reach the higher branches to browse.

First put forward by French biologist Jean-Baptiste de Lamarck, this example is often used to dismiss the idea of inheriting acquired characters. Lamarck, who died blind, poor, and forgotten in 1829, may be one of history's most unjustly treated biologists. His daughter's outcry on his gravestone reads: "The future will remember you, my father." Unfortunately, however, the future to date has not been very generous. When Lamarck published his *Philosophie Zoologique* in 1809, half a century before Darwin's *Origin*, it was largely ignored; however, the few scientists who read it viciously attacked his ideas. The most damaging critic was Georges Cuvier, one of the top French zoologists who was renowned enough to be invited by Napoleon to join his expedition to Egypt in 1798. Cuvier declined because he preferred to work in a laboratory. Ironically, however, the mummified animals brought back from that expedition became one of the main reasons for rejecting the transmutation of species. The embalmed cats, birds, dogs, and monkeys, all of which were 2,000 to 3,000 years old showed no difference when compared with those of today. As someone who embraced the biblical creation story, as most biologists at the time did, Cuvier (1835) considered Lamarck's idea "dangerous enough to merit being attacked." Furthermore, Cuvier's political skills attracted the negative attention of his contemporaries (Burkhardt, 1977). Because Lamarck's reputation has been so tarnished by Cuvier and others, even those who accept the inheritance of acquired characters, among them Robert Chambers, did best to distance themselves from him (Bowler, 1983).

Darwin was introduced Lamarck's idea while studying medicine in Edinburgh from 1825 to 1827 (Løvtrup, 1987). He also came across Lamarck's idea while studying Lyell's *Principle of Geology,* in which a great deal of Lamarck's work was discussed (Løvtrup, 1987). Darwin initially considered Lamarck a "lofty genius," but completely changed his opinion a few years later. In a letter to Hooker in 1844, he wrote: "Heaven forefend me from Lamarck nonsense…" and even called it "veritable rubbish." Barzun (1958) argues:

> The greatest injustice was of course to Lamarck. According to Osborn, 'the disdainful allusions to him by Charles Darwin…long place him in the light of a purely extravagant speculative thinker'. These disdainful allusions were all the more reprehensible that they were misrepresentations, not excusable through ignorance, and made by an unavowed part-disciple. It is clear that Darwin would not have treated a pigeon or a gastropod in this cavalier fashion, and that coming from any other man than an 'intellectual, modest, simple-minded lover of truth' the whole performance would have been damned as shamefully fraudulent.

Such a gross injustice is especially incomprehensible when Darwin himself shifted toward Lamarck's idea. As Darlington (1959) remarks:

> Steadily he began to shift his ground away from natural selection. Retreating in good order he fell back on the assumption that acquired characters were inherited. The action of the environment in changing heredity constituted the 'persisting' cause of variation. It not only *caused* the variation: it now also *directed* it. The environment pushed a soft heredity. And when it had produced the variation the result was hard enough for natural selection to pull it further and to preserve it. The change to this new theory was produced by imperceptible modifications of successive editions.

But according to Darlington (1959), he did not revise his "Historical Sketch" to match the shift;

In the same book we therefore find, on the one hand, statements rejecting Erasmus Darwin, Lamarck and Herbert Spencer and, on the other hand, 'the theory' which now so largely mixed up with theirs."

Darwin's Accomplices and the Birth of the Darwin Industry

Today our duty is to destroy the myth of evolution, considered as a simple, understood, and explained phenomenon which keeps rapidly unfolding before us. Biologists must be encouraged to think about the weaknesses and extrapolation that theoreticians put forward or lay down as established truths. The deceit is sometimes unconscious, but not always, since some people, owing to their sectarianism, purposely overlook reality and refuse to acknowledge the inadequacies and falsity of their beliefs.

-Pierre-Paul Grasse

….we are frequently amazed at how much of what we take for granted emerged from an intellectual battle in which one of an array of alternative theories was the victor. Intellectual victory does not, however, necessarily equate with correctness…it would be foolhardy to cling unreservedly to a particular set of models and hypothesis without at least occasionally questioning their very bases. Unfortunately, however, the urge to defend rather than dissect our intellectual roots is quite strong.

-Jeffrey H. Schwartz

If Darwin's status in science is the result of the "greatest deceit" (Løvtrup, 1987), the question is how could he fool so many people for so long. Darwin could not have created and then maintained the "Darwinian deceit" all by himself, even if he was a master manipulator. In fact, he had friends who would zealously promote him as a great scientist and defend both him and his ideas against all critics.

Arnold C. Brackman, an award-winning author and journalist, published *A Delicate Arrangement* in 1980, three years before his death. This "arrangement," which he calls "the greatest conspiracy in the annals of science," is perhaps the most serious accusation against Darwin.

A Critique of Science

Brackman goes beyond the usual claim that Darwin and his friends (Charles Lyell and Joseph Hooker) conspired to elevate Darwin's part in advancing natural selection and simultaneously minimize the role of Alfred Wallace. He charges that Darwin stole Wallace's idea and concocted an elaborate scheme to cover it. Roy Davies (2008), a former producer of science programs for the BBC, echoes this charge of plagiarism in his *The Darwin Conspiracy: Origins of a Scientific Crime*. This accusation is nothing new; it was first made (although now largely forgotten) by Samuel Butler.

Before analyzing this episode, we need to know a little more about Wallace. This co-discoverer of natural selection, who often merits nothing more than a passing remark in biology textbooks, was born fourteen years after Darwin into a middle class family with modest resources. After finishing his formal education, six years of grammar school, he left to work as an apprentice surveyor for his brother. Like Darwin, this enthusiastic amateur naturalist with no formal training in science taught himself through reading and, more importantly, direct experience. Unlike Darwin, who came from the upper class and had the support of his professor and his father in arranging and paying for his place on the H.M.S. Beagle, Wallace had to self-finance his journeys to the Amazon as a specimen collector (1848-52) and to the Malay Archipelago (1854-62). While he was still engaged in collecting specimens in the jungles, Darwin had long since returned home and published books on his journey. In fact, Darwin's works inspired Wallace. But unlike the conservative Darwin, who remained a creationist throughout his voyage, Wallace soon began to work on the question of how species originated. While in Sarawak (northern Borneo), confined to his bungalow during the rainy season, he developed what he called the "Sarawak Law" and subsequently had it published in *Annals and Magazine of Natural History*. The most important feature of this ten-page article was the "principle of divergence," which posits that species evolve like a branching tree. This was a significant departure from established scientific thought, for until his paper evolution had generally been conceived as a ladder up which the species climbed. Although Wallace's article was stimulating enough to generate correspondence between Darwin and his friends (e.g., Lyell and Blyth), Darwin did not pay much attention to it. This is somewhat baffling, because Darwin also came out with this same tree-branching metaphor on his own.

Having introduced this "law," Wallace sought to explain the underlying mechanism of the branching process. Remarkably, the idea of natural selection came to him while he was lying in bed with a malaria-induced fever. While fighting for his life, he was reminded of Thomas Malthus' essay on population. Malthus' principle emphasizes how cruel nature can be: Animals produce far more offspring than the available resources can support. Thus, their perpetual struggle for survival eliminates the less able, leaving the ones with the better traits. Wallace must have wondered if he belonged to the latter group. The question of his own survivability gave him some insight into the mechanism of evolutionary process. By the end of the day, he had sketched out the draft of an evolutionary process based on the idea of the survival of the fittest. Upon completing it, he sent it to a man he thought would be the most interested in reading it: Charles Darwin. This was truly a fateful decision with a historical significance, for this single act would trigger a chain reaction that would rank as one of the most astonishing episodes in science.

Darwin received this draft, now called the "Ternate paper" in June 1858. Unlike Wallace's Sarawak paper of three years ago, this one caused Darwin a great deal of anguish. In a letter to Lyell, who had previously urged him to publish, Darwin stated: "I never saw a more striking coincidence. If Wallace had my M.S. sketch written out in 1842 he could not have made a better short extract! Even his terms now stand as Heads of my Chapters...So all my originality, whatever it may amount to, will be smashed" (In F. Darwin, 1892). Then Darwin asked Lyell: "Do you not think his having sent me this sketch ties my hand? If I could honorably publish, I would state that I was induced now to publish a sketch...from Wallace having sent me an outline of my general conclusion" (In F. Darwin, 1892).

In effort to "honorably publish," Lyell and Hooker came up with an ingenious scheme that was utterly dishonorable, to say the least: they would use their influential positions as fellows of the Linnean Society, one of the world's most prestigious forums of biological science, to push through their dubious agenda. On June 17, both papers were presented. Both men were given credit for coming out with the theory of evolution "independently and unknown to one another." Then, Lyell states: "neither of them having published his views." This is not true. Darwin had not published, but Wallace had published his Sarawak paper in 1855. Starting out with this false statement, they then tried to establish Darwin's priority (although they claimed to be doing no such thing) by presenting an extract of unpublished work purportedly done in 1844. This extract was supported by two documents: a letter from Lyell and Hooker and an abstract of a letter from Darwin to Asa Gray. Wallace's article was the last item to be presented.

The next unsavory action speaks for itself: "So highly did Mr. Darwin appreciate the value of the views therein set forth, that he proposed, in a letter to Sir Charles Lyell, to obtain Mr Wallace's consent to allow the Essay to be published as soon as possible." Although this sounds like a good-faith effort to obtain Wallace's consent, such consent was, in reality, impossible to obtain given the long time it took mail sent from the Malay Archipelago to reach England and vice versa. If this were a genuine attempt to secure Wallace's consent, they would have arranged both men's joint presentation of their papers at a mutually agreed-upon date. Løvtrup (1987) calls this "a masterpiece of deception."

Wallace was up against what Brackman (1980) terms the Victorian era's "rampant snobbery" on the part of aristocratic scientists. Darwin, Lyell, and Hooker were close friends as well as scientists who enjoyed social and academic status. Wallace, on the other hand, was a relative upstart, an outsider who eked out his living by collecting specimens in the jungle. Darwin's close friends were willing to ensure Darwin's prominent role in evolutionary biology at the expense of both Wallace and academic morality.

Now, let's look at Brackman's more serious charge that Darwin stole Wallace's ideas. One reason for this suspicion is that eight letters from Wallace to Darwin, with the exception of one fragment, are missing. Such carelessness was uncharacteristic of meticulous Darwin. Barbara Beddall (1968) maintains: "It seems surprising that all the material relating to the most dramatic (not to say traumatic) moment in his life should disappear." Another reason is that there are "too many similarities" (Brooks, 1984) between Wallace's idea and Darwin's writing. Brackman suggests that Gray's letter, which was presented to the Linnean Society, was Darwin's opening move. I feel that we should be prudent when accusing someone, especially if the charge is serious charge, because simultaneous discoveries are frequent occurrences.

Orogee Dolsenhe

Darwin's Bulldog Who Didn't Believe in Natural Selection

If Lyell and Hooker planted the seed of the Darwinian myth, it was Thomas Henry Huxley, more than anyone, who is responsible for its growth. Calling himself Darwin's bulldog, he was willing to defend Darwin and his ideas against anyone. He wrote to Darwin: "I am sharpening up my claws and beak in readiness" (Darwin, 1892). But his fervent role in promoting Darwin and *Origin* is quite paradoxical for a number of reasons. First, he used to oppose to any theory of transmutation and attacked anyone for advocating them. In response to the anonymous publication of Robert Chamber's *Vestiges of the Natural History of Creation* in 1844, for instance, Huxley wrote the "cruelest and most cutting review" (Ruse, 1979). Consider some of the expressions he used in the "utmost savagery" ways (Eiseley, 1961): "foolish fancies," "charlatanerie," "pretentious nonsense," and "work of fiction." Even those who had a distinguished status in science could not escape his fierce criticism both in public lectures and in print (White, 2003).

Second, according to Bowler (1988), his generally assumed Darwinian conversion was superficial. Huxley's conversion to evolution came from reading Ernst Haeckel's *Generelle Morphologie* of 1866, not Darwin's *Origin* (Bowler, 1988). He became a full-blown evolutionist only with the newly discovered *Archaeopteryx* found in the Jurassic rocks of Germany, which seem to indicate that birds had evolved from reptiles.

Third, and most importantly, he really did not believe in Darwin's theory of natural selection. Thus Bowler (1997) appropriately calls him a "pseudo-Darwinian." This explains why Huxley almost never discussed lectured on evolution to his students. One student wrote: "One day when I was talking to him, our conversation turned upon evolution. 'There is one thing about you I cannot understand,' I said, 'and I should like a word in explanation. For several months now I have been attending your course, and I have never heard you mention evolution, while in your public lectures everywhere you openly proclaim yourself evolutionist'" (Huxley, 1901). Ruse (1997) eloquently put in this way: "Darwin's bulldog was ever ready to bite; but, in front of the classroom, he had little inclination to bark." In his lecture on the *Origin*'s twenty-first anniversary (1880), Huxley did not even mention the phrase (Løvtrup, 1987).

Fourth, in spite of his misgivings about natural selection, Huxley praised Darwin as much as he could. In his review of *Origin* in *Times*, Huxley (1859) wrote: "so clear is the author's thought, so outspoken his conviction, so honest and fair the candid expression of his doubts." As time went on, his adulation only grew. For instance, in response to Darwin's critics in *Contemporary Review* in 1871 he wrote that it was not only Darwin's "industry, his knowledge, or even the surprising fertility of his inventive genius" that struck the student of nature, "but that unswerving truthfulness and honesty which never permit him to hide a weak place, or gloss over a difficulty." In a memorial printed in *Nature* in 1882, Huxley declared Darwin to have possessed "an intellectual which had no superior...[and] a wonderfully genial, simple, and generous nature...The more one knew of him, the more he seemed the incorporated ideal of a man of science."

Intellectual Neglect in Darwin Worship

> ...if our civilization is to survive, we must break with the habit of deference to great men. Great men may make great mistakes.
> -Karl Popper

> Since Darwin attained sainthood (if not divinity) among evolutionary biologists, and since all sides invoke God's allegiance, Darwin has been depicted as a radical selectionist at heart who invoked other mechanisms only in retreat, and only as a result of his age's own lamented ignorance about the mechanism of heredity. This view is false.
> -Stephen J. Gould

> Education is a method whereby one acquires a higher grade of prejudices.
> -Laurence J. Peter

Human history has abundant examples of elevating individuals to divine stature. For instance, the Sumerian king Gilgamesh was a god-king who possessed superhuman strength. The Pharoahs of ancient Egypt were considered gods on Earth. In a lesser form, the authority and legitimacy of kings and emperors were derived from God's will. In our own time, the personality cults of totalitarian regimes promoted leaders like Hitler, Stalin, Mao, and Kim Il Sung to god-like status. There are two sides to this phenomenon. One significant part is the coercive element at the top. For instance, Stalin's self-glorification was forced upon people by a variety of methods. Yet for some, such veneration represents the genuine fervor of average people. During the Cultural Revolution, millions of "Red Guards" attacked anyone deemed to be counter-revolutionary. Mao's *Little Red Book* became the ultimate authority for his cult. Today, sportsmen, actors, actresses, and singers attract the most hero worshippers. This brings a question to mind: why do people literally worship celebrities as though they were superior to them? Stuart Fischoff, who specializes in the cult of celebrity at UCLA, believes that people, being social animals, are programmed to follow the leader.

Darwin, perhaps more than anyone in science, has attained a divine status in science. His burial in Westminster Abbey following a state funeral in 1882, as well as his picture on the back of the British ten pound note are just some of the visible signs of his stature. In a *New York Times* essay, Carl Safina (2010) argues that we should stop focusing so much on Darwin and more on evolutionary science. Stating that what is going on in biology is the "cult of Darwin," he makes a cogent argument: "Almost everything we understand about evolution came after Darwin, not from him. He knew nothing of heredity or genetics, both crucial to evolution. Evolution wasn't even Darwin's idea." The highly biased version of Darwin can be seen in television programs broadcast by PBS. For instance, on "Charlie Rose" James D. Watson proclaimed that Darwin was the most important person who ever lived on earth. In the same program, Ed Wilson remarked that everything he [Darwin] said was right.

The underlying process for such an extreme elevation of Darwin is, I would argue, is intellectual neglect. To reiterate, we are magnetically drawn to the extreme position whether in praising or denouncing someone. Thus, once the leaning toward the admiration side largely by the effort of Thomas Huxley, Darwin's stature continues to approach divinity as time went on. The indoctrination into this cult starts in the very early years of education, for those who teach it are usually not very knowledgeable in this field. Young students, who learn the consecrated version of both the man and his theory of natural selection, thus acquire the ideologies of modern science by focusing on Darwin rather than evolutionary idea. McComas (1997) cautions:

> ...the teachers must be careful not to rely solely on their textbooks as accurate and complete sources of information. In the case of the discovery of evolution by natural selection, texts are frequently replete with errors of omission and outright inaccuracies.

McComas (1997) adds:

> Not only is the conventional textbook account incomplete and frequently incorrect, the true story is far more interesting than the typical fiction. Texts that omit a description of the fascinating history of the discovery of evolution miss a wonderful opportunity to acquaint students with the human dimension of science...

The Role of Interpersonal Conflict in Science

Adding another bizarre twist to what is certainly the wackiest episode in science involves Huxley yet again. In this case, it concerns the main reason why he backed Darwin's idea so fervently even though he did not believe in it. His pseudo-Darwinian stance is especially ironic, given that he called himself a "man of science," often invoked the "truth" in his debates with theologians, and praised Darwin by comparing him with Newton as "a searcher after truth and interpreter of Nature" (Thomas H. Huxley, *The life and letters of Charles Darwin*).

Huxley's public stance largely stems from his conflict with Richard Owen, a highly regarded anatomist and paleontologist. Owen's particular specialty was the study of dinosaurs, a field in which he was the first to recognize that they were different from modern reptiles. As a respectable scientist and twenty-one years older than Huxley, Owen was in a position to help young Huxley establish a name for himself in the field of anatomy. Huxley, who was having a terrible time securing a grant for his research aboard the HMS Rattlesnake and had been rejected by the Royal Society and the First Sea Lord, turned to Owen for help. After ten days of waiting for Owen's answer, an irritated Huxley send him a reminder. Then one day Huxley came across Owen, and he never forgot or forgave the experience. He revealed his bitterness toward Owen in a letter to Forbes:

> I then on the fourth day afterwards, met him. I was going to walk past, but he [Owen] stopped me, and in the blandest and most gracious manner said, 'I have received your note. I shall grant it.' The phrase and the implied condescension were quite 'touching' – so much that if I stopped a moment longer I must knock him into the gutter. I therefore bowed and walked off.

Huxley's subsequent vindictiveness toward Owen became a driving force in his scientific practice when it came to evolution. For instance, his vicious attack on Chambers' *Vestige* was the best way to tackle Owen's view on species origination, particularly his response to it (Desmond, 1982). Ruse (1997) agrees that "the critique Huxley wrote of Robert Chambers' evolutionary tract *Vestiges of the Natural History of Creation* was in effect being used primarily to harm Owen (some of whose speculations Chambers had used), for the tone is quite exceptional in vehemence and nastiness." Owen and Chambers shared the view that the Creator's power and wisdom are incorporated in explaining the origin of different forms. We will discuss their evolutionary approaches later on. Paradoxically, however, Huxley's own discussion on the subject differed little from Owen's (Desmond, 1982). While Chambers and Owen were the target for settling a score, Huxley made use of Darwin for the purpose of contrast, just as a movie uses the opposite characters of a villain and a hero. White (2003) makes this point:

> In opposition to the figure of Owen as the ambitious and corrupt autocrat, he needed a sympathetic figure of imposing purity and propriety. Thus his personal attacks on Owen proceeded alongside his enthusiastic tribute to Darwin. The elevation of Darwin as a heroic man of science and the denigration of Owen were part of the same process – each was a character witness in a trial staged by Huxley to weigh the merits of the man of science.

Therefore Darwin, the main beneficiary of this long and bitter conflict, was put on a pedestal alongside in Darwin, 1892).

One might ask how just one man could accomplish such an astonishing feat. In short, Huxley was no ordinary man. For over three decades he was at the center of the British scientific establishment (Barr, 1997). As his stature grew, he held such positions as founding editor of *Nature* and president of the Royal Society, which became a platform to advocate for his inclinations. In the refereeing process of journals, Huxley acted as a conduit for the work of which he approved (Ruse, 1997). His influence was felt in key appointments not only in Britain, but also abroad; he made sure that his students or men of whom he approved would occupy most such posts (Barr, 1997). Huxley's students occupied key academic positions in the 1860s and 1870s (Desmond, 1982).

Maupertuis: A Forgotten Genius Who Was a Precursor to Darwin and Mendel

The British forerunners and rivals of Darwin were largely relegated to minor roles in advancing the concept of natural selection. Perhaps the greatest injustice was done to a French biologist named Maupertuis, who was largely forgotten until Bentley Glass (1959), a geneticist at the Johns Hopkins University, rescued him from the oblivion. The sad story of Maupertuis adds another odd chapter to Darwinian absurdity largely orchestrated by Huxley.

Darwin and Mendel are considered the two pillars of modern biology: Darwin for evolution and Mendel for genetics. While Darwin earned a great reputation during his life, Mendel's works on peas were recognized only long after his death. De Vries, a Dutch botanist, presented the theory of mutationism that provided the nature of Darwin's variation of species. Then there is Pierre Louis Moreau de Maupertuis, who lived more than a century before either Darwin and Mendel, yet hardly anyone knows about him. Emery (1988) considers him one of the eighteenth century's greatest scientists, and Glass (1959) sums up his neglected contribution to biology in the following terms:

> In short, virtually every idea of the Mendelian mechanism of heredity and the classical Darwinian reasoning from natural selection and geographical isolation is here combined, together with De Vries' theory of mutations as the origin of species, in a synthesis of such genius that it is not surprising that no contemporary of its author had a true appreciation of it.

Maupertuis' theory of evolution and genetics relied on many types of evidence: anatomy, the study of monsters, the observation of racial differences and family resemblances, animal breeding, microscopy, the natural history of insects, and chemistry (Terrall, 2002). In addition, he was the first to apply the laws of probability to the study of heredity (Glass, 1959). It is beyond the scope of this book to go into the details of his ignored accomplishments; however, a few brief examples must be given. The cornerstone of Mendel's idea, the principle of segregation which states that each parent contributes half of its genetic makeup to its offspring, was foreshadowed by Maupertuis. Maupertuis also proffered another of Mendel's important concept, that of genetic "dominance," but in a somewhat vague manner. In addition, he applied mathematics in calculating the probability of polydactyly (a person having six digits on each hand and each foot) long before Mendel did.

One of the *Origin*'s major components is artificial selection, which became the foundation of natural selection. This seminal idea can also be traced to Maupertuis (1745), who wrote:

> Nature contains the basis of all these variations: but chance or art beings them out. It is thus that those whose industry is applied to satisfying the taste of the curious are, so to say, creators of new species. We see appearing races of dogs, pigeons, canaries, which did not at all exist in Nature before. These were to begin with only fortuitous individuals; art and the repeated generations have made species of them. The famous Lyonnés every year created some new species, and destroyed that which was no longer in fashion. He corrects the forms and varies the colors: he has invented the species of the harlequin, the mopse, etc.

And just as Darwin did, Maupertuis (1745) applied what happens in artificial selection to the natural process:

> What is certain is that all the varieties which can characterize new species of animals and plants, tend to become extinguished: they are the deviations of Nature, in which she preserves only through art or system. Her *works* always tend to resume the upper hand.

It is quite curious how so many seminal concepts could have escaped the attention of so many for so long. Virtually no biology textbook even mentions his name.

This largely forgotten scientist, an original thinker whose contribution to science goes well beyond biology, was the first European to appreciate Newton's law of gravitation. In fact, he led an expedition to Lapland to measure the length and degree of the meridian and confirmed Newton's prediction that Earth is a sphere flattened at the poles. The expedition earned him such a glowing reputation that Frederick II of Prussia personally him to become president of the Berlin Academy of Sciences.

One of his most important theories was the Principle of Least Action, which posits the essentially economical nature of the universe. Simply put, nature uses the smallest possible quantity for any event. This overarching generalization was not fully appreciated until the advent of quantum mechanics in the 20th century (Glass, 1959). Unfortunately and ironically, however, this principle became one of the "most notorious quarrels" in the 18th century and the very reason why Maupertuis faded into oblivion (Terrall, 2002). This regrettable predicament is largely due to two men: Samuel König and Voltaire. Maupertuis and König initially had a friendly relationship. The former opened doors for the latter to the Paris Academy of Sciences and helped him get a job as a mathematics tutor. Interestingly, the onset of mutual enmity can be traced to another notorious quarrel: whether Newton or Leibniz discovered calculus. König argued for Leibniz's priority; Maupertuis, the first European to appreciate Newton, naturally objected. The spirited argument left both men with a sour taste in their mouths. König later published an article in which he charged that Maupertuis' Principle of Least Action was erroneous. An even more serious charge was that Maupertuis knew it had first been proposed by Leibniz. In other words, König accused Maupertuis of plagiarizing Leibniz's idea. In providing the proof of evidence (the Leibniz's letter) that König alleged to have, he had "muddied his claims with so many improbable and unsubstantiated explanations" (Terrall, 2002).

But this was only the first phase of the dispute. The next phase opened when Voltaire, one of the Enlightenment's greatest writers and known for opposing tyranny, bigotry, and cruelty, unexpectedly jumped into the controversy by siding with König. Like his relationship with König, Maupertuis had initially been on friendly terms with Voltaire, for they were comrades in promoting Newtonian ideas. Their relationship began to sour, however, due to Voltaire's jealousy of Emile du Châtelet's affection for Maupertuis and the latter's resentment of Voltaire's sharp wit (Terrall, 2002). Châtelet was a rare female mathematician and physicist in the Enlightenment: not only did she translate Newton's *Principia* into French, but she was also a mistress of Voltaire. What made the situation even worse was that they both became rivals for Frederick's attention, mainly due to their massive egotism (Terrall, 2002). Voltaire portrayed König as a victim of injustice and ridiculed Maupertuis as a tyrant and a buffoon. His persistent attack on Maupertuis turned into a spectacle and became the subject of gossip all over Europe (Terrall, 2002). Totally aware of the power of scorn, Voltaire once said: "Ridicule overcomes almost anything. It is the most powerful of weapons" (quoted in Hellman, 1998). His punchy satires gave the public a false impression of Maupertuis by utterly destroying the latter's reputation (Glass, 1959).

The third phase of injustice to Maupertuis comes from the historical studies of evolution. Huxley (1876), Darwin's bulldog and the first person to investigate Darwin's predecessors, dismisses Maupertuis' works with a single sentence:

> Somewhat later, Maupertuis suggested a *curious* (emphasis added) hypothesis as to the causes of variation, which he thinks may be sufficient to account for the origin of all animals from a single pair.

In reading the sentence, it is quite obvious that Huxley sought to sweep aside Maupertuis' significance as regards the evolutionary idea. Among the many ideas and empirical evidence that this scientist put forward, Huxley chose a single point. His misleading description might have played an even larger role than that of König and Voltaire in ensuring that Maupertuis was forgotten. Quatrefages, Clodd, Packard, Nordenskiøld, Singer and other historians of the theory of evolution have ignored Mauertuis completely (Glass, 1959). Osborn (1894) grouped him with the speculative evolutionists. It is somewhat puzzling why he fails to realize the extensive empirical evidence provided by Maupertuis to support his conclusions. Osborn was a staunch follower of Lamarck and could have used the priority of Maupertuis as a means to reduce Darwin's prestige and promote Lamarckism. Glass (1959) regards this oversight as the chief reason for the "misapprehension" of Maupertuis' significance. It is not that there was no one to recognize Maupertuis's priority over Darwin in the field natural selection. For instance, as far back as 1873 Brewer (1873) had acknowledged: "Having lately had occasion to examine the works of Maupertuis I, like Prof. Jevons, was struck by meeting with anticipatory glimpses of modern theory of Natural Selection." In the 20th century, Guyénot (1941) was one of the first to reassess Maupertuis' contribution: "Already he was reasoning more on the basis of facts than of conceptions of *a priori*. He was one of the founders of Evolution and appears to us, in addition, as the most remarkable precursor of contemporary Mutationism."

While Mayr (1982) contends that Maupertuis was neither an evolutionist nor a founder of the theory of natural selection, he does gives this forgotten scientist credit for being a forerunner of de Vries: "His genetic studies led him to a theory of what we would now call *speciation by mutation* (emphasis added)." The use of this phrase to describe Maupertuis' achievement seems to contradict Mayr's assertion that he was not an evolutionist. In addition, as mentioned earlier, Maupertuis compared artificial selection with natural process. Is Mayr, who is usually not very generous to Darwin's competitors, following in the footsteps of Voltaire, Huxley, and Osborn? Bowler (1988) criticized Mayr for spending a mere half a dozen pages on the alternatives to Darwin and blamed the imbalance on the "Darwin industry."

Religious Intolerance

Like many Greek ideas, the evolutionary concept had to struggle for acceptance against Christianity. Any thought that contradicted the literal interpretation of the Bible was susceptible to being condemned as heresy. In the prevailing Christian theology, God created each and every extinct and existing species. However St. Augustine (354- 430), one of the most important theologians, cautioned against rushing to judgment when interpreting the Bible. For example, the six-day creation of the universe in particular should not be taken as the literal passage of time, but rather a logical framework. He argued that God's creation of all living things as "potentially in their causes, as things that will be in the future," rather than an immediate creation in the beginning (St. Augustine, 1982). In essence, this plasticity of time allowed both creation and evolution. His position is similar to that of Immanuel Kant:

"The creation is never over. It had a beginning but it has no ending. Creation is always busy making new scenes, new things, and new Worlds." Augros & Stanciu (1987) point out that Augustine's view preserves "both the causality of nature and the causality of the Creator." In addition, he addresses the inconsistencies between the Bible and nature in his *The City of God*:

> There is a question raised about all those kinds of beasts which are not domesticated, nor are produced like frogs from the earth, but are propagated by male and female parents, such as wolves and animals of that kind; and it is asked how they could be found in the islands after the deluge, in which all animals not found in the ark perished, unless the breed was restored from those who were preserved in pairs in the ark. It might, indeed, be said that they crossed to the islands by swimming, but this could only be true of those very near the mainland; whereas there are some so distant that we fancy no animal could swim to them. But if men caught them and took them across with themselves, and thus propagated these breeds in their new abodes, this would not imply an incredible fondness for the chase.

Yet his call for prudence has largely been drowned out by not only in the dogma-dominated religious era, but also in science.

Darwin wrote a letter to John Lubbock in 1859 (the same year he published *Origin*) in regards to William Paley's book: "I do not think I hardly ever admired a book more than Paley's "Natural Theology." Paley, an Anglican priest and author of *Natural Theology* (1802), claimed that nature has the "necessity of an intelligent Creator." Although similar arguments were made by John Ray's *Wisdom of God* (1691) and William Derham's *Physico-Theology* (1713), Paley's book was one of the most popular works of its kind. In his highly readable text, he presents persuasive arguments concerning God's handiwork in nature. He starts by reasoning out the basic difference between a stone and a watch found on the ground. Unlike the stone, he argues, the watch's "several parts are framed and put together for a purpose, e.g. that they are so formed and adjusted as to produce motion, and that motion so regulated as to point out the hour of the day." No one would doubt that someone made the watch. He also asserted that a living organism is quite different in a fundamental way: It can replicate itself. Paley contends that the atheist's rejection of the supernatural is as "absurdity."

What was the fate of the design argument during the Enlightenment, when atheism was beginning to gain a foothold and growing rapidly? Ironically, it was not very well received even by theologians. According to Brooke (1991), at least in the eighteenth century, it was not science that undermined the design argument, but rather the religious apologists who were asking too much of it. Thomas Aquinas, Calvin, and other Christian scholars had placed severe limits upon the scope of natural theology, thereby constraining one's comprehension of nature by a literal and strict interpretation of the Bible. Consequently, the freethinking Deists were not well received by the theistic camp. For instance, Blaise Pascal (1977) gave a stern warning:

> All of those who seek God apart from Christ, and who go no further than nature, either find no light to satisfy them or come to devise a means of knowledge and serving God without a mediator, thus falling into either atheism or deism, two things almost equally abhorrent to Christianity.

Like most orthodox Christian apologists during the Enlightenment, Pascal did not see deism as a potential ally in defending Christianity against atheism. This is ironic, because much of the supernatural implication of nature could have been used by either Christian defenders or deistic scholars. However, those with firm Christian convictions would often try to differentiate their statement about the God of nature from those of the deists (Brooke, 1991). The intolerance of devout Christian scholars can be exemplified by the unfortunate fate of Joseph Priestly, the clergyman and theologian who founded Unitarianism in England: He was vilified as an atheist, which he most certainly was not, for his unconventional attempt to combine materialism with theism. He eventually fled to the United States after a mob burned down his church and home.

Thomas Huxley: Science's Extremist

The year in which *Origin* was published was far different from the time when, just a few decades earlier, Lawrence, Matthew, Blyth, and Chambers had published their works. Unlike the early pioneers of evolution who concealed their authorship (Chambers) or abandoned further works (Lawrence), Darwin lived at a time that was far more tolerant of subversive unconventional ideas. Above all, he enjoyed the endorsement of many elite members of the scientific community. Just as the design argument has been a tricky matter for religious scholars, it has also been a divisive issue in the scientific community. Although the rift between both communities has been ongoing for quite some time, the publication of *Origin* may be taken as a convenient point for when their separation became irreconcilable and irreversible. Again, Huxley was a key figure in shaping science's antagonistic stance toward religion, a stance that resulted in the anti-religious and anti-metaphysical ideology among scientists that still dominates western society.

With the greater accumulation of knowledge obtained through comparative anatomical studies of living animals and fossils in the latter part of the nineteenth century, morphologists and paleontologists were in a position to assess the overall historical pattern of life on Earth. In line with most of these specialists, Huxley saw directed evolution producing predetermined trends (Bowler, 1983). Taking such a position, however, meant disagreeing with Darwin's view of the origin of species. As mentioned earlier, Huxley chose to become a "pseudo-Darwinian" at the expense of his own intellectual integrity.

One reason why Huxley played such a dominant role in this religion-science split was his rhetorical skill and effective use of emotionally charged words. He portrayed this conflict as one of good versus evil or hero versus villain. His contemporaries could not miss his religious characteristic in rejecting and ridiculing religion. Known as "Pope Huxley" and "Papal bull" (Hutton, 1870), as well as "the preacher of lay sermons" (White, 2003), Huxley would often debate Church leaders and declare science the enemy of parsondom (White, 2003). His son recalls that he was the "most combatative" of all speakers who participated in these debates (Huxley, 1916). Eiseley (1967) describes Huxley as "a kind of professional duelist's fury." Jensen (1991) adds that "Huxley was a master at launching flank attacks through irony, understatement, and ridicule." In the battle between good versus evil, there was no middle ground and there can be "neither peace nor truce" (Huxley, 1892). One of his favorite rhetorical arguments employed the military metaphors of roadblock: "ecclesiastical barriers," "obstacle to the progress of mankind," and "barricade."

For Huxley, in a world of good versus evil, there was no room for any middle ground and so he attacked those who tried to reconcile the two (Jensen, 1991). In his own words, he confessed that "I have less sympathy with the half-and-half sentimental school...than I have with thoroughgoing orthodoxy" (Huxley, 1900). Jensen (1991) elucidates his Manichean worldview: "This was war, and one is not permitted to wander from one camp to the other or claim partial allegiance to both. Even inaction or indecision was a sign of belonging to the enemy, for those who were not for you were against you." Himmelfarb (1959) calls Huxley "a savage polemicist." Huxley was thus the mirror-image of religious extremism. He was every bit as intolerant and fanatical as the religious dogmatists. He represented science's ultimate two-valued orientation what were these values. His vindictiveness against Owen was the main driving engine for many of his views and was one of the dominant forces that formulated the struggle between science and religion in terms of an actual war (Desmond, 1982). Paradoxically, however, he could present himself as a devoutly religious person in the privacy of his own home (White, 2003) and had an extensive correspondence with Charles Kingsley, a clergyman, on the subject of the soul's immortality (White, 2003).

St. George Mivart: An Independent Thinker Caught in the Middle

St George Mivart was quite an unusual biologist. His conversion to Catholicism at sixteen, despite his father's threat to disown him, is a good example the strong-willed independent mind that would become his trademark. This was an almost foolhardy decision on his part, considering that such a decision would automatically exclude him from Oxford or most other universities in the country. He studied law but never practiced, for his passion was biology. A self-taught biologist "even more than Darwin," as Wallace (1905) acknowledged, he became an expert in the anatomy of carnivores and insectivores. As an anatomist, he had close relationships with the two top British anatomists, Owen and Huxley. Their combined support helped him secure the Chair of Zoology at St Mary's Hospital (Desmond, 1982). He also became acquainted with Darwin and Wallace. Huxley had a major impact on young Mivart; he was not just a friend, but also a "teacher, guide, intellectual confident, and master" (Gruber, 1960). Although initially a Darwinian, as were other disciples of Huxley, his own studies of primate anatomy eventually caused him to doubt parts of the Darwinian explanation, in particular that of "parallel evolution," the prevalence of similar structures rising independently.

Some of the best examples of parallel evolution come from two groups of mammals that are believed to have split about 60 million years ago: placental mammals and marsupial mammals. Most surviving marsupials are concentrated in Australia, which display remarkable similarities of independently evolved structures between both groups of mammals. For instance, the extinct Tasmanian wolf and the placental wolf have similar tail lengths, bone structures, toe nail shapes, as well as a similar ratio of body parts to other body parts. It is almost as if the same body plan were used in the morphology of these two independently evolved mammals. In addition, moles, flying squirrels, and jerboas in Australia basically share the same body plan with placental mammals living in other lands. The Darwinian explanation, which relies on chance occurrences, seems to be unable to account for such parallel outcomes.

Owen offered a completely different explanation: the "archetype" which he defined as a unity among diverse species of mammals. All forms were modeled on a single basic plan that has been modified in various directions. To him, the archetype indicates the Creator's power and wisdom (Bowler, 1983). Such teleological and supernatural position was, of course, strongly rejected by Darwin's circle, including Huxley. Mivart, however, recognized and shared the sensible reasoning underlying Owen's archetype. The conflict between Darwin's natural selection and Owen's archetype was battling within him. He recalls:

> It was in 1868 that difficulties as to the theory of Natural Selection began to take shape in my mind...For the rest of that year and the first half of the next I was perplexed and distressed as to what line I should take in a matter so important, and which more and more appeared to me one I was bound to enter upon controversially. After many painful days and much mediation and discussion my mind was made up, and I felt it my duty first of all to go straight to Professor Huxley and tell him all my thoughts, feelings, and intentions in the matter without the slightest reserve, including what it seemed to me I must do as regarded the theological aspect of the question (Mivart, 1897).

A Critique of Science

Mivart, who approached his mentor in the hope of receiving some sort of guidance, was disappointed by Huxley's response:

> Never before or since have I had a more painful experience than fell to my lot in his room at the School of Mines on that 15th of June, 1869. As soon as I had made my meaning clear, his countenance became transformed as I had never seen it. Yet he looked more sad and surprised than anything else. He was kind and gentle as he said regretfully, but most firmly, that nothing so united or severed men as questions such as those I had spoken of (Mivart, 1897).

This internal conflict exemplifies the intellectual generation gap that occurred in the nineteenth century. Owen was educated in the era when metaphysical concepts were integral parts of the academia, whereas Huxley was one of the dominant members of the new generation who had risen up against the old guards. Once the preoccupation of most scholars, any manifestation of the supernatural in the wonder that is life was being systematically purged from the academy and replaced by the principles upon which science based its prestige and authority: naturalism, positivism, and materialism. In such a climate, it was an internal battle between intellectual integrity versus social expediency. As demonstrated by his conversion to Catholicism, Mivart was not afraid to put himself in a disadvantageous position in order to maintain his intellectual autonomy.

After turning away from Darwinism, Mivart worked on his own view of evolution. His *On the Genesis of Species* (1871) was based on a series of three articles entitled "Difficulties of the Theory of Natural Selection," which had been published eighteen months earlier in *The Month*, a Catholic journal. One of his most powerful arguments against natural selection was the problem of incipient stage: What good is 2 percent of a wing or an eye? Mivart called this "The Incompetency of Natural Selection," and it remains a contentious issue today, 150 years later. The minute changes brought about by chance seem to be too unrealistic to confer any kind of advantage upon the animal in question. Gould (1977) echoes this problem while analyzing the fresh water mussel known as *Lampsilis*. This mussel has a decoy fish that looks and moves like a real one. Its purpose is to attract other fish so that its larvae can be swallowed, because they cannot survive without a host. Gould (1977) uses the word "dilemma" to address the problem confronting Darwinism: What good is 5 percent of an eye? Darwin had to praise Mivart's persuasive argument with words like "admirable art and force" and was forced to respond to his criticism. In an attempt to do so, he added a chapter to a later edition of *Origin*.

After rejecting Darwin's haphazard mechanism, Mivart (1871) introduced an evolutionary idea that emphasized the existence of an internally driven purposeful process. Rather than gradual accumulation, he argued that evolution occurs via discontinuous steps. Rather than viewing new forms as monstrosities, he viewed them as "harmonious self-consistent wholes" (Mivart, 1871). Rather than chance, variability contains an internal guiding principle. Parallel trends in the development of different species were indications of this intrinsic force and were incompatible with chance variation (Bowler, 1997). Mivart's ideas, however, were the exact opposite of those being promoted by adherents of the newly emerging worldview, which was based on materialism and Darwinism. The Darwinians' main strength is their cohesion, and they used it both to attack Mivart (Desmond, 1982) all at once and, as the occupiers of powerful positions, muscled in on this renegade biologist. As Huxley's students captured key academic positions, Mivart's status began to decline.

His reception from the religious side was not any better, despite his attempt to show that one could be a good Catholic and an evolutionist (Brooke, 1991). Although initially welcomed by the Pope Pius IX, the articles he published while at the Catholic University in Belgium antagonized the church hierarchy so much that they were eventually put on the Vatican's Index of Forbidden Reading list. Mivart himself was excommunicated six weeks before his death. And here lies the irony: Mivart, one of the rare biologists in the latter part of the 19th century who did not join either ideological group, ended up being rejected by both of them. Brooke (1991) makes a poignant point: "The career of Mivart was that of mediator and it illustrates the difficulties encountered by one who took that role...Bridges could be built, but powerful interests on either side were apt to detonate them." It seems that the political arena and the scientific community are not so different after all. The only difference may be that the latter is less aware of this problem because of its members' presumed intellectual integrity. Mivart's view, known as "theistic evolution," is quite plausible and still has not been adequately addressed in evolutionary biology.

Group Polarization between Science and Religion

The media regularly informs us that politicians are polarized between the continuum of two ideological positions: "liberal" and "conservative." Such polarization drives those who are in the middle (moderates) toward either end. Either side usually has a set of ideological doctrines to which individuals are expected to conform. People on each side consider themselves good and right, while those on the other side are bad and wrong. In addition, the positions chosen are often mutually exclusive and antagonistic, since they are frequently chosen in order to beat the opponent rather than to devise an optimum solution. Moderates and bridge-builders are typically not welcomed in either camp. For instance, Senator John McCain (R-AZ) has often been criticized being a moderate, the "Democrats' favorite Republican." In the battle of two opposing political camps, many members of the public either join the conservatives or the liberals; the rest choose to be independents. There are more independents than politicians among the average people, because they face far less pressure to conform. Consequently, average people can be more flexible in their views than the politicians in Congress.

As disturbing as it may be for those who believe in science's self-proclaimed emphasis on and respect for logic and evidence, one of the main driving forces behind the intellectual community's adoption of the materialistic agenda during the 18th and the 19th century was this polarization between science and religion. Unfortunately, this polarization represented a significant stumbling block for those people who were trying to select one of the diverse views on the origin of species after subjecting all of them to careful consideration and examination. Broman (2003) addresses this polarization:

> For it was during the Enlightenment that the cultural landscape of Europe was first reshaped in a way that enabled "science" and "religion" to emerge as separate and hostile camps in a long polemical struggle.

He goes on to say:

They did not simply discover what had previously been a latent conflict between "science" and "religion" and bring it to light; to the contrary they created the conflict and trumpeted it as a way of supporting their own notions of truth. Thus, the conflict between science and religion that we believe to have broken out during the eighteenth-century Enlightenment does not necessarily represent an inevitable confrontation between the two antagonistic ways of looking at the world. And to the extent that we continue to believe in a struggle of this kind, we remain trapped by the categories created and bequeathed to use by Voltaire, Gibbon, and the other critics of Christianity.

This mutually hostile polarization had one negative result: scientists went beyond atheism and materialism and became outright anti-religion. Over time, their arguments became science's dominant ideology. Just as in politics or religion, science seems to favor extremists like Thomas Huxley over moderates like St. George Mivart.

The Diversity of Evolutionary Thoughts
The intellectual community's acceptance of evolution was never in doubt after *Origin*'s publication in 1859. Instead, what remained in doubt was its exact mechanism. Biologists in the late nineteenth century were faced two major approaches to life that could not be completely accepted: a naturalistic versus a supernaturalistic account. Darwin's natural selection could not account for all of life's complexity, but the supernatural account of evolution was becoming increasingly untenable. Ironically, this was occurring at the time when the most diverse evolutionary ideas were being offered, because neither approach was dominant. Biblical literalists were active, as were those who wanted to modify the biblical view by suggesting that not every species was created independently. Others sought to combine Christianity and science. Some religious thinkers were even willing to accept evolution as long as the "chance" elements were left out (Brooke, 1991). Agassiz and Owen posited the existence of a mind behind creation, while Spencer emphasized the creative power. Both of them were preceded by St. Augustine, who had suggested centuries ago that evolution was God's method of creation. Lamarck's materialistic explanation differed from Darwin's in that it added the organism's purposes, which were achieved

through a system of trial and error initially set by God. And finally there was the school of orthogenesis, which emphasized predestined direction and cited the horse's evolution as a case in point. For a while, Lamarckism and orthogenesis even eclipsed Darwinism. In fact, both schools remained popular among paleontologists well into the 20th century (Bowler, 1983, 1988).

The geological and paleontological findings made during the 18th and 19th centuries contradicted the biblical literalism and thus forced many intellectuals to sort out the seeming contradiction between science and religion. One approach, "theistic evolution," accommodated the design and divine plan while accepting evolution. Mivart, for instance, argued that the variations among species were predetermined in accordance with a divine plan. Many theistic evolutionists presented the growth process of an embryo to emphasize life's goal-directed and purposeful trend via progressive development. During the latter part of the nineteenth century, however, theistic evolutionists were confronted with the academy's growing, and then finally total, refusal to accept any form of supernaturalism. This increasingly inhospitable climate can be demonstrated by the defection of Edward Drinker Cope. Although he had argued in 1868 that the pattern of growth seen in the fossil record indicated predetermination by the Creator, his later works offer no such argument because he had joined the more politically correct camp of orthogenesis.

The continuous progression of some characteristics might become so disadvantageous that it may destroy the species. Consider the transformation of the *Planorbis levis*, another species of extinct mollusk, which shows a tendency to uncoil, become distorted, and become smaller than its ancestors. Such malformation often takes place just before a species becomes extinct (Hyatt, 1882). Interestingly, such uncoiling and distortion occasionally occurs in individuals that have been injured or are diseased. Why did the process of natural selection allow these characteristics to move beyond what was useful? This question is unavoidable. Wilson (1975) acknowledges: "Phylogenetic inertia...consists of the deeper properties of the population that determine the extent to which its evolution can be deflected in one direction or another, as well as the amount by which its rate can be speeded or slowed." Even Simpson (1949), a cofounder of the synthetic theory, admits: "Yet the fact remains that the sequence cannot be considered as random. The changes involved do have direction and orientation, even though these were not as regular as they have usually been represented. And so, in hundreds or thousands of other cases, it seems clear that there is an orientation of some sort." Recently, Blomberg & Garland (2002) asked: "What are the forces?" Despite the enormous implication of orthogenesis, which contradicts Darwinism, most modern biology textbooks do not even mention it.

Lamarckism was also quite popular on both sides of the Atlantic. As discussed earlier, even Darwin incorporated this school of thought in his *Origin*. In fact, the role of *use and disuse* became more pronounced in his later editions and other works. In addition, Spencer, Ernst Haeckel, and other prominent figures endorsed Lamarck's idea. One of Lamarckism's advantages was that it could present itself simultaneously as materialistic (like Darwinism) and accept vitality and creativity as the driving forces of nature (Bowler, 1983). Among its particularly enthusiastic proponents were Alpheus Packard, Edward Cope, and Henry F. Osborn, the leading American paleontologists of their day. Cope openly accepted the role of the Creator, who worked according to a plan of His own, in the pattern of evolution (Bowler, 1988). American Lamarckists also had strong ties to orthogenesis (Bowler, 1983). Both of these trends were powerful alternatives to Darwinism; the latter's popularity peaked in the 1890s (Bowler, 1983).

For those who accept the materialistic approach to the world, one of the persistent problems is how to explain the emergence of novel properties. How, for instance, could matter give a rise to a completely different entity like the mind? Above all, how could inorganic matter become organic matter? The concept of "emergent evolution" offered a solution to this problem. John Stuart Mill (1843), a British philosopher, pointed out how a specific combination of hydrogen and oxygen could produce a completely different liquid substance: water. Just as these two gaseous chemical elements can produce a liquid, emergent evolution posited that an unpredictable rearrangement of matter could give a rise to a higher level of being like life and consciousness.

At the end of the 1920s (Mayr, 1980), many general college textbooks reflected the diversity of evolutionary ideas by presenting five or six theories. This stands in stark contrast to the information presented in contemporary textbooks: Darwin's idea dominates the discussion. In the next chapter, we will discuss how this happened.

Chapter Eight
The Second Phase of the Evolutionary Absurdity: The Mystique of Incomprehensible Mathematics

> ....Neo-Darwinism, which insists on (the slow accrual of mutations), is a complete funk.
>
> -Lynn Margulis

> To wield influence, a paradigm does not have to be comprehensive, nor does it have to be based on correct information or sound ideas. It can also allow empirical applications of false hypothesis. Only with hindsight might they be seen to have been totally misdirected. What matters most when paradigms are first established is their psychological or epistemological impact.
>
> -Robert G. B. Reid

In 1903, Eberhart Dennett declared the demise of Darwinism with the title *At the Deathbed of Darwinism*. A quarter century later, Nordenskioeld (1928) also stated that Darwin's natural selection has long been rejected in its most vital points. Perhaps Mayr & Provine (1980) summarize the status of Darwin's idea in the late 19th and the early 20th century:

> ...it [Darwinism] was rejected by most biologists and bitterly attacked by many others. It seemed to make some temporary headway in the 1870s, but lost ground again in the 1880s and 1890s and almost received a fatal setback through the mutationist theories of the early Mendelians. In the first decades of the twentieth century the schools of saltationism, orthogenesis, and neo-Lamarckism had decidedly more followers than selectionism. Indeed, only a handful of authors between 1900 and 1920 could be designated as pure selectionists.

In the 1920s and 1930s, nearly all the major books on evolution were more or less antiselectionist (Mayr, 1980). Then, Darwin's idea would be fortified with mathematical equations that hardly anyone could understand. The mystique of incomprehensible mathematics would become the main force of seduction to neo-Darwinism (also known as the synthetic theory). The process of the resurgence of Darwinism in the 1930s and 1940s is every bit as irrational as in the manner that Darwin's idea became established in the latter part of 19th century. Despite its irrationality, neo-Darwinism would acquire an additional absurdity: its intolerance to alternative ideas of evolution.

The Rediscovery of Mendel and Saltationism

> This regular absence of transitional forms is not confined to mammals, but is an almost universal phenomenon, as has long been noted by paleontologists.
>
> -George Gaylord Simpson

The rediscovery of Mendel's work should have forced the different schools of evolutionary thoughts to readjust with the newly found heredity. Yet, the significance of Mendel's finding of inheritance would not be incorporated into the ideas of evolution for some time. Mendel's work was rediscovered by three biologists independently, Hugo De Vries, Carl Correns, and E. von Tschermak, sixteen years after his death in 1900. They have done experiments similar to Mendel's and were searching for prior studies to confirm theirs. For them, there was no doubt in the meaning of Mendel's work. However, when Mendel presented his works 36 years earlier, no one in the room understood the significance of Mendel's report becoming one of the greatest genetic discoveries of all time. The forty people who were attending the meeting included botanists, a chemist, an astronomer, and a geologist (Eiseley, 1959). Carl von Nageli, one of the distinguished botanists at the time, failed to see the importance of Mendel's discovery. Historians long have been puzzled about the overlooking not only at the presentation but subsequent neglect for several decades. Bowler (1989) points out: "the apparent incompatibility between the clarity of Mendel's insight and the failure of his contemporaries to appreciate what is so obvious to the modern reader." One of the most popular explanations is that he was ahead of his time. Barber (1961) offered a different explanation. He suggested that it might be because of Mendel's lack of scientific stature. They didn't expect much from a nameless Augustinian monk and, consequently, they did not see much. This is the exactly the opposite effect of the findings of Peters & Ceci (1982). Those papers with prestigious authors and institutions were published, even though the exact same papers were rejected, because of "serious methodological flaws" without the name and institution as mentioned earlier.

At any rate, what Mendel discovered was that the traits of pea plants such as round seed, wrinkled seed, yellow seed, green seed, tall plant, short plant, etc. maintained throughout generations rather than blending. The hereditary characters of pea plants were determined by discrete units, now called genes, that appear in pairs (alleles) which come from each parent. The obvious implication was that genes seem to be quite stable and it looks like a hindrance for evolution because emergence of new characters would be difficult. A particular difficulty was posed by the Darwinian explanation that relies on variation for evolution. Whereas, one of the main competitors of Darwinism known as saltationism seemed to match better with Mendelism. Francis Galton, Darwin's cousin, was an advocate of saltationism, which considers evolution taking place in discontinuous steps rather than gradually, as Darwin suggested. Galton's support saltationism was based on his observation that extreme characters like height (very tall or very short) in parents are regressed toward the mean to their offspring. In other words, if parents are very tall (or very short), their children tend to be less tall (or less short). This is called "regression to the mean" and Galton thought that natural selection's gradual and incremental change would be neutralized by the reversion. If Galton's regression to the mean is a genuine force, selection would be unimportant in evolution (Gilham, 2001).

Let me briefly discuss the present understanding of Galton's regression to the mean. The generally accepted view is that Galton misunderstood it. Suppose you pick the top ten scorers of one week in, let's say, in the NBA, their scores would likely drop in the following week. For the same token, the bottom ten scorers would likely have a better performance in the following week. In other words, if a player's score is really high, it can only go down. Conversely, if a player has a really low score, it can only go up. Just as an athlete's peak performance can be followed by the decrease of improvement or the worst performance followed by an improved performance, any extreme characters can typically lurch back to the middle. So the conclusion is that regression to the mean is not a causal phenomenon, but a statistical one.

Not so fast. Let's get into a biological phenomenon that somehow has been largely overlooked due to the dominance of Darwinism. Robertson's (1955) experiment is a typical outcome of long-term artificial selection. When animals are selected for large thorax size and small thorax size of fruit flies, it produces an immediate response. However, after a certain number of generations, the response to the selection tapers off. This is not because of a loss of variability but because variation actually increases in later generations. When artificial selection is discontinued, partial or complete reversion occurs. Most importantly, the fruit flies of prolonged artificial selection show a decrease in viability. Ernst Mayr (1984), a staunch Darwinist, admits that there hasn't been a single experiment of artificial selection of a particular characteristic in fruit flies over the last 50 years in which some undesirable side effect has not appeared. The extreme traits do not do well in natural population. For instance, the survival rate of babies with very high or very low birth weight is significantly lower than the average (Karn & Penrose, 1951). The same thing happens to the weights of hatched and unhatched ducks' eggs (Rendel, 1943). James (1962) called it "centripetal selection" to describe the decreased viability of extreme characters. Wesson (1991) borrowed the term "attractor" from chaos theory to argue for a set of permitted states of a system that cohere. Galton's regression to the mean might consist of two separate factors: a statistical phenomenon and a genuine biological phenomenon. Separation of the two effects would be one of the important evolutionary biology's challenges.

At any rate, joining Galton was William Bateson, who would become the most active member of the proponents of saltationism. And Darwin's bulldog, Thomas Huxley, also accepted discontinuous evolution. The saltationists were also later called Mendelians because of the apparent discrete heredity of Mendel's finding. At the same time, biometricians Karl Pearson and W. F. R. Weldon viewed Mendelism as a threat (Provine, 1971). Consequently, the Mendelians would engage in a bitter conflict against biometricians who accepted Darwinian gradualism.

A Critique of Science

The Triumph of Naturalism in Biology

> Physics envy has long been the curse of biology. This curse often expresses itself in the desire for a grand unifying theory with the predictive and explanatory virtues and the testability of Newtonian mechanics.
> -Richard M. Burian

The two camps of saltionism and gradualism started with bitter rivalries that often involved personality conflicts. One changing landscape of the intellectual community that made intensification of the rivalry was that the government began to actively support the useful scientific researches. The camp of Bateson (Mendelians) was able to tap into the new sources of funds by emphasizing experimental studies of heredity that might be beneficial to agricultural production (Bowler, 1989). Whereas, the strength of the Darwinian camp (biometricians) was their demonstration of mathematical prowess. While Mendelians focused on laboratory experiments of a finite number of plants or animals, the main approach of biometricians was toward a whole population using mathematics.

In the middle of conflict between Mendelians and biometricians, a combination of three different events made a significant impact on the direction of biology. The first event was that the Mendelian concept of inheritance would encounter an unexpected backing from the chromosome theory that began to emerge with the rapidly improving microscope in the 1870s by revealing the internal structure of a cell like nucleus. In 1879, Walter Flemming discovered chromosomes in nucleus of salamander eggs. In 1902, two years after the rediscovery of Mendel, Theodor Boveri and William Sutton independently developed the chromosome theory that identified chromosomes as the carrier of genetic material. The pairedness of chromosomes matched with Mendelism's hereditary characters that were determined by discrete paired units from each parents. What was truly significant about the chromosome theory was that the abstract concept of Mendelian inheritance had a real identifiable physical property.

The second event was the advancement of the mutation theory originally proposed by Danish botanist Hugo de Vries. He noticed an appearance of new forms among evening primroses growing wildly in a pasture. He was able to produce many new varieties of evening primroses from the seeds collected from the wild. He suggested that evolution might take place by such large-scale change which he called "mutation."

The third event came from Thomas Henry Morgan at Columbia University, who pioneered the experimental techniques of fruit flies in laboratories. Fruit flies (*Drosophila melanogaster*) became the most popular organisms for genetics studies because of the easily visible four chromosomes. In addition, they have short life cycles and they are easy to take care of. From Morgan's small and cramped "Fly Room," many basic principles of heredity, such as sex-linked inheritance and gene mapping, were born. One of Morgan's students, Herman J. Muller, developed techniques for inducing mutations artificially by exposing fruit flies to an X-ray. He was able to produce mutations several hundred times the normal frequency by exposing fruit flies to radiation (Muller, 1927). Mutations observed in fruit flies had a complete different meaning than what de Vries had initially proposed. The radiation-induced mutation in the lab was considered as micromutation; what de Vries had in mind was macromutation. The micromutations seen in fruit flies could not support the new varieties and species that could appear by saltations (Bowler, 1989). Yet, as we shall see later, the obvious difference between the two types of mutation would be sidestepped.

The three events helped to consolidate the power of naturalism in biology. Evolution was seen as a product of mutations that are driven by spontaneous changes in existing genes, as it was seen in fruit flies (Bowler, 1989). Such an interpretation emboldened naturalism that rejected no need for an intelligent agent to create and maintain the complexity of life. And Darwinism was most suited for incorporations with naturalistic approaches that were strengthened by the new genetics. As naturalism began to dominate, some ambiguous words that might imply supernaturalism and teleology were being replaced by the terminologies borrowed from physical science, like "cause," "factor," and "mechanism" in evolutionary biology in the 1930s (Smocovitis, 1996). Yet, non-Darwinian evolutions like Lamarckism and orthogenesis had remained popular well into the 1930s. The total domination of Darwinism over other evolutionary ideas would come in the late 1930s and onward which was known as "neo-Darwinism."

A Critique of Science

Incomprehensibility as a Form of Value

On March 1, 1919, during the Japanese occupation of Korea, the 33 nationalists signed and read a Korean version of the Declaration of Independence. It emulated the spirit of the American document written143 years earlier. It starts with a paragraph: "We herewith proclaim the independence of Korea and the liberty of the Korean people. We tell it to the world in witness of the equality of all nations and we pass it on to our posterity as their inherent right." There was one crucial difference between the two documents. The Korean version was written in such a difficult language that the average person could not easily understand the content. The reason for the usage of incomprehensible language was because the knowledge of difficult Chinese characters was a symbol of high intellect. The goal of the document was to communicate the declaration of Korea's independence, yet, it was largely compromised by the misguided value system.

Incomprehensibility can become established in the value system and it is especially a big problem in philosophy. Philosophers themselves usually do not recognize or would not admit the problem. A good example is Immanuel Kant's works. Psychologist Thomas Leahey (2004) expresses, "Kant's philosophy is extremely difficult." A Germany-born biologist Ernst Mayr (2004) said that he read or attempted to read Kant's philosophy and didn't understand the writing of Kant, despite it being written in Mayr's native language, German. Stewart (1997) argues, "The striking feature of Kant's work, at first glance, is the vast technical vocabulary and abhorrent prose." Stewart (1997) explains how Kant has garnered his status in modern philosophy:

> First, philosophers generally crave professional respectability. Kant provides the complex vocabulary, the vague idea of a specialized investigative project, and enough smoke and mirrors to convince outsiders that some serious work is being done....In sum, Kant's obscurity is the critical factor in satisfying the urges that have led philosophers to elevate him to greatness. His obtuse distinctions lend an air of professionalism; his convoluted arguments give the impression of profundity; and the resulting inconsistencies provide grist for the debating mills of philosophy.

In a similar tone, Williams (1973) argues: *"The Critique of Pure Reason* is not a systematic, well thought out philosophical work but an agglomeration of philosophical reflections, often contradictory, loosely bound together by Kant's ideas on the nature of reason and reality." This brings a question of how much of Kant's eminence is due to the merit and how much is because of the mystique of incomprehensibility. The problem of incomprehensibility pervades the entire discipline of philosophy. Psychologist Michael Mahoney (1976) bluntly said that philosophers believe that they are on the road to eminence if no one understands what you have said. Similarly, Weinberg (1992) said: "Some of it I found to be written in a jargon so impenetrable that I can only think that it aimed at impressing those who confound obscurity with profundity."

The incorporation of incomprehensibility into philosophy's value system may undermine its important function of bridging between disciplines in the segmented intellectual world. Philosophy's preoccupation of incomprehensibility makes the bridge virtually impassable.

The Mystique of Mathematics and the Meltdown of Wall Street

Math is just a simple tool in decision making, just like computers.
-Nicole El Karoui

As the chair of mathematics at the University of Padua, Galileo was paid only a fourth of leading natural philosophers. The salary disparity gives some indication of undervalued status of mathematics in universities at the time of Galileo. In less than 200 years, mathematics virtually became a requisite of studying science. For instance, David Hartley (1749), an English philosopher, insisted that only in mathematics can one develop theories which can be rigorously demonstrated. Kant also thought psychology cannot become science because it lacks mathematics. The problem nowadays is the exactly opposite of Galileo's time: mathematics might be overvalued. A good example is the 2008 Wall Street meltdown that triggered the greatest global recession since the Great Depression. Felix Salmon (2009) of *Wired Magazine* dubbed it as "the formula that killed the Wall Street." The author of the formula is David Li, a math whiz from China. While working at J. P. Morgan Chase, Li published a paper in *The Journal of Fixed Income* entitled "On

A Critique of Science

Default Correlation: A Copula Functional Approach." His formula is shown below:

$$\Pr[T_A<1, T_B<1] = \Phi_2(\Phi^{-1}(F_A(1)), \Phi^{-1}(F_B(1)), Y)$$

If you don't know what the formula means, don't worry because I have no idea either. I put the formula just to show how incomprehensible some mathematics can be for most people. Even the author of the formula warned that "very few people understand the essence of the model" in a interview with Wall Street Journal in 2005.

Li's formula is essentially a risk management model that aimed at solving a very difficult problem: How are seemingly disparate events related? An example, which he had worked on, is the higher likelihood of a person dying soon after their spouses do, which is known as the "broken heart syndrome." Such a problem may be important to life insurance companies. Li was able to apply the assessment of chance of death by widowed spouses to the Wall Street's nagging problem. What is the likelihood that a number of corporations would default on their bond debt in quick succession? Nobody quite knew how to calculate default correlations. Li's idea was that default of correlated corporation is like the death of a spouse. Li's model was able to assess the likelihood of all correlated companies as pool to default. For investors in Wall Street who abhor uncertainty, Li's equation became quite popular not only from bond investors but also bankers, rating agencies, and even regulators. People were riding on the bandwagon of mathematical mystique. Never mind, hardly anyone understood it. They were making so much money that any warnings were ignored. Li was among those who warned about the limitation of his model in his interview with Wall Street Journal reporter: "The most dangerous part is when people believe everything coming out of it."

The Synthesis, the Compromise, or the Intimidation by Mathematics?

> Some evolutionary biologists....have been focused too much on developing and deploying statistical methods of analysis, ending up glossing over the conceptual underpinnings of the techniques they employ. This emphasis has led to something of a disconnect between empirical approaches and the concept they refer to.

Orogee Dolsenhe

-Massimo Pigliucci & Jonathan Kaplan

[Neo-Darwinism] was from the outset a mathematical discipline, and this tradition has been preserved. Today it has developed into a highly sophisticated subject which, I believe, is accessible only to a small minority of biologists. But although they must have had some premises in common, the mathematical systems developed by the three founders of population genetics are widely different. The outsider with great efforts has penetrated the equations of one of the three should not expect to master simultaneously those developed by others....Nevertheless, what unites the three or, at least, what is considered to unite them, is the belief in the micromutation theory, which claims that it is possible to bring out substantial changes in living organisms through the gradual accumulation of mutations with slight effects on the reproductive success of the organisms in question.

-Sørren Løvtrup

What happened in the meltdown of Wall Street in 2008, oddly enough, has a similarity with neo-Darwinism in one respect: the mystique of incomprehensible mathematics. A major difference between the two is that the flaw in Wall Street was exposed in eight years, whereas the problem in evolutionary biology has been going on for more than a half century and is likely to go on for quite some time unless some drastic change occurs.

In the period between 1936 to 1947, the two competing schools of evolutionary thought, biometry and Mendelism, worked out their differences and synthesized to become the presently dominant evolutionary ideas. The new school of thought that emerged in this period is known as "neo-Darwinism" or the "synthetic theory." Neo-Darwinism is a synthesis of the Mendelian genetics and Darwinian selection with emphasis of mathematization. Some have pointed out that, rather than a true synthesis, it actually was a "truce" (Brooks, 1983) or a "treaty" (Depew & Weber) between the leading population geneticists and systematists of the 1930s and 1940s. They negotiated an interdisciplinary treaty (Ceccarelli, 2001). What happened to the synthesis is in agreement with Collins & Pinch (1993) who stated that scientific disputes are not settled by further experiment but by social negotiations. When a new direction is based on compromise, inconsistencies and problems likely have been disregarded, as is often the case in politics. Dobzhansky (1937) mentions that he was "reluctantly forced" to equate microevolution with macroevolution. Similarly, Stebbins, who proposed the concept of quantum evolution to explain the sudden appearance of certain groups of animals, eventually dropped it from later editions of his book partly under pressure from his gradualist colleagues. Whereas, those who disagreed with the main direction of the Synthesis were excluded, even a founding member of neo-Darwinism like Julian Huxley. Although Julian Huxley, a grandson of Thomas Henry Huxley, coined the term "synthesis," he was considered "disreputable" member of the group because he was not enough a selelectionist and too soft on orthogenesis and other non-Darwinian ideas (Reid, 2007). Likewise, Wright's works on levels of evolution and interdemic selection were removed from the developing synthesis. Wright complained recalls that he wasn't even invited to major meetings. Gould (1980) recalls the meeting with Wright: "'I was out of it' and said that he wasn't invited to major meetings (indeed – and incredibly – he was not invited to attend the major historical conference on the definition, impact and meaning of the synthesis..." In the power game of the synthesis, Fisher was able to stifle many issues from the formation of the new tenets. Schilthuizen (2001) argues that Fisher was the one who put evolution into a "straightjacket of rigorous algebraic formulas" and laid the foundation for neo-Darwinism.

Initially, Mendelism and de Vries's mutation theory appeared to be on the way to victory over Darwinism (Provine, 1971). In the first decade of the 20th century, most Mendelians believed in discontinuous evolution and only a few believed Darwinian selection was an important factor in evolution (Provine, 1971). By 1918, however, many Mendelians began to recognize the effectiveness of selection demonstrated by a number of experiments. William Castle's experiment with guinea pigs and hooded rats might be one of the most crucial factors in accepting Darwinian selection. Castle initially endorsed saltationism largely influenced by reading Bateson's works on Mendelian inheritance. However, he began to see that selection may be effective in generating character beyond the original limits of variation. His selection experiment of guinea pigs produced a population of extra-toed guinea pigs on the left foot. Another experiment was the massive selection experiment with hooded rats that lasted 12 years, breeding about 50,000 rats. The prolonged selection for both increased and decreased size patterns of hooded rats resulted far beyond its original variation. Castle (1905), like other Darwinians, made a conclusion earlier: "there is no essential difference between breeds and species, and that if we can ascertain how breeds originate we can infer much as to the origin of species." Provine (1971) believes that Castle's willingness to reinterpret the results of his selection experiments with hooded rats signaled the final victory for the Darwinian camp.

Some began to see that Mendelian recombination would provide variability, which becomes a raw material for selection. Then, Howard C. Warren (1917), a Princeton psychologist, offered a mathematical argument that Mendelism and Darwinism were compatible. Yule (1902) had pointed out earlier that Mendel's laws were not incompatible with biometrician's continuous variation. He had argued that if a number of Mendelian factors could blend together in the population to give a continuous range of variation. The stage was set for the arrival of the founders of neo-Darwinism: R. A. Fisher, J. B. S. Haldane, and Sewall Wright. They all were mathematical prodigies and their mathematical mystique would play a major role in converting the biological community to neo-Darwinism.

Fisher burst onto the science scene in the 1930s, "Fired up by the impact of melding Mendelism with Darwinism" (Schwartz, 1999). He had a reputation from the years when he was a student in Cambridge that he produced papers that were so sophisticated that accomplished and established mathematicians could not follow his reasoning (Løvtrup, 1987). Provine (1971) explains Fisher's other problems:

> Fisher managed to write papers unintelligible to his seniors and this appears to be a habit which followed him through life, for later in his career 'mathematical statisticians and geneticists were to complain that Fisher's proofs contained intuitive leaps which were not obvious....

Fisher (1958) supported the difficulty of his writings based on incomprehensible mathematics by saying: "No efforts of mine could avail to make the book easy reading." It would be one of the greatest ironies that if biologists bought into Fisher's idea that was mainly based on the mystique of incomprehensible mathematics. Løvtrup (1987) argues that Fisher's problem lies in his "confidence in the supremacy of mathematics as an infallible guide towards the truth has an interesting consequence: he appears to believe that his mathematical demonstrations constitute empirical evidence." Fisher was so confident about his mathematics that he even states that he is "indifferent as to the cause of mutations" in his theory (Fisher, 1934). Løvtrup (1987) believes that Fisher's mathematics was divorced from reality." Løvtrup (1987) expresses his puzzlement over many biologists who have shared Fisher's view.

Unlike the formula that brought down Wall Street, neo-Darwinism is not a singular mathematical model, but is contributed to two others authors, J.B.S. Haldane and Sewell Wright, besides R. A. Fisher. First of all, they differed in how fast evolutionary change would come: Fisher endorses gradual change, Haldane preferred rapid change, and Wright invoked one and at other times the other (Schwartz, 1999). A particular disagreement among them was the issue of how mutation, which is typically recessive, can survive without being swamped. Schwartz (1999) explains the circumstance involving the three mathematical models:

> They did not, however, agree on how that change would be enacted, or on the exclusivity of mutation as the proximate cause of change....They did not even agree on how a mutant, which might introduce the potential for change, would initially be expressed: Would it be in the dominant or the recessive state? Fisher took the former position, expecting the recessive to be converted quickly into the dominant state so that it could be expressed in the heterozygote. Wright and Haldane favored the latter condition, allowing the mutant to spread in the recessive state until it could be expressed in homozygotes...The founders of the evolutionary synthesis—the individuals who formed the core of the Committee on Common Problems of Genetics, Paleontology, and Systematics, which was established in February 6, 1943, to address the problems so clearly emphasized in the committee's title—were obviously faced with three, often *wildly different* (emphasis added) theoretical formulations of the evolutionary process and its perceived elements.

Schwartz (1999) points out the reality that is largely being ignored: "To this date, no one knows how the transformation of a recessive allele into a dominant one occurs." The disagreement over the survival of recessive mutants is the least of the problem.

Even a greater problem of neo-Darwinism is the theoretical base of Darwin's natural selection. Waddington (1961) points out:

> All that is explicit in the theory is that they will leave more offspring. There, you do come to what is, in effect, a vacuous statement. Natural selection is that some things leave more offspring than others, and you ask, which leave more offspring than others; and it is those that leave more offspring; and there is nothing more to it than that. The whole real guts of evolution – which is, how do you come to have horses and tigers and things – is outside the mathematical theory. So when people say that a thing is vacuous, I think they may be thinking of this part of it, this type of statement. The sheer mathematical statement is largely vacuous.

Even one of the founders of neo-Darwinism, Ernst Mayr (1976), questions the validity of mathematics:

> [Fisher, Wright, and Haldane] have worked out an impressive mathematical theory of genetical variation and evolutionary change. But what, precisely, has been the contribution of this mathematical school to evolutionary, if I may be permitted to ask such a provocative question.

Another ignored issue is that Løvtrup (1987) calls the consequences of neo-Darwinism, which are based on sexual reproduction and do not concern asexual organisms, "bizarre". That would limit the explanation of neo-Darwinian being the origin of species on earth.

One of the main problems of almost universal acceptance of neo-Darwinism is the fact that complex mathematics is obscuring the fundamental problems. Hamilton's expression in *Narrow Roads* that university people will listen harder if you first intimidate them with equations can be rephrased: University people can be intimidated by complex equations so that their common sense may be overwhelmed. Rose (2005) makes a crucial statement: "Biology needs to be able to declare its independence from spurious attempts to mathematicize it."

The representatives of some disciplines like developmental biology bitterly opposed the synthesis and stayed out of the process (Mayr, 1993). This may also explain why Simpson (1949, 1953) sometimes was ambiguous, particularly regarding the apparent evolutionary patterns. Other criticisms are: it is "incomplete" (Gould, 1982) and it "never was a single evolutionary theory," unlike the word "synthesis" might seem to imply (Templeton & Giddings, 1981). Fuller (2007) argues that equating microevolution in the laboratory and macroevolution in the distant past based on fossil record is nothing more than an "extended promissory note." Provine (1992) appropriately called the Modern Synthesis as the "evolutionary constriction" of evolutionary theory: "The evolutionary synthesis was not so much a synthesis as it was a vast cut-down of variables considered important in the evolutionary process." Perhaps Reid (2007) sums up best:

> For any kind of synthesis, all the arguments need to be aired, and awkward exceptions confronted at the outset instead of being black-boxed, ignored, or liquidated. The Modern Synthesis has failed this test, and has lost many of the comprehensive features that Huxley gave it. No amount of historical revisionism can make up for that. Yet it persists, and it can still beat the competition.

Despite the complications, the evolutionary synthesis rapidly spread through most countries except France. Once the elite members of evolutionary biology adopted the party line, the synthesis has turned into "unexamined dogma" (Gould, 1983; Eldridge, 1985) and dismiss viable alternatives on the basis of prejudgment rather than hard evidence (Burian, 1988).

A Group of Blind Men Follow the Blind Leader

> We are indeed a blind race, and the next generation, blind to its own blindness, will be amazed at ours.
>
> -L. L. Whyte

As in the case of David Li's formula on Wall Street, for most biologists, the complex mathematical models of population genetics were incomprehensible (Lewontin, 1980). It took someone with a high status to promote their mathematical population genetics. This role fell on Theodosius Dobzhansky. His book *Genetics and the Origin of Species* is generally recognized as the most influential book about genetics, not only by the participants of the evolutionary synthesis, but also among historians (Mayr, 1982). Here is the first irony: Dobzhansky himself was not an accomplished mathematician, as Ceccarelli (2001) stresses:

> At various times in his career, Dobzhansky acknowledged his failure to understand difficult mathematical arguments. He unabashedly admitted that since he could not understand Wright's mathematical treatments, he would invariably read introductions and conclusions of the papers while treating that the mathematical middle accurately connected the two.

He did not, however, completely avoid difficult mathematical equations as Julian Huxley, another promoter of population genetics, did. What Dobzhansky did was to send the part with mathematic equations to Wright for advice, as Ceccarelli (2001) points out:

> Before sending his book to the publisher, Dobzhansky sent chapter 6 to Wright for editorial review and received advice on how to make the account less "simplistic" and more representative of recent "advances." He made the suggested changes without protest.

What Dobzhansky should have done was to have someone else other than the very author of population genetics to critically assess the validity of mathematical population genetics. Here is the second irony: people began to rally around Dobzhansky's book in the acceptance of mathematical population genetics. Gould (1982) remembers:

> ……all the gathered authors looked at Dobzhansky (who clearly enjoyed the accolades) and said that they had drawn primary inspiration from *Genetics and the Origin of Species* (1937).

A Critique of Science

> Dobzhansky had not simply been first, by good for good fortune, in an inevitable line; his book had been the direct instigator of all volumes that followed.

It was like a group of blind men following a blind leader. I suppose they didn't realize they were following a blind man. At any rate, once the elites of the camps of naturalists and geneticists were committed to the synthetic theory, its supremacy in the biological community was a matter of time. One of the greatest ironies in biology is that neo-Darwinism is largely inspired by a single book that offered no new discoveries or theories of its own (Ceccarelli, 2001).

Orogee Dolsenhe

Chapter Nine
The Third Phase of the Evolutionary Absurdity:
The Complexity of Life

Small amounts of philosophy lead to atheism, but larger amounts bring us back to God.

-Francis Bacon

Overwhelming strong proofs of intelligent and benevolent design lie around us.

-Lord Kelvin

In my view the "normal" scientist, as Kuhn describes him, is a person one ought to be sorry for... He has been taught in a dogmatic spirit: he is a victim of indoctrination...

-Karl Popper

Any corporation that fails to recognize and react quickly to the world's constantly changing landscape is susceptible to disaster. Take the well-known case of Polaroid's complacency when confronted with digital technology. A high-flying company that has been a household name for instant film development for decades was caught flat-footed and its products were rendered virtually obsolete overnight. While its rivals Kodak, Hewlett-Packard, Canon and others shifted their gears to adjust to the new technology, by 2002 Polaroid had been forced to declare bankruptcy after struggling to pay off $950 million in debt. Although Polaroid has been reorganized, it never managed to regain its former position. Had the company's executives been more interested in and responsive to the innovation that was sweeping across the world, it likely would have avoided this misfortune.

This also happens in military technology. Consider the development of the rifled musket, which replaced the smooth bore musket. On the surface they appear to be pretty similar. But one simple innovation drastically improved the rifle's accuracy and range: the rifled musket contains grooves on the inside of barrel. These grooves make it possible for a bullet with an elongated cylinder and a cone-shaped head to spin and thus hit a target located up to 500 yards away. The same principle is involved when a quarterback throws a spiral. In comparison, a smooth bore musket's maximum effective range was only 50 yards. The main reason for such high casualty during the American Civil War was the failure of most generals and soldiers to change their traditional tactic of charging in shoulder to shoulder. The last battle of Gettysburg, known as Pickett's Charge, illustrates just how deadly the seemingly minute change could result. Pickett's charge against the Union Army's position on Cemetery Ridge was ordered by General Lee, a brilliant Confederate commander. His 12,500-man division, one of the largest in the Army of Northern Virginia, marched over open fields for three-quarters of a mile under heavy artillery and rifle fire. It was decimated in less than 50 minutes. When General Lee told him to assemble his division, Pickett allegedly replied, "General, I have no division."

This chapter focuses on the biological community's inability to recognize and deal with a drastically changing situation. This third phase of absurdity stems from the gradual realization of enormous complexity of life and the subsequent – and still ongoing – unraveling of traditional views in the second half of the 20th century. The biological community continues to ignore the continuously transforming situation, even though it no longer can be made to fit the basic biological tenets of the 19th century and even those of the first half of the 20th century. The technological leap from the light microscope to the electron microscope and to X-ray crystallography brought about a completely different picture of a cell's structure. The unimpressive looking blob and particles seen during that particular time period has turned out to be far more complex than anyone imagined. Such a realization can be seen by the shift from extreme confidence to a more humble stance by James Watson (1965), co-discoverer of the DNA molecule:

> Complete certainty now exists among essentially all biochemists that the other characteristics of living organisms (for example, selective permeability across all membranes, muscle contraction, and the hearing and memory processes) will all be *completely* (emphasis added) understood in terms of the coordinative interactions of large and small molecules.

Just five short years later, Watson (1970) completely reversed himself:

A Critique of Science

> ...the protein and nucleic acids are very large, and even today their chemical structures are immensely difficult to unravel. Most of these macromolecules are not being actively studied, since their overwhelming complexity has forced chemists to concentrate on relatively few of them. Thus we must immediately admit that the structure of a cell will *never* (emphasis added) be understood in the same way as that of water or glucose molecules. Not only will the exact structure of most macromolecules remain unsolved, but their relative locations within cells can be only vaguely known.

Despite this realization, however, evolutionary biology's basic tenets remain firmly grounded in this now obsolete understanding of a cell's structure. I contend that its practitioners' inability to recognize the significance of this complexity is far worse than what happened at Polaroid in one respect: the biological community's failure to adapt to current technologies and discoveries has been going on for many decades. Ever since the publication of Darwin's *Origin*, evolution has been debated in two vastly different arenas: the scientific community and the court system. We will discuss how each domain has dealt with this contentious issue.

The Limitation of Reductionism

> The truth is that the science of biology has never been weaker; it is in crisis, and that crisis is made worse, not better, by the extraordinary success scientists have had in discovering the structure and basic mechanism of the genetic materials....In so far as genetic science explains living things, biology has no part in the explanation. To put it bluntly, biology has been expelled from its own territory. Instead, the ground is occupied by scientists from alien culture [chemists], who need to know nothing about biology in order to do their work.
> -Richard J. Bird

Reductionism seems to pay insufficient respect to the complexity of the Universe. It appears to some as a curious hybrid of arrogance and intellectual laziness.

Orogee Dolsenhe

-Carl Sagan

The second half of a man's life is made up of nothing but the habits he has acquired during the first half.

-Dostoevsky

For the last two centuries, science's dominant theme has been reductionism. The main purpose for acquiring biological knowledge was to better understand the organism as well as its organs, tissues, cells, subcellular organelles, biological molecules, atoms, and subatomic structure. Williams (1997) summarizes: "This approach, known as reductionism, views nature as something of a Russian doll: Features at one layer are explained by the properties of the layer below. Hence, physicists search for the basic particles and forces; chemists seek to understand chemical bonds; and biologists scrutinize DNA sequences and molecular structures in their efforts to understand organisms." In the case of biology, this approach was developed in the 19th century when its practitioners embraced positivistic materialism. It tried to "model itself as closely as possible on the examples of physics" (Mayr, 1985).

The physics envy can even be seen in *Origin*, where he refers to the laws controlling certain biological processes no fewer than 106 times (Mayr, 1985). This adoption of physics-based terminology also happened in psychology, as discussed earlier. The optimism and enthusiasm with which reductionism was adhered to can be represented by Francis Crick (1966), the co-discoverer of DNA molecule, who stated that "the ultimate aim of the modern movement in biology is in fact to explain all biology in terms of physics and chemistry." Jacques Monod, a Nobel Prize-winning French biologist who is considered one of the founders of molecular biology, confidently stated: "Anything can be reduced to simple, obvious, mechanical interactions. The cell is a machine; the animal is a machine; man is a machine" (Monod, 1971).

Unfortunately for its proponents, reductionism seems to have reached its limitation. The crux of problem is that biological systems are extremely complex and that a large number of higher level properties cannot be explained by studying their individual parts (Van Regenmortel, 2004). A good example is the Human Genome Project completed in 2003. The complete mapping of the human genetic sequence provided neither a better understanding of the biological process nor the medical miracle it had promised. As Smocovitis (1996) explains, the biologists' basic problem is their outdated mindset of "physics envy":

> Because the comparative scale of the sciences weighed so heavily in favor of the physical sciences, and because the physical sciences served as the historical, logical, and epistemic exemplars to the biological sciences, biologists rushed to prove themselves (and to others) that a formerly fragmented and immature science had finally achieved the unity to "rival the unity of physics and chemistry" (in Huxley's words).

Given existing realities, biology needs to go beyond the reductionist models of plastic balls and pegs if it is to remain relevant. This does not mean that we should deny reductionism's accomplishments, but that the biological process seems to be so complex that reductionism can no longer deal with it in a meaningful way. Augros & Stanciu (1987) call for a "new" biology:

> Today, mounting evidence calls for a revision of the most fundamental principles in the life sciences, including the definition of life, how biology relates to the other sciences, the role of evolution in biology, the place of man in nature, what is meant by a scientific explanation, and the very concept of nature.

Ulanowicz (1997), however, sees the difficulty of overcoming a two-centuries-old habit:

> Although physicists over time have come to appreciate the shortcomings of a strictly mechanical approach to natural phenomena, it seems that biologists are reluctant to accept this conclusion. In part their reluctance to loosen their grip on the mechanical worldview is due to their reverence for the originator of the idea of natural selection, Charles Darwin, who was strongly influenced by the Newtonian determinism that characterized thinking during his time.

Orogee Dolsenhe

The Search for Non-Human Intelligence

> Modern scientific theory compels us to think of creator as working outside time and space, which are a part of his creation, just as the artist is outside his canvas.
>
> -James Jeans

> Could there be a more important question in all of human existence than "Is there a God?" And yet there I found myself, with a combination of willful blindness and something that could only be properly described as arrogance, having avoided any serious consideration that God might be a real possibility.
>
> -Francis Collins

> The greatest question in the world to-day: Is there a Living Intelligence behind Nature, or does the great Cosmos somehow run itself, driven by blind force?
>
> -Frances Mason

Are we alone? Are humans the only intelligent entity in the universe? Driven by curiosity, scientists search for electromagnetic transmissions from a civilization that might exist outside our solar system. This is known as SETI (the search for extra-terrestrial intelligence). Astronomers scan the sky and listen for non-random patterns of electromagnetic signals, such as radio or television waves, that might indicate some intelligent origin. If they detect non-random signals, they try to decode and understand their meanings, just as cryptographers try to bread a code. But, what types of signal pattern should be considered as having an intelligent origin? For example, if pulsating electromagnetic waves similar to those given off by a lighthouse signal were detected, would they be interpreted as a potential indication of intelligence? Jocelyn Bell Burnell and Antony Hewish made such discovery in 1967; unfortunately, the source turned out to be a pulsar, a highly magnetized neutron star that rotates and emits a beam of electromagnetic radiation.

One of SETI's most famous advocates is Carl Sagan, an American astronomer best known for the 1980 television series *Cosmos* and author of *Contact,* a novel that became the basis for the 1997 movie with the same title. The novel's main theme is that Ellie Arroway (played by Jodie Foster) works on the SETI program at Puerto Rico's Arecibo Observatory. The project loses its funding because a government scientist considers it to be futile. Fortunately, a billionaire steps in to provide the necessary funding so that Ellie can continue her work at the SETI project in New Mexico. After a while, the team discovers a repeating series of the first 261 prime numbers. Prime numbers can be divided only by themselves and by 1, namely, 2, 3, 5, 7, 11, 13, 17, and so on. Neither 4 nor 6 are prime numbers, because they can be divided by 2 either 2 or 3, respectively. Such a repeating series of prime numbers is considered so complicated that it could not possibly be the result of random chance. Therefore, an intelligent being must be its source. Of course the plot is fictional, for no signals indicating the possible existence of an intelligent source have been discovered thus far.

There is, however, a far more exciting and meaningful search for non-human intelligence. Is there an intelligent agency behind nature? This is one of humanity's most enduring questions, and yet this crucial inquiry is blocked, to a certain degree, by its negative perception within the scientific community. Given naturalism's domination, even raising the possibility of a supernatural agency gradually became stigmatized. In the following sections, we will examine some indications of a supernatural, superintelligent entity behind life and the universe.

Natural Theology, Intelligent Design, and Irreducible Complexity

> Evolution, as a mechanism, can be and must be true. But that says nothing about the nature of its author. For those who believe in God there are reasons now to be more in awe, not less.
> -Francis Collins

> ....science should avoid theories which smack of the supernatural is an artificial restriction on science.
> -Michael Behe

> History shows that the advances of science always have been frustrated by the tyrannical influences of certain preconceived notions that were turned into unassailable dogmas. For that reason alone, every scientist should periodically make a

> profound re-examination of his basic principles.
>
> -Louis de Broglie

> ...whenever scientific men of any age have denied the facts of investigators on *a priori* grounds, they have always been wrong.
>
> -Alfred Russell Wallace

> The most elementary type of cell constitutes a "mechanism" unimaginably more complex any machine yet thought up, let alone constructed by man.
>
> -William Thorpe

As the evidence of evolution began to accumulate during the 18th and 19th centuries, the biologists' main concern was the conflict with religious dogma. Many biologists like Carl Linnaeus, the father of modern taxonomy, never wavered from biblical creationism, whereas others, among them St. George Mivart and Richard Owen, embraced both evolution and creationism. As mentioned earlier, even Darwin's supposedly naturalistic evolutionary approach was quite ambiguous on the role of the supernatural. In his section titled "Organs of extreme perfection and complication," Darwin confesses his doubts about how such a complex organ like the eye could have evolved through natural selection:

> ...to suppose that the eye with all its inimitable contrivances for adjusting the focus to different distances, for admitting amount of light, and for the correction of spherical and chromatic aberration, could have been formed by natural selection, seems, I freely confess, absurd in the highest degree.

The complexity issue has been discussed under the name of "natural theology." Darwin expressed his admiration for its best-known author, William Paley, in his *Natural Theology*. As discussed earlier, Paley compared a stone and a watch found on the ground. The complexity of watch was reflected by the many parts it required to give the correct time. Each part had to have a specific size and shape if the watch as a whole were to perform its designated function. Paley's (1802/2006) argument was that denying the watchmaker's existence was an "absurdity."

A Critique of Science

Ironically, however, it was religious apologists and not scientists who undermined natural theology, at least in the 18th century (Brooke, 1991). Their literal interpretation of the Bible severely restrained natural theology's ability to formulate a design-based argument. Thus, during the 18th and early 19th centuries it was the religious side that was intolerant. In the 20th century, just as the religious side began to show some flexibility, science became increasingly intolerant.

The freethinkers who challenged religious dogma and the dissenters who now disagree with neo-Darwinism share a common fate: they are up against a very formidable resistance. The sad irony is that the intellectual community, both then and now, did not really have a chance to examine the issue of whether the supernatural had a role in nature or not, for it had to deal with pressure coming from those who wanted to maintain the status quo. And then came Michael Denton, a biochemist at New Zealand's University of Otago, whose *Evolution: A Theory in Crisis* (1985) launched the "intelligent design" (ID) controversy. What set his book apart from natural theology was that it incorporated of the enormous complexity that has been accumulating, particularly since the 1950s. Denton (1985) cogently discusses a cell's incredible complexity by enabling us to visualize:

> ...an object resembling an immense automated factory, a factory larger than a city and carrying out almost as many unique functions as all the manufacturing activities of man on earth. However, it would be a factory which would have one capacity not equaled in any of our own most advanced machines, for it would be capable of replicating its entire structure within a matter of a few hours.

He carries this a bit further:

> Altogether a typical cell contains about ten million million atoms. Suppose we choose to build an exact replica to a scale one thousand million times that of the cell so that each atom of the model would be the size of a tennis ball. Constructing such a model at the rate of one atom per minute, it would take fifty million years to finish, and the object we would end up with would be the giant factory, described above, some twenty

kilometers in diameter, with volume thousands of times that of the Great Pyramid.

One of his main critical points was stated in the following terms: "The complexity of the cell, like that of any complex machine, cannot be reduced to any sort of simple pattern, nor can its manufacture be reduced to a simple set of algorithms or programme." Michael Behe (1996), a biochemist at Lehigh University, has termed Denton's point an "irreducible complexity" and defines it as "a single system which is composed of several interacting parts that contribute to the basic functions, and where the removal of any one of the parts causes the system to effectively cease functioning." Behe (1996) proffers the flagellum of bacteria. The flagellum is like an outboard rotary motor that gives bacteria locomotion. And like outboard motor, it is made up of many parts: a rotary motor, a bushing, a drive shaft, a universal joint, a helical propeller, and so on. All parts are required if it is to fulfill its biological function of propelling bacteria.

In an effort to simplify the argument of irreducible complexity, Behe (1996) uses a mousetrap that consists of just five interlocking parts: the base, the catch, the spring, the hammer, and the hold-down bar. All five parts must function in order to catch mice. The crux of irreducible complexity is to explode the core idea of gradual progression through natural selection. In other words, since a partial organ has no survival advantage, just as an incomplete mousetrap cannot catch mice, Darwin's idea of a slow and piecemeal process of evolution is unworkable. In fact, even Darwin (1872) himself saw the potential problem here: "If it could be demonstrated that any complex organ existed which could not possibly have been formed by numerous, successive, slight modifications, my theory would absolutely break down." He then concluded: "But I can find out no such case."

The foremost challenge to Behe's irreducible complexity comes from Kenneth Miller (2008), a biology professor at Brown University who holds that a partial mousetrap could function as something else, such as a spitball catapult or a tie clip. Thus, he dismisses Behe's position on the grounds that "the mousetrap is not irreducibly complex after all." If we use Miller's line of argument, indeed, the various components of a mousetrap can perform many functions. Just imagine yourself stranded on a deserted island with a mousetrap, just like Tom Hanks was stranded with FedEx boxes in the movie *Castaway*. Numerous uses can be found for a mousetrap. The crucial point here is that some agent would have to have an insight as to how an apparatus designed for one purpose could be used for something else, such as in finding many ways to use a pair of skates. Even Miller's argument requires a certain amount of ingenuity.

At any rate, these two men often sit on the opposite side of the ID controversy, whether in college debates or in the courtroom, such as in the Dover trial that will be discussed later. Yet, ironically, they have more in common than most people sitting the opposite sides, for both are Catholics and Behe (2004) does not reject evolution while embracing ID:

> ...a hypothesis of Intelligent Design has no quarrel with evolution per se – that is, evolution understood simply as descent with modification, but leaving the mechanism open...the focus of ID is on the *mechanism of evolution* – how did all this happen."

But Miller (2004), mistakenly, thinks that Behe is against evolution:

> The flagellum owes its status principally to Darwin's Black Box, a book by Michael Behe that employed it in a carefully crafted anti-evolution argument.

Another irony is that Miller, a practicing Catholic, does not reject the supernatural. So, they could be sitting at the same table in the middle, rather than at the opposite ends if the climate of partisanship, which forces scientists to take extreme positions even though their views may not be so characterized, could be reduced.

Given the hostile attitude toward ID, the idea of complexity is being discussed mainly by outsiders – and they are raising some crucial issues that just cannot be ignored. Evolutionists nevertheless characterize their arguments in terms of biblical creationism in an attempt to dismiss them, even though their arguments are quite sophisticated and sensible. J. Scott Turner (2007), a biologist at the State University of New York at Syracuse, sees the phenomenon of design ID as the "most glaring blind spot" in modern biology. The blind spot is not easily recognized by the insiders; however, those on the outsiders can offer some important ideas.

Scientific ideas supposedly compete in an open marketplace (Krantz & Wiggins, 1973). The opponents of ID often mention that its proponents should compete in that arena if they want their theory to be considered a science. However, those who endorse any views that might contradict the status quo quickly realize this ideal of equal opportunity does not apply to them. In fact, the burden of proof is against them. Wells (2006) argues:

> Science journals regularly publish articles *attacking* intelligent design, but they routinely reject articles *defending* intelligent design. For example, Darwinists have criticized Michael Behe's arguments for ID (Chapter Ten) in many peer-reviewed journals, including *Nature*, *Trends in Ecology and Evolution*, and the *Quarterly Review of Biology*. But those journals routinely refuse to publish Behe's responses.

Curtailing the Ridicule, Charges of Disloyalty, and Warmongering

One of democracy's cherished ideas is the freedom of speech and expression. Most democratic nations, however, do not allow absolute freedom of speech. John Stewart Mill once elaborated upon something known as the "harm principle," namely, that which should be excluded from free speech:

> That the only purpose for which power can be rightfully exercised over any member of a civilized community, against his will, is to prevent harm to others.

Joel Feinberg (1985), an American political and social philosopher, argues that this principle is not sufficient, in itself, to protect people against the unjust behavior of others. As a result, he introduced the "offense principle," which states that some speech that may offend others should be censored, or at least strongly discouraged.

This raises an interesting question: Should academic freedom be curtailed in some situations? Consider Huxley's criticism of Chamber's *Vestiges,* in which he used such words as "foolish fancies," "charlatanerie," "pretentious nonsense," and "work of fiction." Should such vicious criticism be covered under academic freedom? I hold that there should be a limitation to academic freedom, as is the case with the freedom of speech, and that this should include the following three practices: ridicule, charges of disloyalty, and warmongering.

A Critique of Science

Lamarck and Maupertuis were ridiculed by Cuvier and Voltaire, respectively. As a result, both of them as well as their theories suffered greatly not only while they were alive, but even after more than a century had passed. Charges of disloyalty are often made by those who use the word "anti-science" to reject someone else's view. This kind of rhetoric was used during the French Revolution. In communist countries it became "counter-revolutionary"; during the McCarthy era it became "un-American," an accusation that terrorized the nation. Ed Murrow's courageous challenge to McCarthy helped to put an end to the anti-communism hysteria. Unfortunately, however, there is no Ed Murrow in science. I would argue that there is no such thing as "anti-science" to begin with, other than the very act of branding another person with this particular term. Warmongering means for one side to "engage in a war" with the other. Portrayng the other side as enemy can easily draw emotional response that can turn into intellectual brawl.

Remarkably, such "militant atheists" as Richard Dawkins, Christopher Hitchens, Sam Harris, and Daniel Dennett possess all three characteristics, so much so that they actually call themselves "the four horsemen." Their penchant for dramatization resembles the rhetoric of the fundamentalist religions they criticize; their combative, abrasive, and vicious language is a mirror-image of the expressions to which those religious extremists who seek extreme solutions give vent. The question we should be asking is whether they are actually benefitting science or society. Do they represent the progress or the degeneration of science?

Systems Biology: The Irreducibility of Systems

> ...biologists seem to be undergoing a paradigm shift right under our noses. Overnight, as it were, biology departments have begun to jump on the bandwagon of Systems Biology.
> -Evelyn Fox Keller

As if irreducible complexity were not complex enough, the complexity problem becomes even greater in systems biology. Since 2000, this has rapidly emerged as one of the life sciences' leading disciplines (Mendoza, 2009). Kitano (2009) believes that systems biology will one day become biology's dominant discipline. The crux of systems biology is the study of the interaction between various levels:

genes, proteins, pathways, subcellular mechanisms, cells, tissues, organs, and organisms. We can compare this discipline with an automobile, for both require the proper functioning of thousands of parts. Each part consists of many subcomponents, each of which has an irreducible complexity of its own. Any automobile is, as a whole, irreducibly complex (with some exceptions like seat belts and the air conditioner).

Karl Ludwig von Bertalanfy (1969), an Austrian biologist, is considered the founder of systems biology. His holistic approach, which he called the "General Systems Theory" (GST), was an exception to the reductionistic paradigm of the 20th century. Like Behe's irreducible complexity, systems biology often draws analogies with engineering systems (e.g., Kitano, 2009) to make its points. Although systems biology does not suffer from the same kind of stigmatization that irreducible complexity is often subjected to, it magnifies biological complexity to a larger degree and scale of irreducible complexity. Paul Nurse (2008), a British geneticist who won the Nobel Prize in 2001, emphasizes the holistic notion of biology:

> Living organisms are complex systems made up of many interacting components, the behavior of which is often difficult to predict and so is prone to unexpected outcomes. Systems analyses of living organisms have used a variety of biochemical and genetic interaction traps with the emphasis on identifying the components and describing how these interact with each other. These approaches are essential but need to be supplemented by more investigation into *how living systems gather, process, store and use information* (emphasis added), as was emphasized at the birth of molecular biology.

As the importance of holism and systems biology start to be recognized, the nature of "information" emphasized by Nurse (2008) becomes critical. The most fundamental component of information in biology is the genetic code, which will be discussed in the next section.

Intangible Complexity and the Genetic Code

> Even the most senior among us has become a student again, working to learn from this rich font of new information...the

> unexpected, unexplored, and explained information in the rapidly expanding genomic database. New terms are suggested, such as "implied information," "metagenome," and "species genome."
> -Lynn H. Caporale

We can compare biological complexity with the most complex machine ever built, such as the space shuttle. We can talk about the tangible side of this complex structure, which consists of thousands of parts. But focusing on this structural intricacy may cause us to overlook a completely different dimension of the $1.7 billion dollar flying machine: that it is an amalgamation of humanity's cultural adaptation, which has been ongoing for the last 10,000 years. Consider the role of the various writing systems that we all take for granted. If they had never existed, our ancestors' knowledge would not have been preserved and passed on to future generations – and thus the space shuttle could not have been built. Educational and library systems, which provide the ways and means to train engineers and scientists, are so essential to modern human culture that the space shuttle could not have been built without them. The material gathering and transportation systems developed over thousands of years are also indispensable parts of building and maintaining this complex structure. The recent innovation of computer systems is another crucial element, as is the fact that each constituent of these intangible systems is fit into the whole by a diligent and intelligent effort based upon the determination to attain a specific goal.

Let's now apply this intangible aspect to life. We often hear phrases like "a product of genes," "has good genes," and "disease-causing genes." Perhaps Richard Dawkins, a British biologist who considers organisms to be the genes' "survival machines" (1976), represents one of the most enthusiastic proponents of a gene-centered view of biology. Brian Goodwin, a Canadian biologist, is an exception to the general rule in that he opposes it. In an interview with David King (1996), Goodwin stated: "A major problem is that in contemporary Darwinism, organisms are actually reduced to genes and their products...organisms have disappeared as real entities from biology, and that, I think, this is a fundamental scientific error."

Another problem associated with jumping on the bandwagon of the gene-centered view is that we do not know exactly what genes do. Pearson (2006) has addressed this point: "Where the meaning of most four-letter words is all too clear, that of gene is not. The more expert scientists become in molecular genetics, the less easy it is to be sure about what, if anything, a gene actually is." Such an admission is a humbling experience for geneticists, considering the high expectations they had when launching the Human Genome Project, headed by Francis Collins: to determine human genome's sequence. This $3-billion project began in 1990 and was finally completed in 2003. The words of President Bill Clinton symbolize the high expectations of most biologists –it would "revolutionize the diagnosis, prevention and treatment of most, if not all, human diseases." But at the end of the project they were disappointed with the limited amount of knowledge that knowing the gene sequence could provide.

The building of a house requires two types of entities: one for the components (like bricks, boards, and beams) and one for the system that guides their assembly (an architectural plan). In biology both types of information are coded within the DNA (Mattick, 2004). The component molecules are fundamentally alike both at the individual and the species level. Then, the differences in animals' forms come from difference in the architectural information (Mattick, 2004). However, one of the major reasons for this disappointment is the lack of understanding of what exactly is meant by "genetic information." Doyle (2006) points out that "the functioning genome challenges us even to examine what we mean by 'information' and by 'code.'" Keller (2000) addresses the large gap between genetic "information" and biological meaning. This might be due to the confusion over the nature of genetic information, specifically the distinction between hardware and software, or due to the lack of progress on software side despite the tremendous amount of growth on the hardware side. The book of life has four basic letters: adenine (A), guanine (G), thymine (T), and cytosine(C). A combination of these three letters forms the basic unit known as "codon." There are 64 (4^3) different combinations of triplet codon. The triplet combination turned out to be an optimum number. There are 20 essential amino acids. If this was double rather triple, it would only give 16 (4^2) possible combinations. If it was single, only four amino acids could be provided. George Gamow, a Nobel physicist, actually predicted that the triplet combination would be discovered.

In order to explain the importance of distinguishing between a gene's software and hardware, let me use a different example. For centuries, many scholars tried to understand the Egyptian hieroglyphic system, one of the world's first writing systems. But the knowledge of how to read it had been lost long ago, no one knew what each symbol represented. During Napoleon's campaign in Egypt in 1798, a French soldier discovered a slab that turned out to be one of archaeology's greatest discoveries: the Rosetta Stone, a tablet upon which the same message had been inscribed in ancient Egyptian hieroglyphs, the Egyptian demotic script, and ancient Greek. By using the Greek section, which scholars already knew how to read, the hieroglyphic system could finally be understood. Once it was deciphered, all other hieroglyphic inscriptions could be understood. Jean-Francois Champollion, a French scholar and expert in six ancient Oriental languages (as well as Greek), broke the hieroglyphic code system. Underneath the seemingly very complex writings were 24 hieroglyphic alphabets. The Rosetta Stone's hardware did not reveal the inscription's meaning until Champollion broke the code. And, the software part is an abstract idea that does not intrinsically exist in the tablet.

The study of genetics shares a similar predicament: we do not know how the 4 billion year old genetic code system works. The main advance made in this field is that experts have now learned how to read the genetic letters; however, the system's overall meaning continues to elude them. In short, no one has yet discovered genetics' Rosetta Stone. The Egyptian hieroglyphs reveal an elegant system that was supposedly created by someone or a group of people who desired to communicate what they knew to others. All types of communication, be they spoken or written, require an arbitrary set of rules that bestow abstract meaning upon tangible symbols that then enable information to be shared and stored. For instance, Morse code has the only two basic elements: the dot and the dash, both of which possess no intrinsic meaning. The codemaker, Samuel Morse and others, developed a system that matched specific combinations of both components with specific letters and numbers. The point here is that developing any communication system requires a conscious agency to create it for a specific purpose. The notion that tangible symbols can somehow acquire meaning on their own is, for me, untenable. This brings to mind Henry Quastler's (1964) proposition that the creation of new information is associated with conscious activity. Eigen (1992) emphasizes the often-neglected issue of DNA: "Our task is to find an algorithm, a natural law, that leads to the origin of information."

The mere existence of the system reflected in the genetic code should be reason enough for us to consider at least the possibility that an intelligent agent created it.

Another crucial issue of DNA is the differing levels of information contained within it. Polanyi (1968) cogently argues for the existence of differing levels that are irreducible by drawing a comparison with language: "You cannot derive a vocabulary from phonetics; you cannot derive grammar from a vocabulary; a correct use of grammar does not account for good style; and a good style does not supply the content of a piece of prose." Meyer (2003) recently compared it with architecture:

> Functional genes or proteins are no more inevitable, given the properties of their "building block," than, for example, the Palace of Versailles was inevitable, given the properties of the stone blocks that were used to construct it. To anthropomorphize, neither bricks and stone, nor letters in a written text, nor nucleotide bases "care" how they are arranged. In each case, the properties of the constituents remain largely indifferent to the many specific configuration or sequences they may adopt, nor do they make any specific structures "inevitable" as self-organizationalist must claim.

DNA's Packaging

People who have engaged in any task that requires handling of a long string, cord, or rope should be familiar with the irritating problem of entanglement. We use spooling and other techniques to prevent this problem. From the evolutionary perspective, the long strands of DNA also should pose a difficult challenge. Just how much DNA is inside the human body? If one were to place all of one's DNA strands end to end, the resulting string would be long enough to reach between the sun and Earth about 100 times (Schroeder, 2001). This blueprint of life is contained in a strand that is about 2 meters long and consists of approximately 3 billion base pairs of DNA in each human cell. Given that a typical cell is only about 30 millionth of a meter in diameter, the fact that it can encompass a 2-meter-long DNA strand is in itself an almost miraculous feat.

Another essential requirement is that any portion of the 3 billion base-long strands must be accessible for copying. Nature solved this problem by wrapping each strand around nucleosomes, spool-like structures made out of histone protein. The spool's grooves are shaped in such a way that they exactly fit the DNA: "The irregularities in the shape of the cord exactly match irregularities in the hollow grooves" (Meyer, 2009). In order to copy itself, the DNA strand of nucleosome must open up. In other words, it must be flexible enough to allow rapid opening but strong enough to prevent any damage to itself, as Schroeder (2001) elaborates: "A combination of strong covalent bonds maintaining the integrity of the basic molecular structure with weaker, more easily broken bonds, holding the strands in the helix, solves the problem." The nucleosomes coil and stack together to form fibers called chromatin, which then loops and coils to form chromosomes. Both Schroeder and Meyer see the clear and unmistakable "face" or "signature" of supernatural agency in such intricate structures that achieve the specific functions that are essential for life to exist.

The Origin of Life

> I would rather believe in fairy tales than in such wild speculation. I have said for years that speculations about the origin of life lead to no useful purpose as even the simplest living system is far too complex to be understood in terms of the extremely primitive chemistry scientists have used in their attempts to explain the unexplainable. God cannot be explained away by such naive thoughts.
> -Sir Ernst B. Chain

> ...there was no such thing as elementary life, in the sense of something that was already life, and yet elementary.
> -Thomas Mann

Aleksandr Oparin proposed that life arose from inorganic compounds existing in the ancient Earth's atmosphere: methane, ammonia, water vapor, and hydrogen. He suggested that when these were jolted with energy from electric storms, they combined to form organic compounds. In 1953, Stanley Lloyd Miller of the University of Chicago simulated these ancient conditions. His mixture of methane, ammonia, water vapor, and hydrogen in a sealed flask produced organic compounds (such as amino acids and hydroxy acids) when electric sparks were discharged. This experiment was accepted as the confirmation of Oparin's theory. Furthermore, since contemporary genetics is the result of random mutations of genes caused by chance events such as cosmic rays, this random formulation must reach back to the origin of life itself (Monod, 1971).

Along with natural selection, randomness is widely accepted as a driving engine for the creation and evolution of life. However, there is a huge gap between the organic compounds produced in a flask and the structure of living organisms. The amino acids produced in the experiment were simple compounds composed of only a few atoms each (Augros & Stanciu, 1987). To form a typical protein molecule, about 300 amino acids must be present in a precise order along a long chain. A single deviation will ruin the proper functioning of the protein. Given that a living organism contains 20 different amino acids, there are 20^{300} possible combinations (or 10^{390}). Out of them, only 1.5×10^{12} different proteins exist (Schroeder, 1997). This means that there are 10^{378} wrong options for each protein type (Schroeder, 1997). Even if a fully-formed protein molecule somehow arose by some miracle, the delicate protein likely would break up in a dynamic environment. Furthermore, a protein molecule alone is a dead end because it cannot reproduce itself. This ability belongs only to RNA and DNA molecules. Paradoxically, however, enzymes are needed to make them, and enzymes are proteins. This is known as the "chicken-egg problem" (Blum, 1951).

Morowitz (1992) also points out that nearly all features of the metabolism and synthesis of proteins and DNA molecules would require the enclosure of membranes, a "barrier to free diffusion" in a watery environment. A single-celled bacterium, life's simplest organism, contains 600 to 800 different kinds of proteins; each cell in a human body contains about 10,000 different kinds of proteins (Augros & Stanciu, 1987). Shapiro (1986) addresses that 90 percent of a bacterial cell's dry weight consists of proteins, nucleic acids, polysaccharides, and lipids. None of these were ever detected in Miller's experiment.

Proteins, however, are just the basic bricks of living organisms. In order for natural selection to take place, a minimal structure for cells—such as a membrane, cytoplasm, enzymes, RNA, and DNA—is necessary. This structure must also enable certain minimal functions, such as the capture of energy, the intake and distribution of water and essential nutrients, the production and repair of the structure, the excretion of energetic and material waste, and reproduction. Morowitz (1968) concludes that forming a bacterium in an equilibrium ensemble would require more time than the universe might ever see if chance combinations were the only driving force behind the creation of its molecules. Crick (1981) concedes that the origin of life seems to be almost a miracle: so many conditions would have to have been satisfied to get it going. Another crucial factor here is that it would be enormously difficult for a single organism to survive all alone. Burkholder (1965) notes: "All organisms are dependent upon the varied activities of other organisms for the supplies of essential stuffs." Similarly, Margulis (1981) argues: "All organisms are dependent on others for the completion of their life cycles." Attributing the origin of life to the random formation of organic molecules in a flask is a gigantic leap of faith.

If we look back and examine some discarded ideas, we might chuckle at their silliness. One such idea is "spontaneous generation" (also called abiogenesis), which was popular from the days of ancient Greece right up to the late 19th century. Aristotle mentioned it numerous times in his *History of Animals*, for instance: "Other insects are not derived from living parentage, but are generated spontaneously: some out of dew falling on leaves..." This theory remained strong even after the controlled experimental results obtained by Redi in 1668, Spallanzani in 1765, and de La Tour in 1837 (see Farley, 1977 for a review of the controversy). It finally collapsed when the French Academy of Sciences sponsored a contest regarding its validity in 1859. Louis Pasteur won the contest by demonstrating that only preexisting microbes could give rise to other microbes.

As offensive as this may sound to the neo-Darwinists, relying upon randomness to account for the origin and the evolution of life is not much better than relying upon spontaneous generation. Both ideas suggest that the complex forms and intricate processes involved in living organisms arise all by themselves: they explain everything without explaining anything and ignore the enormous complexity of living organisms. A fatal mistake of neo-Darwinism is its belief that a larger scale of time and space can overcome the fallacy of spontaneous generation by chance alone. For instance, Wald (1954) calls time the "hero of the plot" and confidently asserts the "impossible becomes possible, the possible, and the probable virtually certain." Using randomness to explain biological complexity is vacuous and misleading: mathematicians cannot define randomness, and no test or formula can determine whether even something as simple as a sequence of numbers is actually random (Chaitin, 1975). Ho & Sounders (1984) argue: "The neo-Darwinism concept of random variation carries with it the major fallacy that everything conceivable is possible." Bird (2003) also contends that "chance is a vague term, a catch-all for what we do not yet understand." Lima-de-Faria (1988) writes, "Randomness is evoked every time a phenomenon has not been clarified in molecular terms," and goes on to state, "Selection and randomness is a morass of concepts, words used to cover ignorance; they must be banished from evolution... These concepts have been the opium of the biologist for over 100 years." Perhaps so many people uncritically buy into the nonsensical concept because it disguises its core with the euphemistic word "randomness."

Fine Tuning the Universe

> A common sense interpretation of the facts suggests that a superintellect has monkeyed with physics, as well as with chemistry and biology, and there are no blind forces worth speaking about nature. The numbers of one calculates from the facts seem to so overwhelming as to put this conclusion almost beyond question.
>
> -Fred Hoyle

Ole Romer, a Danish astronomer and the first person to measure the speed of light (220,000 meters per second), did so by studying the apparent motion of Jupiter's moon Io in 1676. Considering the crude instrument he used, this figure is not too far off the most recent measurement in 1983: 299,792.458 meters per second. This fact is one of many physical constants that physicists have discovered. Perhaps the most surprising finding of the discoveries of physical constants is that they must lie within a very narrow range in order for life to exist. Consider the mass ratio of electrons and protons. A proton's mass is about 1836.1527 times that of an electron. If this ratio were slightly different, DNA molecules could not form. Another example is the constant of strong nuclear force. If it were a little bit larger, no hydrogen could form. If it were smaller, no elements heavier than hydrogen could form. If the electromagnetic force constant were larger or smaller, chemical bonding would be disrupted and life would be impossible. Hugh Ross (2001) lists 93 different types of parameters of fine tuning. If any one of those were just slightly off, the universe would be too inhospitable for life to exist within it. Paul Davies, a British astrophysicist at Arizona State University, has this to say, "There is for me powerful evidence that there is something going on behind it all....It seems as though somebody has fine-tuned nature's numbers to make the Universe....The impression of design is overwhelming" (Davies, 1988) and "The laws [of physics]....seem to be the product of exceedingly ingenious design...." (Davies, 1984).

Some physicists like Frank Tipler, a physicist at Tulane University, converted from atheism to theism because this fine tuning has put life at the center of the universe. Brandon Carter, an Australian physicist, argues for humanity's special place in it based upon what he called the "anthropic principle." Ironically, he put this principle forward during the symposium commemorating the 500th birthday of Copernicus, whose heliocentrism challenged the biblical account of man's special place at the universe's center. Guillermo Gonzalez, a Cuban-American astrophysicist at Grove City College, takes one step further his *The Privileged Planet*, which he co-authored with philosopher Jay Richards. Here, they argue that our planet is the best place from which to observe the universe. For example, they state, consider the relative size of the moon to the sun. The moon is just large enough to block the bright photosphere, but no so large that it obscures the chromtosphere that allow a "perfect eclipse." This event is significant, because it allows astronomers to verify Einstein's general theory of relativity that light would bend according to the sun's gravitational pull. Arthur Eddington indeed corroborated this prediction. This brings to mind Einstein's famous quote: "The most incomprehensible thing about the universe is that it is comprehensible." In the light of Gonzalez's view, it is quite plausible to think that the universe is created with a particular consideration to be comprehensible for human. Such point of view that our species occupy a special place in the universe was put forward by Pierre Teilhard de Chardin, a French Jusuit paleontologist. Unfortunately, his superiors weren't happy with Teilhard's unorthodox worldview and they denied publication of *The Phenomeonon of the Man*. It was posthumously published in 1955. The reception from science wasn't any better. For instance, Medawar (1961) called it "a bag of tricks." Gonzalez would be criticized and eventually denied tenure. We will talk more about Gonzalez's unfortunate experience later.

Lee Smolin, a staunch atheist physicist at the University of Waterloo, has found a way to explain away all of this fine tuning by combining the notions of multiverse with Darwinian evolution. He suggests that a collapsing black hole causes a new universe to emerge on the other side with slightly different constant perimeters. Thus, universes multiply and mutate. In addition, those universes with defective constant parameters will reach heat death and collapse. There is an irony to Smolin's effort: He was able to maintain the positivistic agenda of denying the unobservable supernatural agency only by invoking the idea of multiverse, which is even harder to verify.

The Adaptive Capacities without Natural Selection

What would happen if an organism were inserted into an adverse situation that it had never before encountered? Since it had not gone through the process of natural selection, during which it would have developed the ways to overcome this hostile condition, it therefore should not able to cope with it. For instance, the surgical removal of an eye lens is not a natural occurrence, but Needham (1941) found that a new lens arises from the edge of the iris of a salamander's eye.

The reorganization of the human brain after hemispherectomy poses an even more difficult challenge for the neo-Darwinian paradigm. The human brain has two asymmetrical hemispheres: the right hemisphere normally performs visual-spatial functions, while the left hemisphere is involved in linguistic capacities. Babies who are born with the congenital vascular malformation of a hemisphere must have the affected hemisphere surgically removed in order to avoid possible seizures and mental retardation. This procedure, however, induces some sort of massive reorganization that allows a single hemisphere to accommodate both visual-spatial and linguistic functions. With just half a brain, these babies grow up to function remarkably well in global visual-spatial and linguistic areas (Day & Ulatowska, 1979; Dennis & Kohn, 1975; Byrne & Gates, 1987). This amazing capacity of the brain to reorganize itself humbles us in respect to how little we know. If neuroscientists could ever achieve just a fraction of the brain's capacity to reorganize itself after hemispherectomy, this would constitute a quantum leap in our knowledge of the brain (Dolsenhe, 2005). Our body obviously possesses the faculty to respond to novel situations and to produce a complex and adaptive reorganization. As long as biologists remain fixated on randomness and natural selection as causality, the underlying faculty involved in this extraordinary function as well as others will prove unsolvable.

Teleology and Teleonomy: Science's Verbal Tap Dancing

> If intentionality has been banished from our thinking about adaptation, we are finding difficult, it seems, to write or speak about it without resorting to language that is rife with implicit intentionality.
>
> -J. Scott Turner

The last two sections manifest teleology, defined as the purposive trend in nature. As naturalism toppled the metaphysical throne in the late 19th and 20th centuries, teleology was one of the many subjects that was purged. The purely material-based view of modern science has no room for purposeness and goal-directedness in life, for it contends that the universe runs according to certain physical laws with a mathematical precision and that an unbroken chain of cause and effect determines all events. The rejection of teleology, however, poses difficult challenges to the three major branches of science: psychology, biology, and physics. In this section, we will just focus on biology's response.

Politicians are often faced with a dilemma: their support for one side will upset the other side. For instance, they need financial support from special interest groups but also need votes from the people, or they might have to compromise their personal convictions to deal with the pressure placed upon them by the public or the party. They might find a way to avoid such predicament by presenting themselves as sympathetic to both sides or employ the tactics of evasion, diversion, muddled thinking, or even intellectual laziness when explaining their view on the issues in question.

We can find a similar dilemma in science: the intellectual integrity of scientists who do not conform to accepted views is challenged in order to force them to at least go along with the existing status quo. The word "teleology" comes from the Greek word *telos* which means end or purpose. Teleology is one of necessities of life, as Augros & Stanciu (1987) state:

> Purpose permeates every aspect of life. The metabolism of every cell is ordered to the organism. Growth is aiming at the completeness of form. The organ-tools of animals and plants, the capacity for self-repair. The findings of ethology and ecology, all point to purpose. With elegance and economy, nature subordinates means to end. Matter is for the sake of form, and both are for the sake of operation. Every cell, every tissue, every organ serves a purpose. Every animal, every plant directs its activities to an end. The whole of nature is ordered by purpose.

On the one hand, life's apparent purposiveness makes difficult to simply toss out teleology. Most biologists use teleological language like "the evolution of adaptive strategy," "survival strategy," "sexual strategy," or "arms race," as if the species in question possesses a mental ability to produce plans to accommodate its needs. And yet science continues to "condemn" (Lovelock, 1988) the very notion of teleology. Horn (1988) eloquently describes the teleological paradox in science: "Teleology is the mistress with whom all biologists live, but none like to be seen in the company of."

Mayr's (1976) uses the term "teleonomy" to account for the goal-directedness of living organism without endorsing teleology, or so he thinks. He explains teleonomy as a system that operates on the basis of a program in passive and automatic way, despite the fact that a coded or prearranged information naturally leads to a given end. Augros & Stanciu (1987) have some harsh words for this seemingly clever approach: "One indicative example is the elaborate linguistic subterfuge called teleonomy, invented to deny the evidence of purpose in nature of things."

The Scopes Monkey Trial

Since Darwin's *Origin*, the scientific community has disagreed not so much about evolution, but about the mechanism behind it. In the 1920s, while biologists were debating the mechanisms of evolution, a different type of debate was going on in Tennessee. John Washington Butler, a corn and tobacco farmer and member of the Tennessee House

A Critique of Science

of Representatives, worried that his children and those of his neighbors might become atheists by learning evolution in school. He therefore introduced a bill, the Butler Act, that prohibited its teaching in public schools. The House passed it without any discussion by the vote of 71 to 5, and it was immediately challenged by the ACLU (American Civil Liberty's Union). John Thomas Scopes, a high school sports coach and substitute biology teacher, agreed to serve as the ACLU's front man. Scopes was served a warrant on May 5, 1925 and placed on trial to test the Butler Act. Known the Scopes monkey trial, this even drew international attention.

The prosecuting team included William Jennings Bryan, three-time Democratic presidential candidate. The defense team was led by the famed lawyer Clarence Darrow. The attempt to provide eight expert witnesses for evolution was curtailed by the judge, who ruled that the trial was to determine whether Scopes had violated the act, not to determine the veracity of evolution. Then, the highlight of this eight-day trial unexpectedly turned to Bryan, who volunteered to be an expert witness on the Bible. Thus the examination focused on Bryan's knowledge of biblical stories and religion itself. Although Darrow was able to ridicule his limitation on the subject matter, the jury, composed of ten farmers who did not believe in evolution, found Scopes guilty and sentenced him to pay a $100.00 fine. Bryan may have won the battle but he seemed to have lost the war, for the trial exposed the naiveté of creationism. In fact, this became an ongoing drumbeat among the supporters of evolution and those who advocated the separation of church and state. In short, it was a public relation nightmare for creationists, just as Galileo's trial was a nightmare for the Church.

The Dover Trial

> A secular state, it must be remembered, is not the same as an atheistic or antireligious state. A secular state establishes neither atheism nor religion as its official creed.
>
> -US Supreme Court

It is ironical that, in the very field in which Science has claimed superiority to Theology, for example – in the abandoning of dogma and the granting of absolute freedom to criticism – the

> positions are reversed. Science will not tolerate criticism of special relativity, while Theology talks freely about the death of God, religionless Christianity, and so on.
>
> -Herbert Dingle

In 2005 in Dover, PA, the teaching of evolution in a public school became another court case that drew national headlines. And like the Scopes trial, the heavy hitters for both sides, lawyers and expert witnesses, gathered around the Dover case. Besides those common factors, however, the two cases bore almost no resemblance to each other. In fact, the 80 year separating the two cases revealed just how much times had changed. First, the roles of plaintiff and defendant were reversed: the advocate for evolution was the plaintiff, not the defendant. Second, the Dover schools were already teaching evolution. The board members wanted to change the status quo, not ban the teaching of evolution. As we shall see later it was not even about teaching creationism alongside evolution, although the board members would have preferred this arrangement. The controversy stemmed from the board's decision to require teachers to read a short statement in the 9th grade biology class.

In order to explain how the Dover Area School District became the defendant, we need to understand the decentralized American public educational system. Unlike most countries, American public schools are governed by self-directing school boards who are mostly elected by local voters. The board members have the power to decide the direction of the schools, including the purchase of textbooks, which would later become the main source of the infamous clash. The storm began to take a shape when deeply religious members became the majority after the 2003 board elections. Two board members, Alan Bonsell and William Buckingham, were particularly vocal and became the driving force behind the controversy. Bonsell, who lacked any science training and believed that Earth is somewhere around 10,000 years old (Irons, 2007), wanted to bring prayer and faith back into the schools. Buckingham saw the threats posed by secular liberalism and the arrogance of science, especially evolutionary biology (Slack, 2007). The board decided not to purchase biology textbooks, because they were "laced with Darwinism," unless they were *balanced* with the teaching of creationism (Slack, 2007). Thus, these textbooks became hostage to the school board (Irons, 2007).

A Critique of Science

The very next day, the board's decision was reported in the *York Daily Report* and it was notified by the ACLU that it would take legal action if they tried to teach creationism in the class. In the next board meeting about a week later, Buckingham had considerably shifted his extreme position after talking with Richard Thompson, president of the Thomas More Law Center. For example, he started using the phrase "intelligent design" instead of "creationism." In addition, the board toned down his demand for purchasing biology textbooks that balanced evolution with creationism. He suggested that *Of Pandas and People*, an intelligent design biology textbook, should be used as a "curricular supplement," rather than the main textbook. About four months later, the board further retreated by using *Of Pandas and People* for reference. On November 9, 2004, it issued a press release stating that teachers would be required to read the following statement to all students in the 9th grade biology class starting in January 2005:

> The Pennsylvania Academic Standards require student to learn about Darwin's theory of evolution and eventually to take a standardized test of which evolution is a part. Because Darwin's Theory is a theory, it is still being tested as new evidence is discovered. The Theory is not a fact. Gaps in the Theory exist for which there is no evidence. A theory is defined as a well-tested explanation that unifies a broad range of observations.
>
> Intelligent design is an explanation of the origin of life that differs from Darwin's view. The reference book, *Of Pandas and People*, is available for students to see if they would like to explore this view in an effort to gain an understanding of what intelligent design actually involves.
>
> As is true with any theory, students are encouraged to keep an open mind. The school leaves the discussion of the origin of life to individual students and their families. As a standard-driven district, class instruction focuses upon preparing students to achieve proficiency on standards-based assessment.

Judge for yourself. This does not sound like a statement made by religious fanatics. Reading the statement, I cannot help but think that they have presented a thoughtful, reasonable, and humbling statement. There is no sign of rejecting evolution, given that it asks for people to have an "open mind." Considering the one-party rule in biology despite many difficulties, it is not a bad idea for students to be reminded that biology does have to contend with a specific predicament. In fact, such a statement should have been required reading not only for every student, but also for every biologist as well. In particular, the first paragraph might have been written by evolutionary biologists who also see gaps in neo-Darwinism, as some critics of neo-Darwinism have used much harsher words like "funk" (Margulis, 1991). Virtually no biology textbooks mention the existence of dissenting voices. Evolutionary biologists have placed all of their eggs in one basket; ironically, it is the board members who were trying to present an alternative basket.

At any rate, the three board members who voted against this statement resigned in protest. In addition, the teachers refused to read it to their students; a school administrator had to do so. The ACLU, acting on the behalf of eleven parents, filed suit on December 14, 2004. The plaintiff team eventually dwarfed that of the defendants. In addition to the ACLU lawyers, the team added representatives from Americans United for Separation of Church and State (AU) and the National Center for Science Education (NCSE). Slack (2007) remarks that "that's a lot of names to keep track of, let alone big egos to keep from bumping against others." On the defendants' side, Richard Thompson of Thomas More Law Center was the chief council. The publisher of *Of Pandas and People*, the Foundation for Thought and Ethics (FTE), was not allowed to participate, for the judge ruled that it had failed to demonstrate that it had a significant protectable interest in the case.

As the trial progressed, the defendants faced a serious setback: the "wedge document," which had been drafted in 1998 by the Discovery Institute. Marked "Top Secret," it became a liability in the attempt to promote the contention that intelligent design is not a form of religion. The sloppy artwork of the document's cover portrayed Michelangelo's *The Creation of Adam*, which portrays God's hand as reaching out to Adam. Why was this such a big deal?

Let's now discuss the circumstances surrounding this particular document. One of the coveted ideas of the American Constitution is its purported separation of church and state, as spelled out in the First Amendment: "Congress shall make no law respecting an establishment

A Critique of Science

of religion, or prohibiting the free exercise thereof..." This constitutional provision to sever organized religion's connection with the state was truly a breakthrough. As much as the concepts of the separation of powers and the system of checks and balances, separating church and state symbolizes the audacity and wisdom of the Founding Fathers. Notwithstanding the First Amendment, achieving this separation did not come immediately. As demonstrated in the Butler Act, elected politicians were able to ban the teaching of evolution in public schools. The ACLU deserves a lot of credit for their passionate effort to fight to achieve this ideal via the Scopes trial. The Butler Act, however, was not finally repealed by both houses of Tennessee legislature until 1967. This abrupt change was a consequence of the "Sputnik moment" that occurred on October 4, 1957, when the Soviet Union successfully launched Sputnik 1, the world's first artificial satellite. Americans, who believed that Soviet technology threatened their security, decided to upgrade of their country's educational system to improve its technological and scientific capabilities. Following Sputnik, the teaching of evolution in public schools became widespread.

In the early 1980s, several states tried to bring creationism into the classroom. For instance, in 1981 Arkansas' Act 590 mandated equal time for creation science and evolution in public schools. In *McLean v. Arkansas Board of Education*, Judge Overton ruled that this act sought to advance religion and therefore violated the separation of church and state. A similar case in Louisiana, *Edwards v. Aguillard,* went all the way to the Supreme Court in 1987. In a seven to two majority, the Court ruled that a Louisiana law requiring that creation science be taught in public schools along with evolution was unconstitutional.

These setbacks convinced the religiously inclined intellectual community that it was time to devise a new strategy, one that would strip away any traces of religion so that they could insert creationism into the public school curriculum without being accused of violation the separation of church and state. Their solution, known as "intelligent design" (ID), is defined as follows: "The theory of intelligent design holds that certain features of the universe and of living things are best explained by an intelligent cause, not an undirected process such as natural selection" (from the Center for Science & Culture website). The Center for Science & Culture is a subsidiary of the Discovery Institute that, in turn, is a Seattle-based nonprofit organization founded in 1990 to promote ID. The most ambitious and aggressive part of the Discovery Institute's plan was explicitly stated in the wedge document: "The Center seeks nothing less than the overthrow of materialism and its cultural legacies...and to replace it with a science consonant with Christian and theistic convictions." Seeing scientific materialism as a "giant tree," their strategy was to formulate a wedge that, while relatively small, could split the trunk when applied at the tree's weakest points. Naturally, this document was viewed as the smoking gun, proof that the Discovery Institute was trying to force religion into the public school system. Forrest & Gross (2004) dubbed ID "creationism's Trojan horse."

The Dover trial began on September 26, 2005 in the Harrisburg courtroom as a bench trial without a jury. District Judge John E. Jones III presided and would be the only one to hand down a ruling. In contrast to the Scopes trial, the Dover trial was more like a graduate-school seminar in which both sides offered expert witnesses from biology, history, theology, education, and philosophy (Irons, 2007). The centerpiece in the Scopes trial was the confrontation between Darrow and Bryan on biblical issues; in the Dover trial, the seminar-like atmosphere engendered an intellectual clash between two eminent scientists, Michael Behe and Kenneth Miller. Ironically, both of them are Catholics.

Judge Jones ruled that ID is a form of creationism and, therefore, the school board's requiring the statement to be read in the classroom was unconstitutional. Adding insult to injury, all eight school board members who had voted for the teachers to read the statement were defeated in the next election. Thus ID's proponents were rejected both by the court and the public. The Dover trial demonstrated the changed landscape as regards the public's acceptance of the principle of separation of church and state. The lopsidedness in favor of evolution is just one of many facets of the secular environment in the 21st century.

Nova, the highly respectable documentary program of the Public Broadcasting Service (PBS), produced a two-hour program entitled "Judgment Day: Intelligent Design on Trial,"
which depicts the turmoil created by the ID debate in Dover. This was yet another blow to its proponents, not to mention another public relations nightmare. The program outlined the lopsidedness of the overall trends in how this supernatural approach is used to understand life and the universe. While the majority of academics might have cheered evolution's triumph over creationism, some concerns have begun to be

aired. Thomas Nagel (2008), an atheist philosopher at New York University, for instance, believes that this is an "intellectually unhealthy situation" because the evidence does not rule out the possibility of intelligent design despite the biologists' confidence that random mutations in DNA are a sufficient explanation. He makes the following point:

> Skepticism about the standard evolutionary model is not limited to defenders of ID. The skeptics may be right or they may be wrong. But even if one merely regards the randomness of the source of variation as an open question, it seems to call for the consideration of alternatives.

While Nagel may be an open-minded atheist, Anthony Flew, a British philosopher and life-long atheist, reversed his former rejection of anything supernatural. A long-standing proponent of the standard version of naturalism, namely, that there is no need to believe in any sort of Creator to explain nature, his embrace of theism was big news. After all, he is considered the world's foremost philosophical atheist. His debates and dialogue, starting in 1985 with Gary Habermas, a philosopher and theologian at Liberty University, made him reconsider his atheism. In 2004, Flew informed Habermas that he had become a theist because, in his own words, he "had to go where the evidence leads."

One sign of intellectual maturity, whether in an individual or a society, is the capacity to entertain alternative views. Even when we disagree with them, we should show them at least a minimum amount of respect. Most people cherish this democratic ideal. Besides the democratic principle underlying this basic right, mutual respect facilitates the provision of an intellectual space in which all ideas can be presented and discussed in a civil manner. As we all know, however, one of history's continuing problems is that some people might try to justify their denial of a particular person's or group's basic rights. Here I cite the ACLU's alliance with the plaintiff in the Dover trial, for this arrangement seems to go against their stated principles. In fact, I would argue that in this case, the defendant seemed to be more in need of having his constitutional rights protected. This is exactly the opposite of what happened in the Scopes trial.

Orogee Dolsenhe

Common Errors Made during Evolutionist-Creationist Debates

> If tyranny and oppression come to this land, it will be in the guise of fighting a foreign enemy.
>
> -James Madison

    The lopsided landscape that favors naturalism and Darwinian evolution makes very difficult to have impartial and fair-minded discussion in science. There is one neutral space in which the issue of supernaturalism can still be more or less debated. Various television programs like Nightline, on networks like CNN, or in large college and university auditoriums provide both sides with equal footing and equal time to make their case. Perhaps the first evolutionist-creationist debate occurred between Thomas Huxley and Bishop Wilberforce in 1860, just a few months after the publication of Darwin's *Origin*. The highlight of this event was a single exchange: Hoping to deliver a knockout punch, Wilberforce asked Huxley whether he would prefer to think of himself as descended from an ape on his grandfather's or his grandmother's side. Huxley's alleged legendary counterpunch was that he was not ashamed of having an ape as his ancestor; instead, he would be ashamed to be connected with a man who used his great gifts to obscure the truth. The story has repeated ever since as an example of a confrontation in which science's reasoning triumphs over religion's ignorance. Gould (1991), however, discounts the widely held view of who won, saying that both men thought they had won the debate. This contest was inherited by their sons, both of whom thought their fathers had won. Caudill (1997) summarizes: "The debate myth shows how history is built, at times, more upon foundations of belief and hindsight than upon historical evidence." I suppose that spin doctors used by Huxley and his followers did a far better job than those of Wilberforce.

    Many modern evolution-creation debates can be seen on YouTube, where both sides repeat some basic errors. The first one is that of mutual exclusiveness. Evolutionists try to reject any supernatural agent by presenting evidence for evolution; creationists, on the other hand, attempt to disprove evolution by employing the design argument. As St. Augustine argued, creation and evolution can be complementary. Known as "theistic evolution," a number of 19th-century biologists embraced it. The second common error is as follows: evolutionists interpret the relevant evidence as proving Darwinian evolution, whereas creationists assert that the difficulty of Darwinian natural selection negates evolution itself. The fact is that there are many evolutionary ideas and Darwin's idea is just one of them.

The third common error is that if everything complex had to be created by some agent, then who or what created God? The concept of an omnipotent God is not so different from that of an omnipotent materialistic universe that possesses the properties and conditions needed to create life. In fact, both ideas rely on the involvement of some "miraculous" entities. The idea that the universe has always existed and does not need a creator is not that different from the argument that God has always existed. Aristotle's term was the "Unmoved Mover." The notions of a self-existing God or a self-existing universe are essentially the same type of deep philosophical question regarding its origin: Why there is anything at all, rather than nothing? It is also the problem formulated by Parmenides: *Ex nihilo nihil fit* (Nothing can come from nothing). However, in order to for anything to exist, something had to exist before it. this could be either the universe or God.

The fourth common error is that both sides portray the other side as the enemy. The wedge document, for instance, portrays scientific materialism and Darwinism as destructive forces that have "infected" virtually every area of life. Evolutionists, on the other hand, view creationists as a threat to science and the public educational system. Furthermore, the debate has often turned into a conflict between science and religion. We need to break out from this "us versus them" mindset and start seeing them as fellow seekers of knowledge.

The fifth common error is what happens afterward: Each side's supporters try to spin the event so it looks like their side won, as was the case after the Kennedy-Nixon debate. Rather than admitting that their opponent had some good points and that their own side had some weak points, the highly partisan support for each side is the rule. Just take look at some of the vulgar language that is now, unfortunately, quite common on YouTube.

Three Devious Ways to Reject Supernaturalism

> Among us ridicule is a deadly weapon; all its blows are mortal.
> - François-Vincent Raspai

Three powerful weapons are used to reject God's existence of God: demarcation, the double standard, and ridicule. Demarcation means that certain ideas and issues supposedly do not meet the standard required to be considered as science. In essence, this is the institution-sanctioned discrimination practiced against anything metaphysical. It seems that "ungodly" has been replaced with "unscientific" as a way of justifying intolerance. As for the double standard, this is applied to creationism's "God of the gaps." Evolutionists argue that whenever difficulties arise in the naturalistic accounts of phenomena, creationists turn to God's omnipotent power. Apparently, they do not realize that a similar argument can be used against them when they proclaim a "randomness of the gaps" or a "natural selection of the gaps." The use of ridicule can perhaps be traced to Bertrand Russell, a British philosopher who showed a "marked hostility" toward the concept of God (Ross, 1990). In fact, his famous metaphor for the rejection of God was "Russell's teapot," which appeared in his article entitled "Is There God?" (Russell, 1952). He compared belief God to a china teapot revolving around the sun. Due to its small size, however, no one can observe it or disprove its existence. This metaphor is a rhetorical argument in the form of ridicule. Others joined in this ridicule: the "invisible pink unicorn" (Dawkins, 2006), the "flying spaghetti monster" (Henderson, 2005), and the "dragon in my garage" (Sagan, 1995). It seems that scientists are joining Russell's intellectual bullying of those who do not agree with them. Notwithstanding its unjust rhetorical argument, ridicule is very effective.

Evidence for God's existence can be discussed in three ways: positive evidence, negative evidence, and the absence of evidence. These should be discussed without a priori rejection for any reason. The scientific community must provide a neutral space in which all ideas can be discussed and thoroughly examined. The burden of proof should apply to each approach equally. Yet the call for fairness and equality, unfortunately, is an unrealistic request in the dogma-dominated scientific community, for the proponents of supernaturalism are guilty until proven innocent. Above all, ridicule must not be used to undermine scientific debate.

Building Bridges between Science and Religion

Long before Darwin's *Origin*, St. Augustine suggested what is now known as theistic evolution: that evolution might be the way of God's creation. This idea has largely fallen victim to the polemical atmosphere surrounding the confrontation between science and religion. In the 18th and 19th centuries, as mentioned earlier, the strict and literal interpretation of the Bible severely limited both theistic evolution and natural theology. In the 20th century, as neo-Darwinism became the dominant school of thought, the tables have been reversed. This has been the case particularly since the Synthesis, for biologists now display the same degree of intolerance that characterized the Christian apologists who were active during the Enlightenment. It is certainly ironic that in our own day the religious side actually displays a greater degree of diversity and intellectual freedom than does the scientific side. Consider Pope John Paul II's 1996 message to the Pontifical Academy of Sciences:

> And to tell the truth, rather than speaking about the theory of evolution, it is more accurate to speak of the *theories* (emphasis added) of evolution. The use of the plural is required here—in part because of the diversity of explanations regarding the mechanism of evolution, and in part because of the diversity of philosophies involved. There are materialist and reductionist theories, as well as spiritualist theories. Here the final judgment is within the competence of philosophy and, beyond that, of theology.

Not only did he endorse evolution, but he also used "theories" rather than "theory." This demonstrates his personal open-mindedness as regards a variety of options, an attitude that is lacking among contemporary scientists. John Haught (2000), a theology professor at the Georgetown University, suggests that an engagement of theology with evolution will be beneficial for both science and religion. He admits that the knowledge acquired in Sunday school is hardly expansive enough to incorporate the nuances of evolutionary thought and suggests to break through an "obsessive and restrictive" notion of God.

Two years before his death Theodosius Dobzhansky (1973), the most important bridge builder in the Synthesis, published a short article in the *American Biology Teacher* with the oft-quoted title of "Nothing in biology makes sense except in the light of evolution." The title is somewhat misleading, however, because the article actually seeks to build a bridge between science and religion. His view of evolution reminds one of St. Augustine's:

> The organic diversity becomes, however, reasonable and understandable if the Creator has created the living world not by caprice but by evolution propelled by natural selection. It is wrong to hold creation and evolution as mutually exclusive alternatives. I am a creationist *and* an evolutionist. Evolution is God's, or Nature's method of creation. Creation is not an event that happened in 4004 BC; it is a process that began some 10 billion years ago and is still under way.

There is an irony to Dobzhansky's role in building bridges between naturalists and geneticists. The neo-Darwinian bridge has become so restrictive that his subsequent effort to build bridges between science and religion was doomed from the start.

The Problem of Unanimity

> Most of the great controversies and conceptual oppositions of the nineteenth century are still present at the beginning of the twenty-first century: religion and vitalism versus evolution and materialism, structuralism versus functionalism, reductionistic versus holism, gradualism versus saltationism, selectionism versus nonadaptationism, the inheritance of acquired characteristics, and nurture versus nature. What has changed is not so much the nature of the ideas but the evidence supporting them and the intensity of the debates.
>
> -Jan Sapp

In February 2008, Fidel Castro announced his retirement from his nearly 50-year reign as Cuba's leader. In the following week, the country's National Assembly voted for a new leader. The sole candidate on the ballot, Fidel's brother Raul, received the votes of all 597 assembly members. Those who support Cuba's political system might say that everybody loves Raul. For instance, assembly member Luis Felipe Simon Cabreza says: "In my opinion, Raul is the only option." The lack of dissent, however, could be interpreted as the lack of freedom due to

A Critique of Science

the likelihood that everybody is forced to conform. Any indication of free thinking, be it in a country, organization, institution, or corporation, is dissent. While it might be time-consuming and chaotic at times, it gives people the chance to examine a range of viewpoints. Just imagine what might have happened if the Kennedy administration had not analyzed differing options during the Cuban missile crisis as we shall see later. The real source of communism's problematic doctrine is its inability – if not outright refusal – to consider a wide range of options. Beware of unanimity, for a large problem might be lurking underneath the apparent conformity.

If diversity were used to measure the degree of intellectual freedom, modern biology would reflect a very low degree of diversity indeed! Even a cursory study of commonly used biology textbooks reveals that hardly any alternatives to neo-Darwinism are even mentioned. If these alternatives are mentioned, they appear as no more than passing remarks or as objects of ridicule. This culture of intolerance has become so well entrenched that people no longer see anything wrong with the picture. Ever since the onset of the Synthesis, modern biology has become a one-party intellectual dictatorship.

One might say this reflects the field's unanimous agreement, because virtually all biologists have concluded that the neo-Darwinian approach is correct. But some biologists continue to challenge this supposed unanimity openly. For example, Goldschmidt (1940) writes: "I cannot agree with the viewpoint of the textbooks that the problem of evolution has been solved as far as the genetic basis is concerned." Goldschmidt (1960) recalls their reaction to his dissenting view: "Neo-Darwinians reacted savagely. This time I was not only crazy but almost a criminal." Sewell Wright, a key member of the Synthesis, was not invited to major meetings – no doubt a "suitable punishment" for daring to disagree with various neo-Darwinian ideas. When it comes to diversity and tolerance, science and religion have been marching in the opposite direction: science has moved toward intolerance and conformity, while religion has moved toward tolerance and diversity. Hunter's (2007) argument in *Science's blind spot* that Darwinism is religious and ID is empirical may well be justified in numerous ways.

Orogee Dolsenhe

Academia's Hostile and Intolerant Intellectual Landscape

> Unfortunately, in our blinkered times, to be against the scientific establishment is to be against science itself.
> —Steve Fuller

> The test of courage comes from when we are in the minority. The test of tolerance comes when we are in the majority.
> —Ralph W. Sockman

> Science is supposed to give the full range of possible explanations a fair chance to succeed....Evolutionary biology, by limiting itself exclusively to materialistic mechanism, has settled in advance the question of which biological explanations are true, apart from any consideration of empirical evidence.
> —Willliam A. Dembski

On the website of Michael Behe, one of ID's leading proponents, one notices the following disclaimer:

> My ideas about irreducible complexity and intelligent design are entirely my own. They certainly are not in any sense endorsed by either Lehigh University in general or the Department of Biological Sciences in particular. In fact, most of my colleagues in the Department strongly disagree with them.

I cannot help but wonder why he had to write this statement. I can only assume that he did so to protect the reputation of his department and his colleagues. His department's concern over ID is clearly displayed on his website:

> The faculty in the Department of Biological Sciences is committed to the highest standards of scientific integrity and academic function. This commitment carries with it *unwavering support for academic freedom* (emphasis added) and the free exchange of ideas. It also demands the utmost respect for the scientific method, integrity in the conduct of research, and recognition that the validity of any scientific model comes only

as a result of rational hypothesis testing, sound experimentation, and findings that can be replicated by others.

The department faculty, then, are unequivocal in their support of evolutionary theory, which has its roots in the seminal work of Charles Darwin and has been supported by findings accumulated over 140 years. The sole dissenter from this position, Prof. Michael Behe, is a well-known proponent of "intelligent design." While we respect Prof. Behe's right to express his views, they are his alone and are in no way endorsed by the department. It is our collective position that intelligent design has *no basis in science* (emphasis added), has not been tested experimentally, and should not be regarded as scientific.

These two paragraphs seem to be mutually contradictory: the phrases "unwavering support for academic freedom" and "no basis in science" seem to cancel each other out. Despite this, however, Behe's situation at the Lehigh University is far better than that of other ID proponents. Having embraced an alternate theory after obtaining tenure, he escaped the regrettable fate of other ID adherents.

Consider William Dembski, an ID proponent who studied mathematics, psychology, and philosophy. In 1999 he established the Michael Polanyi Center, the first ID university-level research center at Baylor University, with the support of its president, Robert B. Sloan. One of its high points was a 2000 conference, "The Nature of Nature," which attracted some major figures in science and philosophy, including Nobel physicist Steven Weinberg. The conference also brought out an unwanted consequence: the Faculty Senate voted 27-2 to abolish the center. The center was eventually closed in 2003 and Dembski moved to Southwestern Baptist Theological Seminary.

Another case of intolerance can be seen in the treatment of Guillermo Gonzalez, a Cuban-American astrophysicist at Iowa State University. He and the philosopher Jay Richards published *The Privileged Planet* (2004), in which they claim that a supernatural agency fine tuned fundamental physical properties in order to provide the conditions to support life. In addition, they suggest that the universe was designed for us to discover its mysteries. For instance, they note, the relative sizes of the moon and the sun permit a perfect eclipse so that the theory of relativity could be tested. Thus, they offer double dosages of the purposive design type. The book immediately received a hostile review from William H. Jefferys, an astronomer at the University of Texas at Austin, who ended it by saying: "I hope that he [Gonzalez] does not throw away his career on such nonsense" (Jefferys, 2005).

An irony is that Jefferys might actually have instigated the process of getting him dismissed from the university. The most damaging attack, however, comes from Hector Avalos, a faculty advisor to the campus Atheist and Agnostic Society. A professor of religious studies and former Pentecostal preacher, Avalos co-authored a petition urging all faculties to uphold the university's integrity and reject ID; about 130 ISU faculty members signed it. Similar petitions appeared at the University of Iowa and the University of Northern Iowa. In 2007, the university rejected Gonzalez's tenure and President Gregory Geoffroy rejected the subsequent appeal. Geoffroy stated the rejection had nothing to do with Gonzalez's support for ID; rather, he simply did not show the trajectory of excellence expected of a candidate seeking tenure. Curiously, Gonzalez has authored 68 peer-reviewed articles, more than four times the 15 papers required by the department's own tenure procedure. At any rate, after the Iowa Board of Regents rejected Gonzalez's final appeal, he accepted a job at Grove City College, a Christian school located in Pennsylvania.

During the long history of man's quest for knowledge, it was usually the religious or political establishments who impinged upon scholars' academic freedom. Such external censorship gradually diminished when universities began to appear, for their autonomous status enabled the scholars they employed to govern themselves. But something unexpected happened in the 19th century: censorship began to build from within. This phenomenon was the kind of danger against which James Madison, the godfather of the American Constitution, had warned: "The great danger is that the majority may not sufficiently respect rights of the minority." Madison's caution proved to be prophetic not only in America, but in the rest of the world as well. The Bill of Rights conferred upon the minority the basic rights to which all citizens are entitled. While most parts of the free world have instituted some ways to protect the dignity and rights of all people, academia has developed a sophisticated set of rationalizations to deny these basic human rights. In the midst of the modern world's near universal acceptance of the freedom of expression, the academy has, unfortunately, acquired the very traits of the religious or political tyranny against which freethinkers have always fought.

What happened at Baylor University and Iowa State University was an example of discrimination sanctioned by the academy, which has, over time, turned into the intolerant and hostile defender of positivistic materialism. In other words, the oppressed has become the oppressor. Well aware of the academy's hostility toward ID, why did these three professors risk their reputations and job security to openly support it? They also have something else in common: they are all senior fellows of the Center for Science & Culture, a subsidiary of the Discovery Institute, and have a strong religious faith. This may explain why they would take such unwise, if not reckless, path. One could even say that they might be appealing to a "higher standard," one that transcends their academic status. Brian Martin (1993) seems to describe quite well the present state of science:

> Textbooks present science as noble search for truth, in which progress depends on questioning established ideas. But for many scientists, this is a cruel myth. They know from bitter experience that disagreeing with the dominant view is dangerous...Rather than being a search for the truth, science is closely bound up with the exercise of power. This is normally acknowledged for totalitarian regimes and for military dictatorships, where intellectual suppression is overt. But the same sorts of processes occur, usually in a more subtle fashion...

The Ironies of Multiverse

The undetermined nature of an unobserved event is quantum mechanics' crux of weirdness. In 1935, Erwin Schrodinger used a simple thought experiment to demonstrate the paradox of a quantum event. Known as "Schordinger's cat," it is explained in the following terms. Suppose a cat is placed in a steel chamber along with a glass flask that contains a poison gas. The cat's fate depends upon a Geiger counter that has a 50 percent chance of decay. When that point is reached, a hammer will strike the flask, which will cause the poison gas to be released and kill the cat. If it does not decay, the hammer will not strike the flask. In such a quantum state, the cat is simultaneously alive and dead until someone opens the box and looks inside. The crux of this paradox is that it takes consciousness to determine event.

From the point of orthodox physicists, Schrodinger's cat is certainly a thorny issue that simply will not go away. In his 1956 dissertation at Princeton University, Hugh Everett proposed the notion of multiverse to account for this wacky situation in quantum mechanics. Although he was not the first propose the idea, since William James had come out with it as early as 1895, what he did with it was incredibly important. When the observer opens the box, the observer becomes entangled with the cat, both dead and alive. In other words, the universe branches out into two universes that cannot interact with each other. Steven Weinberg, one of the most enthusiastic supporters of the multiverse concept, said that he has enough confidence in it to bet his dog's life on it. Andre Linde at Sanford University is willing to bet his own life (quoted by Trimble, 2008).

Although his idea of multiple universes seems to have rescued itself from the dreaded implication of quantum mechanics, namely, an observer-created universe and mind as the ultimate reality, it comes with an obvious drawback: it creates an almost infinite number of branching universes. The crux of Everett's approach is that in order to save a tiny quantum event, an entire universe is created. This seems to go against the principle of Occam's razor. An even greater irony is that in his attempt to preserve the positivistic particle concept, Everett relied on parallel universes – the existence of which can never be proven.

If consciousness is the ultimate reality, as Wigner has said, then psychology's major role is to explore the mental world. And yet psychology is the most insecure discipline, because mind-stuff has been mostly purged from the academy for more than a century.

Relativity, Quantum Mechanics, and the Fixation of Physicists

> If quantum mechanics hasn't profoundly shocked you, you haven't understood it yet.
>
> -Niels Bohr

> But soul is not in the universe, on the contrary the universe is in the soul.
>
> -Plotinus

A Critique of Science

Previously we discussed some historical examples of what happens when people are unable to recognize that circumstances are changing circumstances, such as Polaroid's executives and Civil War generals. In this section, we will discuss the inflexibility found among physicists. By the end of the 19th century, so many major discoveries had been made in the field of physics that Lord Kelvin declared that there was nothing new to discover. Yet only a few years later, in the early years of the 20th century, two ground-breaking discoveries overturned Newtonian physics: relativity and quantum mechanics.

One of materialism's key features is the particle concept. The ancient Greek atomists Leucippus and Democritus believed that things consist of an almost infinite number of particles (or atoms). Basing himself upon their work, Newton considered particles as "massy, hard, impenetrable, moveable particles of sizes and figures." Another key feature is the notion of absolute space and time. In his *Principia*, Newton describes space: "Absolute space, in its own nature, without regard to anything external, remains always similar and immovable." As regards the nature of time, he says: "Absolute, true and mathematical time, of itself, and from its own nature flows equably without regard to anything external...."

In the late 18th century, Antoine Lavoisier and Joseph Louis Proust inaugurated the actual empirical approach to investigating fundamental particles by studying chemical reactions. In the early 19th century, John Dalton revived the concept of the atom, which literally means "indivisible" in Greek. But J. J. Thompson had already demonstrated in 1897 that the atom is divisible: by using a cathode ray tube he detected the "electron," a negatively charged subatomic particle. Ernest Rutherford, his former student, went on to discover the "proton," a positively charged particle at the center of the atom. Rutherford proposed that the planetary model be applied to the atom: a cloud of electrons circles the nucleus, which is tightly packed with protons and neutrons.   And then in 1905 came Einstein's theory of relativity, which toppled the commonsense notion of absolute space and time. In this world-shattering approach, an observer becomes central to space and time. The observer's role becomes even greater in quantum mechanics: an observer is elevated to a participator. Quantum physicists used phrases like "the observer-created universe" and "the participatory universe." What do they mean?

Quantum physicists often use the double-slit experiment to demonstrate the sheer weirdness of this concept. Suppose that photons have passed through two holes (A and B). As they pass through these holes, the waves emanating from each hole interfere with each other and thus produce a patterned dark and bright band on the screen. This resembles the two approaching and eventually overlapping waves caused by tossing two pebbles into a pond. Suppose that the photons pass through either hole one at a time and that, after doing so, they leave behind a single dot of imprint on the screen as a proof of their existence. So far so good. But when we continuously fire photons one at a time for a given period of time, an interference pattern emerges on the screen. It seems as if a single photon was interfering itself.

Here is even a greater surprise. Suppose a detector is placed next to the hole to see which hole the photon went through. For some reason this causes the interference pattern to disappear, as if it somehow knows that it is being measured. The strange story gets even stranger when "delayed-choice" becomes involved. Originally proposed by John Wheeler, a Princeton physicist, a delayed-choice setting means that the decision on how the photons are to be measured is made after they have passed though the slits but before they have arrived on the screen. In other words, the physicist reaches a conclusion only after the event has taken place. Kim et al. (2000) reported that they have confirmed Wheeler's proposal on the ground that their experiment revealed that the past event can be erased. Another way of saying this is that the past event or the identity of a particle is not determined until the actual observation has been made. We can go even further and say that there is no event in which an observer is not involved.

Einstein was mystified by quantum mechanics: "I have spent at least one hundred times more time in thinking about quantum mechanics than about general relativity" (quoted by Ghirdi, 2005). The founding members of quantum mechanics were also deeply affected by the strangeness of this new phenomena, so much so that most of them wrote about its philosophical implications. The comment made by Eugene Wigner (1962), a Nobel physicist, captures an irony of quantum mechanics: "The very study of the external world led to the conclusion that the content of the consciousness is an ultimate reality." Arthur Eddington, a British astrophysicist who confirmed the prediction of Einstein's relativity in 1919, had this to say:

A Critique of Science

> The mind-stuff of the world is, of course, something more general than our individual conscious minds....It is difficult for the matter-of-fact physicist to accept the view that the substratum of everything is of mental character. But no one can deny that mind is the first and most direct thing in our experience, and all else is remote inference (Eddington, 1928).

James Jean, a British physicist and astronomer, made a similar point: "The Universe begins to look more like a great thought than like a great machine. Mind no longer appears to be an accidental intruder into the realm of matter....we ought rather hail it as the creator and governor of the realm of matter." French physicist Bernard d'Espagnat (1979) agrees with Eddington and Jean: "The doctrine that the world is made up of objects whose existence is independent of human consciousness turns out to be conflict with quantum mechanics and with facts established by experiment."

Notwithstanding this evidence, which clearly contradicts the reigning materialistic worldview, most physicists behave as though nothing has happened. One reason for this is that, as German physicist Carl Friedrich von Weizsäcker (1952) argues, the new physics is against the commonsense worldview:

> The physical world view of the nineteenth century...took the forms of our perception, in far as they correspond to classical physics, as absolute, and therefore we thought that a process which was not perceptible to the senses had been understood only after it had been reduced to a model after the patterns of the perceptible. We recognize how this conception, too, derives from the thought of a unified picture of the world. The picture was a grandiose attempt, and it was natural that physics should allow it as far as possible. But the advance of our knowledge has decided against it.

Heinz Pagels, a physicist at Rockefeller University, makes a similar point: "Grasping quantum reality requires changing from a reality that can be seen and felt to an instrumentally detected reality that can be *perceived only intellectually* (italics mine).

Perhaps even a greater reason may be that the new physics challenges the status quo, for many people have a vested interest in the particle concept. Among the physicists who have won the Nobel Prize since 1901, a high percentage of them received this award for their direct or indirect elaborations upon this very concept. Despite their field's very limited practical function, particle physicists enjoy generous funding as compared to other fields. As Davies (2003) argues: "I do not know any who considers that the most fundamental branches of physics will have *any* technological relevance to our lives..." A similar point was made earlier by Ben-David (1971):

> It is true that since World War II the research of subatomic particles has been greatly furthered by financing that became available because of the original practical application of atomic research. But the results of this expensive research have had no practical application, and this shows how tenuous is the relationship between practical purpose and scientific theory.

Consider the European Organization for Nuclear Research's (CERN) Large Hadron Collider (LHC) located along the Franco-Swiss border. The world's largest high energy particle accelerator, it is one of the most expensive scientific instruments ever built. This structure, which has a circumference of 27 kilometers (17 miles), cost $6 billion dollars to build and was funded by 20 European countries. It is far larger than Fermilab, which has a circumference of 6.3 kilometers (3.9 miles) circumference, which was established in 1967 near Chicago. Particle accelerators, which began to be built during the 1930s, were originally no larger than a lunch box. But then the competition to build ever-larger colliders took on a life of its own based on the rationale that a larger collider would generate a greater energy for colliding particles. They also used plenty of rhetoric, all of which was designed for outdoing others: "the quest for advancing the deepest laws of nature," "unravel the deepest secret of the Universe," "God particles," "creating mini-Big Bangs," and so on.

Not to be outdone by the Europeans, American physicists wanted to construct an even larger one: the Superconducting Super Collider (SSC), to be located near Dallas. Its circumference of 87.1 kilometers (54.1 miles) dwarfs that of CERN's LHC. The estimated budget for this project rose from $4 billion in 1987 to $8 billion in 1992. Congress finally cancelled it in 1993 for a variety of reasons, among them cost overruns and the end of the cold war. Above all, both scientists and politicians well understood that practical results were neither the object nor to be expected (Greenberg, 2001).

Notwithstanding the ample grounds for this cancellation, there should have been a greater reason: a very careful examination of the particle concept before committing a significant amount of dollars to it. The implications of quantum mechanics should force us to consider an alternative view that can, to a certain degree, be justified: the world is made up of mind-stuff rather than particles. For politicians, however, the deep philosophical implications of quantum mechanics are not exactly at the top of their list for why such a project should be cancelled. As others have done in the past, those physicists who seek an outlet for their aggravation are attacking not the politicians and the funders, but those who state that the world is made out of mind-stuff as Lynch (2001) mentions:

> "Scientists" and the "defenders of science and reason" are accused of being frustrated by recent cuts in basic research budgets, so that they turn their rage upon the "academic left" rather than the fiscal conservatives who made the cuts....it is worth dwelling for a moment on arguments to the effect that scientists frustrated by the dwindling status and resources in, say, particle physics are scapegoating "sociologists."

Orogee Dolsenhe

Chapter Ten
The Demise of Randomness as the Cause of Biological Complexity: Gaps in the Neo-Darwinian Paradigm

> Nothing happens at random; everything happens out of reason and by necessity.
>
> -Leucippus

> To improve a living organism by random mutation is like saying you could improve a Swiss watch by dropping it and bending one of its wheels or axis. Improving life by random mutations has the probability of zero.
>
> -Albert Szent-Gyorgi

> Some genuinely testable theories, when found to be false, are still upheld by their admirers – for example by introducing *ad hoc* some auxiliary assumption, or by re-interpreting the theory *ad hoc* in such a way that it escapes refutation. Such a procedure is always possible, but it rescues the theory from refutation only at the price of destroying, or at least lowering its scientific status...
>
> -Karl Popper

In determining causality, we ascribe everything in the world either to chance or to necessity (Riedl, 1978). Chance represents uncertainty, unpredictability, and unmanageability, while necessity usually is associated with certainty, predictability, and control. We often use expressions like "chance encounter" and "accident" to explain unexpected events or to evade responsibility when accounting for mishaps. The latter is associated with an understanding of causality; the former renounces causality. We certainly prefer the latter over the former. Our inborn curiosity to understand the underlying cause of events or processes is noticeable from a very early age. Undoubtedly, much of the content of mythology, superstition, and religion throughout the world derived from the human inclination for the latter. Philosophy is a more elaborate form of inquiry into causality, which has been developed in many historical civilizations. Yet the scientific pursuit of causality is set apart from the rest by its emphasis on empiricism and the application of mathematics. Alhazen, Galileo Galilei, and Isaac Newton led the way to progress toward the present scientific practice based on necessity. The application of empiricism and mathematics allowed the causality of nature to be explained with precision and predictability, rather than mere general descriptions. This gave a new breed of scholars their special status; scientists proudly broke away from philosophers. With the newfound physical laws, the universe is understood as a giant clockwork that obeys mathematical principles. Prior conditions supposedly establish the outcomes of events; an unbroken chain of cause and effect determines all activities. The machineries and the technological progress we use and enjoy are the demonstration of the power and success of the mastery of causality.

An obvious difficulty in accounting causality of biological organization with Newtonian mechanics is that the structural organization of living organism exceedingly complex. In facing the dilemma, the synthetic theory accounted for biological complexity—which could not be explained by the laws of Newtonian physics—with the concept of randomness. Neo-Darwinism's random mutation is the occasional accident or mistake in the replication of genes that allows offspring to differ from parents. It provided the mechanism for what Darwin called the "tendency to vary," which allowed natural selection to take place. In the domination of neo-Darwinism, something very peculiar has happened. Biological complexity that cannot be described by "necessity" has been pushed toward "chance." Paradoxically, however, the crucially important concept has been left out in the synthesis as Mayr (1980) admits:

> The role that chance plays in evolution has been controversial ever since 1859. Yet this argument has not played a very important role either in delaying or in furthering the synthesis.

Thus, the word usually associated with the forsaking of causality has been turned into causality itself. Monod's (1971) *Chance and Necessity* characterizes the derivation of the latter from the former. Although he won the Nobel Prize by discovering the complex controls involved in the manufacture of proteins through a sophisticated feedback mechanism similar to that of a thermostat, he considered that same regulatory mechanism to be a product of chance:

> Chance *alone* is at the source of every innovation, of all creation in the biosphere. Pure chance, only chance, absolute but blind liberty is at the root of the prodigious edifice that is evolution... It today is the *sole* conceivable hypothesis, the only one that squares with observed and tested fact (Monod,1971).

Monod even spelled out the philosophical implication: "The ancient covenant is in pieces; man knows at last that he is alone in the universe's unfeeling immensity, out of which he emerged only by chance." Steven Weinberg (1977), a Nobel physicist at the University of Texas, echoes: "The more the universe seems comprehensible, the more it seems pointless."

Because "chance" has the potential for an unlimited range of application, every evolutionary outcome can be explained away, with no need for any sort of necessity explanations. If the concepts of randomness and chance are given such nearly unlimited potential, there ought to be a concerted effort to understand them. Yet, there is lack of even a thorough analysis of the history of the gradual acceptance of chance (Mayr, 2004). The accidental view of the world has become the norm in biology. This has created a number of gaps and numerous inconsistencies that could not be corrected by making minor modifications.

The Gap with Accidental Evolution

The neo-Darwinian views errors in copying DNA as the source of random mutation. This provides the mechanism for variation so that the selective force can direct the species to evolve. However, the neo-Darwinian mechanism must travel through a highly improbable sequence in order to work. First, the errors occur with a probability of between 10^{-10} and 10^{-8} in prokaryotes per generation and between 10^{-11} to 10^{-9} in eukaryotes per generation (Spetner, 1996). Such low probability of errors is due to the proofreading mechanism that corrects them (Hanawalt, 1989). Second, mutations tend to be selectively neutral (Kimura, 1968; Kimura & Ohta, 1971; King & Jukes, 1969). Third, because most proteins in mammals are represented by multiple genes, a single point mutation does not necessarily confer a change (Hood et al.,

1975). Fourth, even if there were a number of mutations that ultimately brought about a change, it is highly unlikely that the change would provide an adaptive advantage. A minimum number of point mutations toward a specific direction are required to confer any sort of advantage.

The Gap with Evolutionary Patterns

Webster's dictionary defines random as "lacking a definite plan, purpose, or pattern." Yet, contradicting the neo-Darwinian principle, numerous unmistakable evolutionary patterns exist. Paleontologists have long used various names to discuss the tendencies or trends (e.g., Cope, 1896; Osborn, 1893; Watson, 1926; Simpson, 1949): orthogenesis, orthoevolution, aristogenesis, inertia, evolutionary momentum, rectilinear evolution, and phylogenetic inertia (see Burt, 2001 for review). An undeviating evolutionary direction often proceeds even when there is little or no adaptive value. The evolutionary course of *Gryphaea* (an extinct mollusk related to the oyster) provides an example. This mollusk had two valves or shells; one side of the shell was larger than the other side, resulting in coiling. Fossils show that the coiling became progressively tighter and tighter, to the point that it was unlikely that the shell could be opened at all; at that point, the animal must have died. In some characteristics, continuous progression actually becomes so disadvantageous that it may cause the extinction of the species. For instance, the transformation of *Planorbis levis*, another species of extinct mollusk, shows a tendency to uncoil, to become distorted, and to be smaller than ancestors. Furthermore, malformation often takes place just before the extinction of a species (Hyatt, 1882). Interestingly, the uncoiling and distortion can occasionally occur in individuals with wounds or diseases.

Why didn't the process of natural selection prevent the characteristics from moving beyond usefulness? This question is unavoidable. Wilson (1975) acknowledges: "Phylogenetic inertia...consists of the deeper properties of the population that determine the extent to which its evolution can be deflected in one direction or another, as well as the amount by which its rate can be speeded or slowed." Even a cofounder of the synthetic theory, Simpson (1949), admits: "Yet the fact remains that the sequence cannot be considered as random. The changes involved do have direction and orientation, even though these were not as regular as they have usually been represented. And so, in hundreds or thousands of other cases, it seems clear that there is an orientation of some sort." Recently, Blomberg & Garland (2002) echoed: "What are the forces?"

The Gap with Endosymbiosis

During the 1960s, Margulis proposed the radical idea that mitochondria and plastids were once separate organisms incorporated into a host cell. Margulis (1970) arugued that some of the small aerobic bacteria became symbionts and functioned as mitochondria of the larger host cell. The intricacy of the newly established symbiotic relationship was serendipitously discovered by Jeon (1972). Though he initially was distraught over his valuable amoeba getting sick and dying with a bacterial infection, a few amoeba somehow survived the infection. He found that the infectious bacteria were incorporated into the hosts. Even more surprisingly, amoeba were so dependent on the new symbionts that they could no longer survive without them. Obviously, some vital function had been relegated to the newly incorporated member which would require a massive reorganization of the internal system. Some symbiotic organisms like lichens (association of algae and fungi) can survive if they are artificially separated (Ahmadjian, 1966). Taylor (1974) was the first to point out that hereditary symbiosis in the hosts clearly involves a form of Lamarckian inheritance. Unlike most non-Darwinian ideas, the idea of symbiotic evolution has been accepted without much controversy.

The Gap with Directed Mutation

Neo-Darwinism understands organisms to be passive systems driven by random processes, with no control over their evolutionary fate. However, Cairns, Overbaugh, & Miller (1988) have found that adaptive change in bacteria may be due to nonrandom mutation or directed mutation. In their study, strains of bacteria unable to metabolize lactose (hence Lac^-) were placed in a medium containing lactose as the sole energy source. Because the bacteria are starved, they cannot give rise to a colony. Cairns, Overbaugh, & Miller (1988) discovered that mutants that can metabolize lactose (Lac^+) arose within 48 hours. But, remarkably, the mutations arose only if lactose was present on the plates. They proposed that bacteria can choose which mutations occur. Hall (1988, 1991) not only confirmed these results but also showed that a sequence of two mutations can occur, giving a benefit to an organism, while a single point mutation would offer no advantage. Caporale (2006) points out that genomes often find repeated paths to the same solution in spite of the expansive landscape of possible random changes. Although mainstream biologists remain unconvinced of directed mutation, some geneticists recently have begun to challenge the dogmatic view. For instance, McClintock (1984), a Nobel Laureate, considers the genome as a highly sensitive organ that can sense unusual and unexpected events and make necessary changes to overcome problematic situations. Goodwin (1976) suggests that organisms are "cognitive" systems that take in information. Campbell (1985) proposed that mechanisms for projective evolution have evolved, consisting of five attributes:

1) access to multiple sources of information about self and the environment.
2) ability to analyze this information to determine which physiological responses will be adaptive.
3) access to genes.
4) access to the germline or a reproductive system.
5) ability to induce anticipatory physiological adaptations.

Shapiro (1992) contends that genomes are complex interactive systems that can function as true smart systems like computers, making adjustments as conditions require:

> Instead of the 'constant genome' we now have the 'fluid genome'. Rather than a rigidly defined storage system that changes only by occasional copying errors and physicochemical accidents, we now think of a dynamic storage system subject to constant monitoring, correction and change by dedicated biochemical complexes. Instead of a collection of individual isolated genes whose information is utilized in an automatic/mechanical fashion, we now think about integrated, multigenic systems that can be turned on and off in a coordinated fashion according to the needs of the organism.

McClintock, Goodwin, Campbell, and Shapiro represent a new breed of biologists who defy orthodoxy. They consider genetics and the endocrine and/or nervous system to be the source of the adaptive capacity. Campbell (1985) declares the present state of evolutionary biology this way: "The crucial question for evolutionists now is not whether sensory evolution occurs, but how extensively prokaryotes and eukaryotes have recruited their sophisticated adaptive sensory systems to aid their evolution."

Another astonishing finding that challenges the orthodox view is the smart interactive system is enable to incorporate genes from another species, phylum, or domain lines known as "lateral gene transfer" (LGT). For instance, Stroun et al. (1970) reported that many bacteria release their DNA into the plant cells they occupy and the host picks up and incorporates the DNA for its own benefit. It seems the lowly organisms long have mastered the sophistication of genetic engineering that scientists only began to understand a short time ago. Another important implication of LGT is this: it renders life more as the "web" of genomes rather than the branching "tree" that Darwin originally portrayed (Dolittle, 2005). Jeon & Danielli (1971) purported the holistic approach more than three decades ago: "Organisms and genomes may thus be regarded as compartments of the biosphere through which genes in general circulate at various rates and in which individual genes and operons may be incorporated if of sufficient advantage."

The Gap with Epigenetic Change

Yet another class of heritable change refuses to fit the neo-Darwinian scheme: so-called "epigenetic change." One of the main assumptions of neo-Darwinism is that heritable change arrives through genetic transformation. However, this assumption is absolutely incorrect (Jablonka & Lamb, 1995). Many phenotypical changes from one generation to the next do not involve any alteration in DNA sequence. For instance, when *Unica*, a type of pea plant, is raised under a constant temperature (10, 17, or 20 degree Celcius), the size of the plant and the number and viability of seeds continuously decreases over six generations, until it becomes unviable (Highkin, 1958). Interestingly, if the plant is put back into an environment with fluctuating temperatures, it regains its vigor. Brink (1960) called such change "paramutation," while Cullis (1988) preferred to implicate the Lamarckian dimension. A more widely-used term is "epigenetic inheritance system" (EIS). The main characteristics of EIS are: 1) It is usually less stable than the genetic system, 2) The induced changes are more often directed than those in the genetic system and are more predictable and repeatable, 3) The variations resulting from EIS are likely to show only a limited number of alternative states (Jablonka & Lamb, 1995). In recent years, EIS has become one of buzz words in biology (Jablonka & Raz, 2009). In a review, Jablonka & Raz (2009) listed over a hundred cases of nongenetic informational transfer between generations. What is involved in the heritable changes that don't require genes? This is an obvious challenge that can overthrow neo-Darwinian dogma.

The Gap with Parallel Evolution

Suppose two persons toss a coin ten times. These two random events should be completely independent; thus, if these supposedly unconnected events produced identical outcomes, we would be perplexed by the unexpected matching results. We would be even more puzzled if even larger numbers of people produced parallel outcomes. Our common sense and curiosity would lead us to search for the source (necessity) other than randomness (chance).

As implausible as it may sound, the proponents of neo-Darwinism explain away parallel biological forms with the latter than pursuing the former. Evolutionary biology is plagued by cases of parallel morphology that serve no adaptive advantage. For instance, *Pieris occidentallis* of North America and *Tatochila vanvolxemii* of Argentina share similar seasonal wing patterns (Shapiro, 1976). Ho (1988) dubbed it as "pseudomimicry." Mimicry is the superficial resemblance of one species to another, such as the viceroy mimicking the wing pattern of the unpalatable monarch butterfly. Pseudomimicry, on the other hand, seems to have no adaptive functions.

Another interesting case of parallel evolution is the regressive osteological reorganization of a population of blind cave fishes follows the pattern of another (Breder, 1944). There are two caves (La Cueva Chica and Cueva de los Sabinos) in a particular region of Mexico, about 15 miles apart. Both have populations of the blind fishes, *Astyanax mexicanus*, that were isolated from the surface population. When fishes and other animals live in an environment of perpetual darkness for many generations, their eyes degenerate. The degenerative pattern of ocular anatomy follows "some master plan that controls the architectural changes" (Breder, 1944).

The Gap with Isomorphism

Even more intriguingly, parallel forms can be shared between living organisms and non-living. One of the most well-known examples is the striking similarity between the self-organizing structure known as Belousov-Zhabotinsky reaction (BZ reaction for short) and the spiral waves of slime mold aggregation (Winfree, 1987). BZ reaction is named after two Russian scientists who discovered it in 1958. In a typical BZ reaction, rotating spirals or concentric rings are spontaneously formed when the oxidation of malonic acid by bromate ions catalyzed by ferroin. Each spiral and ring expands and collides with neighboring waves without penetrating each others. The most puzzling aspect of BZ reaction is that the wave patterns are remarkably similar to a well known biological pattern, the wave pattern of the aggregating slime mold amoebas.

Another puzzling case is the equiangular spirals of plants (known as phyllotaxis) being obtained in physical laboratory experiments (Douaday & Couder, 1992). Phyllotaxis of flower typically has two sets of equiangular spirals that are superimposed, one set in a clockwise direction and the other in a counterclockwise direction. For example, sunflower has 21 spirals in one direction and 34 in the opposite direction. The two sets of spirals like sunflower can be produced by dropping drops of ferrofluid, that are magnetically polarized so that they repel each other, to fall at the center of disk with a film of oil (Douaday & Couder, 1992).

Thompson (1941) did pioneering work in identifying isomorphism between physical and biological systems. Von Bertalanffy (1960) put forward a "general system theory" that investigates the similar structures in various disciplines. Whyte (1949) considered the disconnected state between physics and biology a "crisis" and recommended: "...both sciences must work together on to a new common foundation." Presently, the field known as "complexity" or "self-organization" takes on the common features (or isomorphism) in physics, biology, chemistry, physiology, and sociology (see Goodwin, 1994; Solé & Goodwin, 2000; Bird, 2003 for review). Furthermore, quantum mechanics demands a unification of psychology, physics, and biology (Morowitz, 1984). A cross-disciplinary theory has been put forward by Dolsenhe (2000, 2005).

Orogee Dolsenhe

The Gap with Mathematic Equations in Morphology

Many decades before Newton's discovery of the mathematic equations of physical laws, René Descartes found the beauty of a simple mathematical equation in the morphology of mollusks. Mollusk shells are essentially three-dimensional spirals (or logarithmic spirals, in technical terms) that can be written in the mathematical equation $r = ae^{\theta}$. Each species' spiral may show a different curve and expansion rate. In addition, shells have distinctive patterns of stripes, parallel and oblique lines, dots, and triangles that can be written in mathematic equations (Meinhardt, 2003). Computer generated graphics can simulate morphology and decorative patterns that are virtually identical to those of mollusks. The most perplexing aspect of the various patterns is that they serve no survival advantage because they are buried in the shallow sediments of the sea floor and are covered with an opaque membrane that hides the shell decoration (Sole & Goodwin, 2000). We must ask: How does the humble mollusk possess a mathematic equation that humans discovered only recently? Without natural selection, what force was involved in the mathematical morphology? The classic work by Thompson (1942) reveals morphological patterns that simply cannot be accounted for via neo-Darwinian principles of random mutation and natural selection. Even Simpson (1950) admitted that the various forms of horns of different species were not adaptively significant.

The Gaps Within

Taking a strict neo-Darwinian approach has created three major contradictions between what scientists preach and what they do (Dolsenhe, 2005). First, consider the issue of free will. Most of us believe we have some control over our future and can make a difference for ourselves and for others. However, this lies in disagreement with the prevailing scientific view that considers the outcomes of biological and mental events to be nothing more than the results of matter obeying physical laws and random process. Sinnot (1950) addressed this paradox: "A number of biological philosophers... have endeavored in various ways to avoid the extremes of both materialism and vitalism by finding some middle position which can accept the results of physiology and genetics but still find room for freedom, purpose, and value."

Second, bear in mind the strong belief in science that everything somehow should make sense, summed up astutely by Lindley (1993):

"They (scientists) simply assume that there must be a rational answer to the questions they are asking and a logically coherent theory that will tie together the phenomena they are studying. If scientists did not believe this—if they thought that at bottom the physical world might turn out to be irrational, to have some irreducible 'fortuitous' (emphasis added) element in it—there would be no point in doing science." Such a belief, however, is antithetical to the accidental view of nature; if nature's driving engine is random processes, a mere roll of the dice, why should nature be coherent? These paradoxes force biologists to live double lives; they are self-contradictions and intellectual laziness.

The Gaps with Society

Monod (1970) sanctimoniously admits the inevitable conclusion of the present "accidental" scientific paradigm: "The ancient alliance has been destroyed; man knows at last that he is alone in the universe's indifferent immensity out of which he emerged only by chance." Monod was quoted in saying: "Man has to understand that he is a mere accident" (quoted in Judson, 1979). If such a cold and impersonal view creates a dilemma that forces scientists to live double lives, how does an average person deal with the implication? While scientists pretend the problems don't exist, most laymen have opted, unsurprisingly, for alternatives that provide spiritual satisfaction, alternatives like religion and the New Age movement. Harman & Sahtouris (1998) perhaps summed it up best: "Our current neo-Darwinst account of evolution is intellectually and spiritually so unsatisfying that one can easily see why people flock to creationist alternative…"

The Evasive Champ and the Dawn of Revolution

> The synthetic theory possesses a remarkable elasticity: when under attack it expands to exclude almost anything, but once the pressure is relaxed it contracts.
> -Peter Saunders & Mae Wan Ho

> The synthesis that emerged was, by stark contrast, largely intolerant of criticism and resistant to change.
> -Jeffrey Schwartz

> As the synthesis hardened, "could be explained" turned into "are explained," yielding outright dismissal of competing theories. To some this seemed to turn the synthetic theory into an *a priori* rather than an empirical doctrine.
>
> -Richard M. Burian

A consistent number of dissenting voices has pointed out many discrepancies ever since the formulation of neo-Darwinism (e.g., Waddington, 1953, 1960; Koestler, 1970; Hardy, 1965; Thompson, 1942; Needham, 1942). However, they have been, for the most part, ignored and rejected. Neo-Darwinism is like an aged boxing champion surrounded by an aura of mythic invincibility. He is so revered that no challenger is considered worthy of a fight. The myth is maintained by the loyal followers like Dawkins (2001) who accuses those who disagree with Darwin's view being "ignorant, stupid, or insane" or Dennett (1995) who even believes Darwinism as the greatest idea ever conceived, superior to the ideas of Newton and Einstein. The truth, though, is that he is old, weak, vulnerable, and defenseless. But even more surprising, he won the title not by knocking out an opponent but by declaring a "truce." The only way he could maintain the comfortable status quo was by evading fights. Smocovitis (1996) describes their maneuvers as taking on "more and more of an enigmatic and elusive quality" which Burian (1988) calls it a "moving target."

However, the idea of punctuated equilibrium by Eldridge & Gould (1972) and the neutralist theory by Kimura & Ohta (1971) and King & Jukes (1969) mounted a challenge that neo-Darwinians could not evade. According to neo-Darwinian principles, the source of variability is point mutation or micromutation. And the direction of evolutionary change is determined by natural selection operating in a gradual shift of gene frequency. But rather than reveal gradual and smooth transitions, the paleontological data display an abrupt appearance of new groups (e.g., Williamson, 1981, Greenwood, 1981). Eldridge & Gould (1972) proposed that the gaps in the fossil record are real and the results of rapid speciation. Gould (1980) declared: "That theory (the synthetic theory) as a general proposition, is effectively dead, despite its persistence as textbook orthodoxy." The neutralist theory (Kimura & Ohta, 1971; King & Jukes, 1969) suggests that most mutations do not alter biochemical products and consequently are selectively neutral. Punctuated equilibrium and the neutralist theory were incorporated into neo-Darwinism even though they were originally viewed as non-Darwinian (Brooks, 1983; Goldsmith, 1990). For instance, Mayr (1996) states that the theory of punctuated equalibria is an "elaboration" of his theory of speciational evolution. A popular neo-Darwinian slogan seems to be: if you can't beat'em, join'em. Eldridge (1995) appropriately call the adjustment as the "reinventing Darwin." Although punctuated equilibrium and the neutralist theory did not bring about a revolution, they perhaps effected one very important change in evolutionary biology: neo-Darwinism was no longer sacred ground. From the 1970s and onward, a growing body of literature has challenged neo-Darwinism (see Ho & Saunders, 1979; Brooks, 1983; Riddiford & Penny, 1984 for review). Depew & Weber (1985) address the present predicament of neo-Darwinism: an intense debate has arisen about whether to re-articulated neo-Darwinian principles to accommodate the seeming inconsistencies or to break with neo-Darwinian conceptualization. Some biologists have invoked the word "crisis" to depict the present situation in biology (Ho & Saunders, 1984; Bird, 2003; Denton, 1986; Reid, 2007). Lima-de-Faria (1988) considers neo-Darwinism a "hindrance" to the progress of science and, similarly, Reid (2007) calls it an "obstacle." Provine (2001) states: "Now I see these same theoretical models of the early 1930s, still widely used today, as an *impediment* (emphasis added) to understanding evolutionary biology, and their amazing persistence in textbooks and classrooms as a great topic for other historians."

Kuhn (1962) draws a parallel between political revolution and scientific revolution: "an existing paradigm has ceased to function adequately in the exploration of an aspect of nature to which that paradigm itself had previously led the way." Aforementioned are the examples. And just as Kuhn described, in the face of anomalies neo-Darwinians have devised numerous ad hoc modifications of their theory intended to eliminate apparent conflicts such as those indicated by punctuated equilibrium and the neutralist theory. More commonly, however, they simply ignore anomalies; whatever does not fit into the box is essentially invisible. The responses (or non-response) of neo-Darwinists measure up with Kuhn's argument that revolution does not take place by logically convincing the orthodoxy.

According to Kuhn (1962), scientific progress is not smooth but alternates between periods of stasis and periods of rapid development.

Perhaps the transition from behaviorism's domination to the cognitive revolution demonstrates the Kuhnian model of transformation quite well (see Palermo, 1971; Weimer & Palermo, 1973). During the heyday of behaviorism, psychology was dominated by a single mode of practice; living organisms, including humans, were reduced to stimulus-response. John B. Watson, the founder of behaviorism, rejected consciousness by stating that it is neither definable nor a usable concept in science. Behaviorists turned a narrow slice into a "whole pie" (Wilson, 1999) and undermined their opposition with unusual vigor and success (Toates, 2004). Scientific progress stalls during the period of stasis because the domineering view is able to censure alternatives. At any rate, with the demise of behaviorism, many dormant subjects have been revived.

Like behaviorists in psychology, a group of prominent biologists took a position that all biological complexities could be explained by random mutation and natural selection. And like behaviorists, neo-Darwinists have become intolerant of alternative views. Any phenomena that seem to contradict the prevailing principles are ignored, rejected, or manipulated into an uncomfortable incorporation. However, some cases will not be so easily defeated. Perhaps Whitehead (2004) is right in his contention that Kuhn did not go far enough: "Anomalies tend to get swept under the carpet until there are so many of them that the furniture start to fall over." The neo-Darwinian paradigm's current plight is much worse than psychology's crisis preceding the cognitive revolution. One more straw might break its back—if, that is, we do not view it as already broken. Perhaps it currently stands only because of the crutches of loyal followers. When it finally falls, we should expect a spurt of growth in many different fields. And just as in the cognitive revolution, many ideas that have been barred in the past might be resurrected. After the dust settles, people will puzzle over the *extent* of the irrationalities science had embraced.

Kuhn (1962) suggests that revolution will be brought on by those who learn to see science and the world differently and by those whose attention has been intensely concentrated upon the "crisis-provoking problems." There certainly are many such problems that attract some innovative investigators such as in directed mutation, isomorphism, epigenetic changes, etc. Obviously, we have to examine alternative explanations to account for those that cannot be explicated by neo-Darwinian approaches (see Morowitz, 2002; Dolsenhe, 2005 for examples). What is happening in science presently may be the dawn of a scientific revolution.

Orogee Dolsenhe

Chapter Eleven
The Demarcating Wall of Science

....the sciences are simply 'organized' or 'classified' common sense...

-Thomas Nagel

.....the whole of science is nothing more than a refinement of everyday thinking.

-Albert Einstein

....there is no reason to think that [science] is in possession of a special method of inquiry unavailable to historians or detectives or the rest of us.

-Susan Haack

To reiterate, if a person or a group achieves something greater than others, an innate tendency is to elevate oneself or the group and at the same time to put down others. Some of the effects of the zero-sum impulse may be long lasting and some may become established in society. The age-old institution of the Indian caste system is an example of how devastating the consequences of the zero-sum impulse can be. Science beginning in the 19th century was born under the zero-sum impulse with a set of justifications for excluding some subjects. The discussion of the rules and criteria for exclusionary practice is referred to as the "demarcation problem." In plain speaking, it tries to draw the boundaries around science so that its elite status is preserved. The problem is that people who were raised in a culture are not aware of its unjust rules. Scientists may not be aware of their own caste system while they are dismayed by the Indian caste system; they are blinded by a normalized pathology. Those subjects and scholars who fall outside the demarcation are forever segregated from science. You often hear comments such as: "It's not science." "But is it science?" The worse is the derogatory terms "pseudoscience," "nonscience," "antiscience," "unscientific," "junk science," "pathological science," etc. As much as insiders, outsiders fail to recognize the profound unjust practice perhaps because of the mystique of science.

The obsession of the Indian caste system is purity, whereas the preoccupation of science is positivism. Based on a positivistic agenda, science rose above religion and philosophy. A sad irony is that there was hardly any opposition to the discriminatory rule. I suppose that they were busy trying to demonstrate that they were worthy enough to join the elite club. A prime example is psychology that was to barter away its own soul for entry into the upper class. We shouldn't be fooled by anyone's attempt to justify discriminatory practices. The main goal of this chapter is to argue against the irrationality and injustice of excluding ideas or disciplines from science.

The Discriminatory Canon of Science

> Real equality is immensely difficult to achieve, it needs continual revision and monitoring of distribution.
> -Mary Douglas

> Human progress is neither automatic nor inevitable....Every step toward the goal of justice requires sacrifice, suffering, and struggle; the tireless exertions and passionate concern of dedicated individuals.
> -Martin Luther King, Jr.

In ancient times, the greatest obstacle to the progress of knowledge was the separation between two modes of knowledge-seeking—experimentation and reasoning. These modes had been separated by two classes of people: white collar workers and blue collar workers. And there are three Islamic scientists (Alhazen, al-Biruni, and Avicenna) who played a critical role in breaking down the wall. They possessed artisan skills and were able to produce experimental apparatus to test their ideas, as Galileo and Newton did in Europe in the Renaissance. Alhazen's adoption of the strict approach involved avoiding what cannot be investigated directly through experimentation and mathematics. However, something odd happened in the process. In essence, the positivists in Islam and Europe were not only boosting the value of experimentation but also devaluing phenomena beyond direct experience. Like once-persecuted Christianity becoming the persecutor,

A Critique of Science

those disciplines with empirically-based knowledge-seeking practices began to put down those who do not rely on empiricism especially in the 18th and 19th century. Philosophy and theology were the main casualties of the transformed value system. Ironically, however, philosophers were eager to embrace the newly emerging caste system. For instance, David Hume argued that any statements not based on direct experience should be rejected as worthless discourse. A French philosopher, Auguste Comte, took it one step further by identifying three stages of knowledge-seeking: the theological, the metaphysical, and the positive. Comte's depiction of the three stages represents the official adoption of the ideology of setting boundary in science. The zero-sum impulse has taken over and science's caste system began to solidify in the academia that became the institutionally-sanctioned practice of discrimination.

There was one problem with the positivistic agenda. Although empiricism was considered the foundation of science, physicists in particular recognized that some theoretical terms, such as "force," "mass," and "field," referring to what was not directly observable, were necessary. Even entities like electrons and quarks may not meet the positivistic criteria because they are unobservable. Thus, the predicament was to find a way to include the unobservable entities without negating the lofty aspiration to reject metaphysics. They went around the predicament by adopting what is known as "logical positivism," which Leahey (1987) elucidates:

> The problem, though, was how to admit science's theoretical vocabulary as legitimate, while excluding metaphysical and religious nonsense. The solution the logical positivists arrived at was to closely tie theoretical terms to bedrock observation terms, thereby guaranteeing their meaningfulness. The logical positivists argued that the meaning of a theoretical term should be understood to consist in procedures linking it to observational terms. So, for example, "mass" would be defined as an object's weight at sea level. A term that could not be so defined could be dismissed as metaphysical nonsense. Such definitions were called "operational definitions," following the usage of Percy Bridgman, a physicist who had independently proposed the same idea in 1927.

Leahey (1987) adds:

> To the logical positivist it did not matter if there were atoms or forces in reality; what counted was whether or not such concepts could be systematically related to observation. Logical positivists were thus, for all their apparently tough-minded insistence on only believing what one observes, really romantic idealists (Brush, 1980), for whom ideas—sensations, observation terms—were the only ultimate reality.

Thus, they were able to justify the apparent double-standard of excluding metaphysics while upholding positivistic ideals.

To emphasize the inherent injustice of logical positivism, let me compare it with similar kinds of things in society. In America, the southern states after the Civil War faced an unwelcome predicament: The Fifth Amendment ratified in 1870 prohibited denying a citizen the right to vote regardless of race, color, or previous condition of servitude (i.e., slavery). At the same time, some policy makers wanted to find a way to exclude or reduce the numbers of eligible African American voters, getting around the Fifteenth Amendment. As they say, where there is a will, there is a way. African Americans had two major disadvantages: poverty and lack of education. The poll tax and the literacy test were effective in preventing eligible African American voters from exercising the Fifteenth Amendment.

Let me give you another example of something that happened recently. In any sports, whether in golf, tennis, football, baseball, or Olympic sports, one of the most fundamental aims for those who set up the system is to provide a neutral space. In other words, everyone should have equal requirements and criteria for entry and judging. Players' value should be based on their performance. Some regulations, however, may be designed to exclude or reduce the targeted group under the guise of equal treatment by the majority. Consider LPGA's new rule in 2008. LPGA announced that "all players" must pass an oral evaluation of their English skills and those who failed would be suspended from the Tour. The justification for such policy was cleverly described by Libba Galloway, the Deputy Commissioner of LPGA: "There more fans, more media and more sponsors. We want to help our athletes as best we can succeed off the golf course as well as on it." And she uses the typical rhetoric that comes with disguised discriminatory effort: LPGA is "a global tour and is not targeting any specific player or country." Yet, it is not difficult to notice the fact that the players from Asia, particularly from Korea, would be the ones that would be most susceptible to suspension. There were forty-five Korean players in LPGA and seven of them were in the top twenty. Mike Walker, Senior Editor of *Golf Magazine,* expresses his concern: "The world of sports is supposed to be a true meritocracy. You should be measured by your skill, not your personality or parents or linguistic prowess. If Steve Ballesteros was subject to a rule like this one, he never would have won the 1980 Masters." Under the threat of lawsuit, LPGA eventually backed out of the English proficiency requirement.

It took many decades before some discriminatory rules like poll taxes and literacy tests were outlawed. LPGA's attempt to require players to pass an English proficiency test, fortunately, was shot down even before the actual implementation. Yet one of the most enduring discriminatory rules in the modern world still exists in science. The problem is that scientists don't even realize it is an unjust and unfair rule. In fact, it is viewed as something very positive, rather than negative. That is the reason why the discriminatory rule—positivism—could have endured for over a century and still endures.

Karl Popper's Falsificationism

> There are two ways to slide easily through life: to believe everything or to doubt everything; both ways save us from thinking.
>
> -Alfred Korzybski

> ...why have philosophers of science been so keen to present themselves and their history as *pro*-science? Under the circumstances, one might reasonably expect that theirs would be a history resentment, perhaps of genius spurned or suppressed... Instead, philosophers relate to practicing scientists as *underlabourers,* a term John Locke coined to characterize his own relationship to his friend Isaac Newton, the scientific 'master builder'. As underlabourer, Locke's job was to clear the rubbish in the way of the master builder's work.

Orogee Dolsenhe

-Steve Fuller

The next discriminatory justification for excluding metaphysics came from Karl Popper, an Austria-born British philosopher. It is known as falsificationism. Like logical positivism, falsificationism arose from the difficulty of upholding the positivistic principle in the face of the theory of relativity and quantum mechanics. In the 19th century, Newtonian mechanics had ruled supreme. Lord Kelvin's declaration exemplifies the confidence of physicists at the time: "There is nothing new to be discovered in physics now. All that remains is more and more precise measurement." The certainty in Newtonian physics came from evidence accumulated for two centuries. Then, at the turn of the century, relativity and quantum mechanics overturned fundamental assumptions of time, space, and energy; simply put, they were black swan events. Thousands of observations don't necessarily prove the assumption that all swans are white. It was a two-pronged assault on Newtonian physics. Einstein acknowledged the inherent weakness of relying on empirical evidence by saying that no amount of experimentation can ever prove him right but a single experiment can prove him wrong.

Popper devised a new way of drawing a boundary between science and nonscience. His new criterion for defining science was "falsifiability" or "refutability," rather than verifiability of positivism. Those ideas that can be tested for falsification are considered science and those that cannot are outside science. Popperian falsificationism attempts to put all ideas in one of two categories: those falsified and those not yet falsified. Thus, it was an antithesis of positivism; nothing is positive. It was a reactionary, extreme answer to positivism. Among the ideas rejected by Popper's falsificationism were Freud's psychoanalysis, Marx's theory of communism, and even Darwin's theory of natural selection. Paradoxically, however, Popper (1974) himself uses the very Darwinian notion of the survival of the fittest that he rejects to put forward an ideal of what scientific process should be like: "where ideas compete in the marketplace and where, after rigorous selection, the best one survive—best for a day always, and with new rivals ever on the horizon."

Scientists are usually quite receptive to the demarcation problem because it promotes the idea that what they are doing is exceptional. Here is the irony of the long saga of building a wall between science and

A Critique of Science

non-science. First, the wall was mostly designed and built by the members of the excluded class. Second, whenever the wall seemed to be in danger of being breached, the excluded members rushed in and renovated it. Simply put, philosophers kept finding ways to justify science's discriminatory practice against others, including philosophers themselves. It is like members of the lower class of the Indian caste system justifying the unjust system directed against them. Thus, the demarcating wall of science was designed, built, maintained, and fortified mostly by outsiders while the insiders enjoyed their prestigious position based on the discriminatory rule.

Philosophers eventually did come to realize the untenability of the demarcating wall. The problem now is that scientists are reluctant to recognize philosophers' recantation. It seems that scientists are rather happy with preserving the wall even though the justification and rationale of it had been largely rejected by philosophers. Mahoney (1976) addresses this ironic situation:

> Thus, attempts to erect a single criterion for demarcation have failed miserably—even to the point of public concession by their originators. Nevertheless, the contemporary scientist goes on waving the banners of positivism and logical empiricism as if they were the undisputed champions of scientific process. Were he to take a moment to look behind him, the scientist would see that he has been long since abandoned by his philosophical comrades...

Humans are plagued by a pair of opposite traits in certainty judgment: overconfidence and insecurity. One of the main causes for overestimating or underestimating one's estimation of certainty is past success or failure. We swing from extreme certainty to extreme insecurity. Overly optimistic and excessively pessimistic certainty judgment can also happen in different schools of thought in the philosophy of science. Positivism is a version of an overconfident approach to knowledge-seeking. Positivism confines itself to the types of knowledge that can be confidently assessed by verification through observation and experimentation. In the wake of relativity and quantum mechanics that overturned Newtonian mechanics, however, Karl Popper came to the conclusion that nothing can be proven with certainty. Popper completely about-faced from positivism's certainty to the extreme uncertainty stance: Knowledge can only be falsified, not verified.

Another form of the highly insecure approach to knowledge-seeking is "relativism." It stresses that knowledge has no absolute truth or validity, only relative truth. Thomas Kuhn is one of the most well-known advocates of relativism. Kuhn believed that most scientists are narrowly-trained specialists who try to work entirely within their paradigm until too many unsolved puzzles accumulate (Fuller, 2004). He compares a paradigm shift with a conversion experience in religion. A paradigm dictates what and how scientists see the world. Kuhn's commitment to relativism lessened in his late years perhaps due to criticism.

At any rate, overconfidence and insecurity are the innate traits that hinder optimal decision-making process. Positivism, falsificationism, and relativism further complicate our inherent flaws of overconfidence and insecurity. The role of the philosophy of science, as the intellectual watchdog, is to find ways to compensate for our innate flaws, not to make worse.

Crude Stupidities and the Streetlight Effect

> We do not have a fixed scientific method to rally around and defend....I think it does no good for scientists to pretend that we have a clear a priori idea of the scientific method.
> -Steven Weinberg

Wolfgang Kohler (1931) had a chimpanzee named "Sultan" who was smart enough to figure out how to reach a banana that was suspended from the ceiling. He was able to have an insight to stack boxes on one another to reach the food. After four weeks of the same kind of training, Kohler switched to a different mode of getting food. Kohler put the food outside the cage and left a stick to retrieve the banana within reach. However, Sultan dragged boxes and began to build rather than use a stick to reach the banana. Apparently, Sultan didn't realize that he had to use a different approach to reach the banana. Kohler (1931) called it "crude stupidities arising from habit." We might chuckle at Sultan's inability to adopt a new method of getting food, but the fact is that positivism is science's crude stupidity, as offensive as that may sound.

A Critique of Science

A positivistic scientific methodology may have been quite useful in physics. However, other disciplines shouldn't rely on the exact same method that physicists used successfully. By the same token, biology doesn't have to mathematize evolutionary biology just because mathematics was useful in Newtonian mechanics. Above all, psychology must find its own method of investigating the introspective entity called consciousness.

Since my blunt comparison with Kohler's chimpanzee would likely offend some scientists, I might as well add another one. A policeman sees a drunkard searching for something under a streetlight and asks him what he lost. The drunken man says he lost his keys and they both start to look for the keys under the streetlight. After a while, the policeman asks him if he lost his keys underneath the streetlight. The drunkard answers, "no, but this is where the light is." This is called the streetlight effect and Freeman (2010) discusses the human flaw of looking in places where it is easiest to look. The streetlight effect is a form of bias that can result in the failure of experts like finance wizards, doctors, relationship gurus, celebrity CEOs, high-powered consultants, health officials, and scientists. Bossomaier & Green (1998) unequivocally state: "Science tends to be a little bit like the drunk who is looking for his wallet under a street lamp at night.

Positivism is an example of the streetlight effect. But it has an additional complication. Those who look under the streetlight become the status quo whereas those who look away from the streetlight are rejected and stigmatized. Positivism is the drunkard-established value system. The inevitable consequence of the drunkard's dominance is that scientists are forced to explain everything within a confined realm that meets positivistic agenda regardless of how inadequate or illogical it may be. At the same time, those items that fall outside the streetlight are dismissed no matter how useful they may be. Investigators must look wherever and use whatever the method to meet the need of particular subjects.

Feyerabend (1988) has argued that scientific methodology impeded the course of scientific progress. I believe the problem is not in the scientific method itself but in confining to a specific method to study a variety of conditions and characteristics that may be inadequate or inappropriate. The scientific method is a tool to be used in accomplishing certain functions and experimentation and mathematics should help determine the veracity of theory. We shouldn't restrict ourselves from investigation based on specific tool that might have been previously useful for some other situation.

I would argue that the great advancement of science in the 21st century won't be made by some new discoveries; instead it will be made when scientists are free to roam around the dark area and use any kind of method (or tool) that the investigator decides to use. In order for that to happen, the mindset of scientists must change from intolerance to tolerance, intellectual straightjacket to intellectual freedom, anti-religious to secular, and from anti-metaphysics to metaphysics-friendly.

Michel de Montaigne and Self-Knowledge

> The most important product of knowledge is ignorance.
> -David Gross

> Thinking is a process of conversation with one's self when the individual takes the attitude of the other...
> -George H. Mead

> To philosophize is to doubt.
> -Michel de Montaigne

> Skepticism is the first step on the road to philosophy.
> -Denis Diderot

In the Christian-dominated culture of Europe in the Middle Ages, there were two major factors responsible for intellectual independence from religion. One was, as discussed earlier, the creation of institutions of higher learning, or universities. Another was the spirit of rational inquiry that is supposedly the signature brand of the Enlightenment. While the establishment of autonomous universities provided a neutral space for learning, the spirit of free inquiry became the driving engine for reviving old ideas and stimulating new thoughts. They complemented each other and became a two-pronged assault in breaking down the mental trap that for centuries plagued the Dark Age in Europe. In historical analysis, it is easier to see and examine the causes of the external changes in the establishment of universities. It is not so evident,

A Critique of Science

however, to notice the subtle transformation taking place in the mindset and culture of intellectuals. Yet the renovation of mind was every bit as important as the renovation taking place on the outside. The question is: how did the spirit of free inquiry arise in the midst of religious intolerance and dogma?

There is no doubt that the creation of institutions of higher learning greatly contributed to the rising spirit of rational inquiry. At the same time, we can pinpoint one individual who made a greater impact than anyone else in promoting the new intellectual movement. His name is Michel de Montaigne, a retired politician in France. Why do I put so much emphasis on this one man and what did he do? He was born a member of a privileged family in 1533, in a town near Bordeaux, thirty-one years before Galileo's birth, sixty-three years before Descartes's, and 109 years before Newton's. After Montaigne retired at quite a young age from a rather successful political life, he retreated to the library in his chateau and began to write about himself. He produced a sort of autobiography dominated by his reflections on thought, or more precisely his thoughts on thoughts. Writing autography is rather common in modern days, particularly for those with fame. However, in Montaigne's time, such an effort was considered self-indulgent unless it was made in the context of religion, as in Augustine's *Confessions*. Montaigne began to work on *Essays* (*essai* literally means trial or attempt in French) in 1571 and published it in 1580. He continued working on it and published the second edition in 1588, the third edition in 1595 (posthumous).

If I have to pick a single paragraph that best represents the significance of *Essays*, it would be a page from the chapter "On Repentance" that contains three major points. First, in a plain-spoken manner, as if someone else is inside his mind, Montaigne describes the internal state of his mind by using the pronouns "he" and "him."

> *He* (emphasis added) is always restless, and reels with a natural intoxication. I catch *him* (emphasis added) here, as *he* (emphasis added) is at the moment when I turn my attention to *him* (emphasis added). Cohen, On repentance, 235

Then, he depicts the dynamics and inner conflict within himself:

> It is a record of various and variable occurrences, an account of

> thoughts that are unsettled and, as chance will have it, at times contradictory, either because I am then another self, or because I approach my subject under different circumstances and with other considerations. Cohen, On repentance, 235

He then takes a humble position about the degree of confidence he has in his knowledge:

> Could my mind find a firm footing, I should not be making essays, but coming to conclusions; it is, however, always in its apprenticeship and on trial. Cohen, On repentance, 235

Although he was describing his own thought, the statements are deeply thought-proving to anyone who has encountered them, especially to those who encountered them in the 16th century. What he was doing was observing the observer and judging the judge. This is known as "self-knowledge" and it is the mental process of looking at our inside selves and examining our own thoughts. It is as if we become a watchdog of our own thoughts. This capacity to self-examine or be self-vigilant is one of the most important attributes that sets us apart from other animals, as we will discuss further later. Its enormous importance, however, has usually been a neglected subject, despite the fact that numerous sages of the ancient world emphasized it. This is because of our tendency to focus on what goes on outside us. In China, Lao Tzu said, "Knowing others is wisdom, knowing yourself is Enlightenment." In a similar vein, Chuang Tzu said, "A man who knows he is a fool is not a great fool." In Greece, Thales said, "The most difficult thing in life is to know yourself."

But the most famous reference to self-knowledge comes from Socrates: "Know thyself." Among ancient philosophers who emphasized self-knowledge, Socrates deserves the honor of the top dog title because he greatly elaborated the significance of self-examination in Plato's *Phaedrus*, essentially a report of a long dialogue between Socrates and Phaedrus. Socrates exposes the multiplicity within a person by comparing a person with a charioteer and a pair of winged horses. The cooperation and conflict among the charioteer and two winged horses is something we all can identify with in our decision-making process. The success and failure of our actions largely depend on managing the different parts of our mind. Socrates encouraged his students to reason critically rather than to appeal to authority. For that, he was tried for heresy and for corrupting the mind of youths. In his trial, he went as far as saying that the unexamined life is not worth living. Socrates was renowned for shaking others' confidence in their knowledge. He claimed to hold no views on his own, but only to submit the views of others to rational criticism. This is critical for the development of "skepticism" (Popkins & Neto, 2007). The word "skepticism" comes from the Greek word *skepsis,* meaning "enquiry," but it evolved into different meanings among laypersons. It often means "distrustful" of particular kinds of beliefs, such as political views or religious convictions (Popkins & Neto, 2007). For the early Christian theologian Tertullian, the skeptical approach to knowledge was something unacceptable, and a philosopher was a "patriarch of the heretics."

The process of critically appraising one's knowledge was crucial to acquiring independence from the overwhelming influence of religious authority in Europe. Like other philosophers who promoted self-knowledge, Montaigne's *Essays* provided stimulation to scholars to be cautious about the degree of confidence they had in their knowledge: "Nothing is so firmly believed as that which we least know." He considered sureness as a "plague on Man." What Montaigne did in the 16th century was to re-address and elaborate upon the self-observational effort. Here was a man who never considered himself a philosopher, yet he may have influenced European philosophy more than anyone else in the Renaissance and the Enlightenment. He is called an accidental philosopher (Hartle, 2003). The renowned philosophers after him like Bacon, Hobbes, and Locke, to name a few, all cite him. Ironically, however, his *Essays* have been mostly acknowledged as the origin of a new literary genre, rather than as a philosophical masterpiece (Hartle, 2003). One particular philosopher who owes a great deal to Montaigne is Descartes. Mehey (1997) accuses Descartes of borrowing heavily from Montaigne:

> Much of Montaigne's text turns up in his [Descartes's] writing, reworked and refigured, those aspects that would assault the cogito undergoing a repression: one may even see reinscribed, unacknowledged citations, marking precisely what needs to be delimited, subordinated—excluded through an interior confinement—interwoven in Descartes's texts.

Langer (2005) sums up Montaigne's legacy: "After Montaigne, they became more doubtful, more free-spirited, more open." In the resurrection of self-knowledge and skepticism, no one perhaps played a bigger role than Montaigne. Thus, European intellectual endeavors were optimally equipped in both tangible and intangible ways. The tangible feature was the establishment of autonomous universities and the intangible dimension was the focus on self-knowledge and skepticism that were essential to advance knowledge. The stage was set for the spirit of free enquiry and the advancement of knowledge with a pace and depth unseen before.

Metacognition, Metaexecutive, and Metaknowledge

> One of the painful things about our time is that those who feel certainty are stupid, and those with any imagination and understanding are filled with doubt and indecision.
> -Bertrand Russell

> Ignorance more frequently begets confidence than does knowledge.
> -Charles Darwin

One of the most crucial features in human evolution is the capacity to look inward. It has been variously called "knowing of knowing," "cognition of cognition," "self-monitoring," "cognitive monitoring," "thinking about thinking," "internal observer," and "self-perception." In psychological jargon, the term "metacogntion" is used. Simply put, a part of our mind constantly engages in giving us a critique or feedback on our mental process. Metacogntion is an indispensable part of human intelligence that sets us apart from other animals because it gives us the capacity to evaluate our ideas. Flavell's (1976) definition illustrates the benefit of having the self-reflective capacity to the decision-making process:

> Metacognition refers to one's knowledge concerning one's own cognitive processes or anything related to them, e.g., the learning-relevant properties of information or data. For example,

A Critique of Science

> I am engaging in metacogntion if I notice that I am having more trouble learning A than B; if it strikes me that I should double check C before accepting it as fact.

What would happen if we didn't have a metacogntive capacity? The obvious problem is that whatever part of us that is involved in performing a certain task is so busy with operating the function that there is very little mental resource left for evaluation.

What goes on within a society is similar to the processes of metacognition within a person's mind. Often, an organization or a group that is performing a certain activity is so focused on performing the work that it doesn't have time to self-evaluate. Among the many virtues of the democratic form of government is the system of monitoring and critiquing what the executive branch is doing. The ideals of the separation of powers and the institutionalizing of checks and balances provide a way of monitoring the governing system outside of the executive branch. Freedom of press established by the First Amendment provides an additional assessment of the government outside the government. The ideals envisioned by James Madison and others in creating the American Constitution were essentially an implementation of a meta-executive system. Among the many reasons why we should oppose a tyrant, then, is that the dictatorial system, like an animal, is seriously disadvantaged without the metacognitive capacity. It is intolerant to dissent and criticism. Thus, the self-monitoring system within a person or within a country serves the same vital function and enables them to make intelligent decisions through self-monitoring.

Now, let's get into a discussion of the self-monitoring system in knowledge-seeking. The higher learning institutions—madrasas in the Islamic world and universities in Europe—provided intellectual space for scholars to assess and critique. What constantly interrupted scholars' intellectual freedom and self-monitoring processes in both Islamic and European sciences was religious interference, but the paths of Islam and Europe were exactly opposite regarding the effects of religious intervention. Islamic scholars enjoyed a greater intellectual freedom in the first few centuries after the rise of the Islamic Empire, but that freedom declined with the Islamicization of madrasas. European scholars, on the other hand, after the adoption of Christianity as the state religion by the Roman Empire, lost their intellectual freedom and diversity of thought, inherited from the Greeks. The closing of the Athens Academy represents the reversion of the Greek intellectual legacy, and Europe was plunged into an intellectual abyss for many centuries. What brought Europe out of the intellectual dark ages was the systemic creation of institutions of higher learning. The autonomous university permitted by corporate law began to gradually drive out religious interference. In the 18th and 19th centuries the value system in which religion was viewed flipped, and religion came to be viewed unfavorably. The notion of supernatural agency became a collateral casualty. At the same time, a positivistic agenda established the intellectual caste system. Science dealing with empirical evidence and mathematics moved to the top of the pecking order. As mentioned earlier, the establishment of the caste system, ironically, was enthusiastically endorsed by the members of the lower class. The upper class, on the other hand, reveled in the prestige that positivists bestowed upon it. Then, in 1962, Thomas Kuhn's *The Structure of Scientific Revolution* challenged the traditional view. He rejected the view that science's knowledge-building process is a gradual and accumulative progression. Instead, scientists work under what he call "paradigms," which determine the scope and types of scientific research. He argues that paradigms force scientists to investigate some esoteric problems in detail, but at the same time cause them to reject novel and anomalous findings:

> No part of the aim of normal science is to call forth new sorts of phenomena; indeed that [which] will not fit the box are often not seen at all. Nor do scientists normally aim to invent new theories, and they are often intolerant of those invented by others. Instead, normal-scientific research is directed to the articulation of those phenomena and theories that the paradigm already supplies.

Kuhn (1962) draws a parallel between political revolution and scientific revolution:
> Political revolutions are inaugurated by a growing sense, often restricted to a segment of the political community, that existing institutions have ceased adequately to meet the problems posed by an environment that they have in part created. In much the same way, scientific revolutions are inaugurated by a growing sense, again often restricted to a narrow subdivision of the

scientific community, that an existing paradigm has ceased to function adequately in the exploration of an aspect of nature to which that paradigm itself had previously led the way. In both political and scientific development the sense of malfunction that can lead to crisis is prerequisite to revolution.

Kuhn (1962) repudiated science's mystique. Simply put, scientists, being human, are like anyone else and they are susceptible to the same errors that laypeople are prone to. Others joined in assessing science critically and launched new disciplines, like "philosophy of science," "sociology of scientific knowledge," "science and technology studies," and "history and philosophy of science." Regardless of the different terms, they shared a common goal of using metaknowledge to assess and critique science. What they were attempting to do was similar to what metacognition within a person or the meta-executive system in a country would do.

The problem was that some members of the upper caste, who had been spoiled by constant praise, didn't appreciate scrutiny by the lower class. Particularly, many physicists belittled and still belittle the significance of the philosophy of science, as Steven Weinberg, a Nobel Prize-winning physicist at the University of Texas, expresses: "We [scientists] learn about the philosophy of science by doing science, not the other way around." Richard Feyman, a Nobel Prize-winning physicist at Caltech, exemplifies the contempt that physicists typically have for the philosopher's critique: "Philosophy of science is about as useful to scientists as ornithology is to birds." It's not just the philosophy of science that he is disdainful but also the philosophy of other fields:

> Because of the success of science, there is, I think, a kind of pseudoscience. Social science is an example of a science which is not a science; they don't do [things] scientifically, they follow the forms—or you gather data, you do so-and-so and so forth but they don't get any laws, they haven't found out anything. They haven't got anywhere yet—maybe someday they will, but it's not very well developed, but what happens is on an even more mundane level (Feynman, 1999).

Perhaps he looks down on other disciplines because he doesn't know much about them, as he admits:

> I didn't have time to learn and I didn't have much patience with what's called humanities...I'm really still a very one-sided person and I don't know a great deal. I have a limited intelligence and I use it in a particular direction (Feynman, 1999).

Ironically, Feynman (1999) seems to know one can easily fool oneself:

> .....you must not fool yourself—and you are the easiest person to fool. So you have to be very careful about that. After you've not fooled yourself, it's easy not to fool other scientists. You just have to be honest in a conventional way after that.

It's almost as if he himself is exempted from his own ruling. I call it the "Brahman delusion."

While they themselves don't know much about the philosophy of science, they confidently reject the arguments raised by the philosophy of science and justify that dismissal on the grounds that philosophers don't know much about physics. A startling comment made by Alan Sokal, a physicist at New York University, reveals physicists' entrenched attitude toward the philosophy of science: "I'm merely a theoretical physicist with an amateur interest in the philosophy of science and perhaps some modest skill at thinking clearly" (Sokal, 2001). He feels that his lack of knowledge in the philosophy of science can be made up by his "modest skill at thinking clearly." I don't know any statement that shows more clearly the lack of clear thinking than Sokal's. Imagine any scientist saying that he doesn't know much about the topic he is discussing but that his ignorance is made up for by his clear thinking capacity. It would have made more sense if he had stated that his IQ score could make up for his ignorance in the subject. Compare Sokal's comment with Nobel Prize-winning physicist Brian Josephson's: "We think that we think clearly, but that's only because we don't think clearly" (Josephson, 2003).

The attitude of these physicists is reminiscent of the "Dunning-Kruger effect," named after Cornell University psychologists Justin Kruger and David Dunning. They argue that unskilled people suffer from an illusory superiority in which they rate their ability much higher than it actually is. Feynman, Weinberg, and Sokal probably are much more thoughtful and prudent in their own field. However, when they confront an area they hardly know, their confidence soars.

At any rate, dismissing the importance of the monitoring and critiquing capacity in knowledge-seeking is as flawed as is dictatorship without freedom of the press or a person without the metacognitive capacity. We should start to see the metacognitive or overseeing capacity as the buddy system that recognizes blind spots. We should be tolerant, receptive, and appreciate criticism and input from others.

The Mutual Dependence Principle in Science

> Science often behaves like a rebellious teenager who thinks he knows everything about everything.
>
> -Piet Hut

The long history of human progress consists of two dimensions. In one dimension are informational and technological advances. In the other is the moral and ethical progression that emphasizes everyone should be given equality, justice, and respect. A remarkable thing is that they are usually tied together. Most democratic nations that respect individual rights are more prosperous than countries run by tyrants. I would call it the "mutual dependence principle." The question is, then, why they should come together in a package. The answer is quite simple. If society respects individual rights, people are free to express their views, and this allows all options to be examined and discussed.

How can we relate scientific knowledge to the mutual dependence principle? If scientific knowledge is compared to the mutual dependence principle, science seems to have the two sides. On the one hand, science has achieved an enormous growth in our understanding of nature that enables us to enjoy modern comforts. On the other hand, a physicist William R. Corliss compiled over 40,000 anomalies in the Source Book Project, indicating that there may be some major defects in our understanding. The mutual dependence principle would suggest that science has the dual personality in the progress of morality and ethics. The good side can be seen in scientific conferences that each participant is treated with respect. The Jekyll side typically reveals in dealing with discussions that meet the party line. However, when it comes to the ideas that contradict the prevailing paradigm, science turns into the Hyde side.

Orogee Dolsenhe

The entire world has been progressing in morality and ethics. One of the most well-known expressions of equality is "all men are created equal" in the Declaration of Independence. It took many decades to achieve the ideal that Thomas Jefferson laid down in the document. The Equal Protection Clause of the Fourteenth Amendment right after the Civil War that states "no state shall....deny to any person within its jurisdiction the equal protection of the laws" was an effort to fulfill the equality principle. Yet, the actual realization of the equality principle came with the Civil Right Act of 1964 that outlawed the discrimination against blacks and women. On the other side of the Pacific, the long lasting Indian caste system has been made illegal.

The scientific community sadly has been marching backward when it comes to the ethical progress. Under the misguided banner of positivism, scientists practice the institutionally sanctioned discrimination. Yet, the discriminatory practice is seen as something positive and prerequisite for science. Science is a rare community in modern era without diversity, flexibility, and mutual-respect when it comes to the crucial subjects. The subjects like supernatural agency and immaterial concept of mind have been so stigmatized that hardly anyone would dare to openly discuss and endorse the condemned concepts. Perhaps one of the main reasons for the discriminatory practice to continuously exist in science is because it is geared toward ideas rather than person. However, ideas do not exist by themselves. Only conscious agents—scientists, philosophers, or scholars—can interpret, embrace, and utilize ideas. In that sense, discriminating ideas is really against people. This brings the contention raised by Charles Wright Mills (1956), a sociologist who specializes in the structure of power, in *The Power Elite*:

> The power of ordinary men are circumscribed by the everyday worlds in which they live, yet even in these rounds of job, family, and neighborhood they often seem driven by forces they can neither understand nor govern. 'Great changes' are beyond their control, but affect their conduct and outlook none the less.

I hope the great changes of science are not beyond control of individuals. I hope the scientific community could and should self-examine and take the necessary corrective actions.

A Critique of Science

The Risk Management of Decision-Making by Individuals and Groups

> Diversity is a way of coping with the possible. It acts as a kind of insurance for the future.
>
> -François Jacob

> It is the mark of an educated mind to be able to entertain a thought without accepting it.
>
> -Aristotle

> The central problem of our age is how to act decisively in the absence of certainty.
>
> -Bertrand Russell

> Absence of evidence is not evidence of absence.
>
> -Martin Rees

You often hear the proverb: "Don't put all your eggs in one basket." In essence, these words of wisdom reflect an approach to risk management in an uncertain situation. Perhaps the financial world is the most devout follower of this age-old maxim, for diversifying one's financial portfolio reduces risk by investing in a number of different assets. This risk management principle can also apply to the basic decision-making process in science. Let's consider one of science's most contentious issues: the existence of supernatural agency. Various options are available: the naturalistic account, which rejects the existence of supernatural agency, and theism, which advocates for the notion of a personal God as seen in monotheistic religions. Deism accepts a supreme being but denies a personal God, prophecies, or miracles, whereas agnosticism states that it is not possible to know for sure whether God exists. The existence of a supreme being cannot be easily proven either way, as even an ardent natural physicist like Edis (2006) admits: "Practically no physicist would be rash enough to say that the present state of play in physics absolutely rules out any supernatural agent....No one can construct a god-detector, walk into the lab, and report that they have found none." Given the present stage of science, which is full of uncertainties, it is best to have diverse approaches. Unfortunately, however, all of our eggs seem to have been placed in one basket: naturalism. Why should scientists take such an irrational and illogical stance when there are so many uncertainties? The main reason is that the contemporary science's value system vilifies the very notion of supernatural agency.

People continue to confuse two different types of decision-making: the conclusive and the non-conclusive. The first one typically involves a time-constraint in which a person or group has to reach a decision within a specific time. For example, one must decide to turn left or right to avoid an obstacle while walking, running, driving, or navigating. In the case of a group, one example is when a jury has to determine a defendant's guilt or innocence after weighing the evidence presented. Most decisions in government and corporation are geared toward drawing conclusions to determine which policy or course of action should be implemented. For instance, during the Cuban missile crisis members of the Kennedy administration had to choose among several possible courses of action within a limited time.

In a non-conclusive setting, there is no urgency for an individual, a group, or even the scientific community to arrive at a single choice. We do not have to choose immediately in situations, such as accepting or rejecting supernatural agency; rather, we can remain open-mind until some experience helps us decide. The knowledge-seeking process does not typically require the emergence of a conclusion. Just as financial investors diversify their wealth, individuals within the scientific community should seek out and investigate various options. Allowing dissent and diversity is a form of risk management that helps the group as a whole to remain open-minded.

There are three reasons why we should tolerate "wrong" ideas. First, it may turn out to be correct. A case in point is the "apparent superstition of uneducated people" (Goldstein & Goldstein, 1978) that the mild disease of cowpox might confer immunity against smallpox. Jenner tested this superstition, even though it required taking the risk of deliberately infecting people with a disease. Second, even a wrong idea may make a great contribution. The "EPR paradox" (or Einstein-Podolsky-Rosen paradox) is one of the best examples of this, for it reveals the fundamental nature of the world. Einstein and his colleagues designed this thought experiment to reveal the incomplete nature of quantum mechanics. Alain Aspect, a French physicist, proved that a quantum event occurring at one location can instantaneously affect an event at another location despite the lack of any connecting mechanism. Thus, what Einstein had called "spooky action at a distance" in the EPR paradox paradigm turned out to be a statement about reality. The effect is sometimes referred to as "nonlocal behavior" or "quantum weirdness." What essentially happened was that Einstein made one of the greatest contributions to science by being wrong. Third is the golden rule of ethics, which says that you should treat others the way you want to be treated. Mutual respect should be a norm applied to all aspects of society. The only exception should be toward those who are intolerant to other idea, person, and group. As Buddha said, we should only be intolerant to intolerance.

This brings us to the question of whether modern science might be too preoccupied with finding truth and thus declaring the "winner" and the "loser" prematurely. Truth should not be forced upon people, but should be a byproduct of the knowledge-seeking process. Therefore, scientific investigation should be more like an exploration of uncharted territory in which there is no right or wrong, like an explorer's experience that can benefit others who take the same path. Positivism, which essentially declares an entire area to be false and prohibits scientists from exploring it, is an intellectual tyranny.

Leon Festinger (1954), an American social psychologist, posits that holding conflicting ideas simultaneously will cause tension and uncomfortable feelings, a phenomenon known as "cognitive dissonance." It is like we have an innate impulse to put all eggs in one basket. This can also be seen among groups or societies that are working their ways through various disagreements. In a group discussion, individuals seek to make their ideas the dominant ones. Mind can be viewed as a fragmentary entity composed of competing impulses and identities, rather than as a unitary entity (Moldoveanu & Stevenson, 2001). The idea of a "divided self" goes back to Plato, who proposed that the soul had three parts. Freud also came out with a tripartite division of mind: the id, the ego, and the superego. One's intellectual laziness may "self-censor" (Bem & McConnell, 1970) an uncomfortable idea, just as a group entrusted with reaching a decision might ignore or reject an inconvenient truth before carefully considering it.

My argument is this: People who respect their own ideas will respect the ideas of others, and those who are intolerant of their own thought will be intolerant to that of others. Simply put, intrapersonal and interpersonal conflict resolutions spring forth from the same mental habit. Here is why: When we discuss an issue with someone else, we are not really having a direct debate with him or her, since our thoughts are private and no one can get into our personal realm. What actually happens during a debate is that a part of our mind represents our understanding of the other person's thought. Thus, interpersonal dialogue is a form of intrapersonal dialogue for they are the same processes. I call this the "internal representative dialogue" (or IRD). In essence, such notions as debate, argument, dispute, disagreement, and other techniques of interpersonal communication are illusions. The question here is whether one's mind is tolerant and open-minded enough to provide a neutral space in which the part of mind that represents someone else's idea can have an equal footing in the internal discussion taking place. The source of some leaders' intolerance toward others is really the consequence of IRD's failure to accommodate an alternative view. The rigidity of Hitler is a good example, for he was just as intolerant of himself as he was of others.

The Black Swan and the Expectancy Effect

In 1697, a Dutch explorer named Willem de Vlamingh discovered black swans in western Australia. Once used as an expression of impossibility, however, after this event the phrase "black swan" was transformed into what Nassim Nicholas Taleb (2007) calls the "Black Swan Events," those rare and surprising events that have a major impact. In short, no matter how obvious and certain a given view may seem, there is always a chance the anticipated view could be wrong. This is another reason for not putting all our eggs in one basket.

A potentially greater problem, however, is that some people may not even notice that the swans are black. As they say, we see what we expect or want to see. As discussed earlier, our expectations can cause our perception and judgment to become distorted (Bruner & Postman, 1949; Asch, 1951). One of the primary reasons for this is that nothing is totally black and white, for various degrees of subtlety are involved. Unless we are mentally prepared to recognize an unanticipated possibility, what is actually happening may not even register in our perceptual process. One way to minimize this is to utilize the fragmentary aspect of our mind. Simply put, we should not put all of our eggs in one basket.

A Critique of Science

We can let our mind hold differing views, just as a democratic society allows dissenting views to exist. Diversity and dissent are virtues not only in a society, but also within a person. Having to deal with many internal voices, both within an individual and a democratic society, can be time-consuming and annoying at times. But there is one major benefit of this: the ability to better prepare for possible "black swan" events by allowing all options to be discussed and compared. Of course, this should be done without isolation and compartmentalization of differing thoughts, within an individual or a society.

Gullibility and Rigidity: The "Not Me" Myth

> Not be absolutely certain is, I think, one of the essential things in rationality.
>
> -Bertrand Russell

> Remember that changing your mind in the light of new evidence is a sign of strength not weakness.
>
> -Stuart Sutherland

> Doubt or believe everything: these are two equally convenient strategies. With either we dispense with the need for reflection.
>
> -Henri Poincare

The human mind is susceptible to two paradoxical failures in decision-making: gullibility and rigidity. We are easily persuaded to believe in things that often seem to be somehow unreasonable and, once we accept certain view, are unable to change it even when there is sufficient reason to do so. We like to think of ourselves as a sound decision-makers, as intellectually autonomous individuals, and as being immune to manipulated into accepting illogical and irrational views. The main problem here is that we are not very thoughtful when it comes to examining our own thought process because we are so busy judging others. In other words, the lessons derived from one or more historical failures apply only to others, not to me. Margaret Singer, a clinical psychologist at the University of California, Berkeley, calls this the "not me" myth. People who are in a position to examine and critique others are especially vulnerable to this myth. For instance, the Inquisitors of the

Middle Age saw all of the wrongs in others people and yet exempted their own actions from any scrutiny. Driven by self-righteousness and the two-valued impulse, they saw themselves above and beyond the "heretics" they were examining, judging, and condemning. They never realized that they were committing immoral and reprehensible acts in the name of God.

This "not me" myth can strike us when we least expect. Consider the Beltway snipers who terrorized the nation's capital in the fall of 2002. Oscar Wilde, an Irish playwright, said life imitates art far more than art imitates life. His words eerily came alive when the snipers, John Muhammad and Lee Boyd Malvo, demanded $10 million dollars, a hundred times more than the amount demanded by the serial killer portrayed in the Hollywood blockbuster *Dirty Harry*; "Scorpio" only demanded $100,000. The diabolical scheme of the snipers in the movie and in the real-life drama were the same: Show me the money, or people will continue to be shot. The Beltway snipers engendered one of the largest investigations in the history of American law enforcement. Over 2,000 law enforcement officers, police and sheriff's deputies from the county, as well as state and federal agents from the FBI and the ATF, were involved. The person who led this investigation was Montgomery County Police Chief Charles A. Moose and there was a sad irony behind this massive manhunt.

The irony can be said to start with Chief Moose's caution expressed to the media: "We want to stress that when we release that information that we not release it in a way that somehow now eliminates people and causes us to get too focused on a path and potentially miss the suspect." The information he was referring to was a "white truck." In the early morning of October 3, 2002, Sarah Ramos, a 34-year-old woman from El Salvador, was sitting on a bench in Silver Spring, MD, reading while waiting for her ride when a .233 rifle bullet entered the top of her skull. When the dispatcher radioed for an ambulance, two medic engines in the vicinity could not respond because they were tied up with two other sniper attacks that had occurred less than an hour ago. A little more than an hour later, Lori Lewis Riviera was fatally shot while vacuuming her minivan at a Shell gas station, only five miles away from the site of the Ramos shooting. The final victim that day was Pascal Charlot, a Haitian carpenter who was about to cross the street in Washington, DC, not far away from the earlier shootings. Seven shootings occurred during the two days; six people were killed.

Detectives were frustrated by the lack of bullet casings, a pattern among the victims, and other substantial pieces of evidence. One person, however, did give a detailed description of the Ramos shooting. Juan Carlos Villeda, a Guatemalan immigrant working with a landscaping crew at the shopping center, had noticed Ramos walking past a nearby post office, picking up a booklet on top of one of the mailboxes, and reading it while sitting on the bench. He heard a loud explosion. When he looked back at Ramos, he saw that she had slumped to her right and that the booklet was soaked in blood. He then saw a white truck with a small cab and box-type rear speed in front of him, turn onto a side street, and head toward Georgia Avenue. Two men were sitting inside, and the truck had a dent in the rear and a damaged right rear bumper. Police were told to look out for a beat-up white box truck and even released a composite graphic of it. In the following days, such trucks were stopped indiscriminately, often at gunpoint, and their drivers were handcuffed.

Witnesses also reported the presence of a dark blue Chevrolet Caprice at the various sites, including those where both Ramos and Charlot were shot. In fact, detective Tony Patterson of the Washington, DC, police department had put out a teletype for all of those involved to be on the lookout for the Chevrolet Caprice. Sadly, his announcement did not capture the attention of the detectives on Moose's investigation team. On October 13, following a two-sentence mention of the Chevrolet Caprice in the *Washington Post*, CNN's Wolf Blitzer asked Moose about it during a television interview. Moose played down its importance, saying that there was "not a big push for public feedback on that." Perhaps one of the reasons for doing so was that they already had four promising suspects, all of whom had guns, strange histories, and access to white trucks. The investigators were so sure about the suspects that they were expecting the "we got 'em" phone call any minute. Thus they were shocked when Linda Franklin was killed in the parking lot of a Home Depot in Falls Church, VA. As the suspects were under surveillance and could not be in two places at the same time, their names had to be erased from the board located next to the command table.

Yet the fixation on the white van continued until the perpetrators were captured. In other words, even while Moose was expressing caution on the possibility of being too "focused" on a specific path, he himself and the rest of the detectives were fixated on the white truck. Even though several witnesses had seen the dark blue Chevy Caprice, not to mention that it had actually been stopped numerous times for a license-plate check, its significance was drowned out by everyone's overwhelming preoccupation with the white van. At the end, the snipers literally gave themselves up by bragging about committing another murder in Alabama in order to convince the authority they were the ones who had demanded the $10 million.

The outlines of the Beltway sniper debacle are obvious. Among many leads, for some reason the descriptions of the white truck stood out and, once the investigators began chasing it, made the leads on other vehicles seem irrelevant. Moreover, any information that contradicted the white truck approach was brushed aside. I believe there are three reasons for this. First, Villeda's detailed and vivid account caused the vague and general descriptions of the Chevy Caprice by other witnesses to be considered less. The white truck just happened to be at a wrong place at the wrong time, and Villeda happened to notice it at that moment. Second, those involved indulged in the human tendency of "confirmation bias," which causes us to interpret information in a way that confirms our preconceptions and filter out contradictory information. During the 17th century, Francis Bacon (1960/1620) cautioned: "It is the peculiar and perpetual error of the human understanding to be more moved and excited by affirmatives than negatives." Thus, we have inherent biases in our decision-making process to lead us to favor certain types of information. While Chief Moose cautioned others not "to get too focused on a path and potentially miss the suspect," he and his investigating team did exactly that. This is a good example of the "not me" myth: they were gullible (accepting one lead among many). At the same time, they were rigid (they dismissed evidence that "didn't fit in" with their theory) which is the third factor. It is the difficulty of changing one's mind, particularly if one has been strongly committed to it.

Rigidity also happens in science. On August 10, 1990, B. F. Skinner, one of the world's leading behaviorists, made his final public speech at the APA's annual convention in Boston upon the occasion of receiving an Outstanding Lifetime Contribution to Psychology award. His 20-minute speech was largely an insistence that psychology could never be a science of mind, only a science of behavior. He died seven days later. Despite the facts that the cognitive revolution overthrew behaviorism nearly a half a century ago and that virtually all contemporary psychologists reject radical behaviorism, Skinner went to his grave firmly and stubbornly insisting that consciousness could not be studied scientifically. This perhaps shows how difficult it is for someone to change one's mind when confronted with the possibility that certain longstanding and deeply held commitments might be incorrect. This also reminds us of Kuhn's (1962) articulation that revolution does not take place by logically convincing opponents. New science has to take over and dominate while the old way of thinking dies out. Perhaps we should not expect to convince opponents through rational discussion. Even if one realizes his or her mistake, admitting it remains extremely difficult. Judicial history is filled with cases of judges and prosecutors refusing to admit that they have caused innocent people to be imprisoned and even executed. We may never know whether Skinner or skeptics were aware of their errors or not. The only thing we can be sure is that it would be almost impossible for them to admit their mistakes.

Any one of us could face the same situation. Given this, we should be mindful that this phenomenon can also occur within groups, organizations, and other bodies. We should be asking whether the scientific community might be suffering from the "not me" myth and might be committing a cardinal sin in the name of science. Might scientists have become caught in the trap of rigid conformity and group thinking? We should be open to the possibility that we have been raised or indoctrinated into a culture that legitimizes intolerance and discrimination.

Orogee Dolsenhe

Chapter Twelve
Kennedy's Failure and Success

> ...sound decisions are those that are made well, with full consideration of all the relevant aspects of the problem. Good decision makers carry out each of the essential steps of effective decision making carefully and completely. Unsound decisions are marked by flaws in the process, such as ignoring available factual information about known risks or failing to consider some important alternatives.
>
> -Daniel D. Wheeler & Irving L. Janis

The demarcation problem, in my view, is an unnecessary complication that is utterly counter-productive in the knowledge-seeking process. A simple guideline in any decision-making process, whether those of individuals or groups, should be thoughtful consideration of all available options. We shouldn't reject options based on some arbitrary dogmatic principles put forward to stratify or discriminate against anyone. Ideas should be given an equal base and the minimum value just as all individuals deserve a base value of respect. Conversely, no idea should be given preferential treatment because of the status, reputation, or past achievement of its originators or proponents. The assessment of an idea should be based on its own merit, not some irrelevant extrinsic factors. As simple as this may be, it is easier said than done. In this chapter, we will examine the decision-making process in the Kennedy administration to show how easy it is to fail and, paradoxically, how simple it is to remedy this failure.

The national government is, theoretically, where the country's brightest gather. The leader of the country is surrounded by advisors with extensive knowledge who have proven themselves effective decision-makers in their various fields. These also have access to extensive networks of information gathering system to aid them in comprehending constantly changing situations in the country and around the world and facilitate intelligent decision-making. Yet, this elite group of people are susceptible to lapses of judgment just like anyone else. The only difference from average people is that their decisions have much bigger consequences. History often depends on decisions made by this small group of people in the government.

What makes one idea a failure and another a success? One way to evaluate the factors involved in making good or bad decision is to examine and compare the events stemming from what we consider failures with ones that we view as successful. In this chapter, we will examine two famous cases with completely opposite characteristics, both made by the same group of exceptionally talented men in the Kennedy administration: the Bay of Pigs and the Cuban Missile Crisis. The former has been classified a failure resulting from poor decision-making quality while the latter is called a success from the good qualities of the same process - yet, both were made by the nearly same group of people. A closer examination of these episodes of the Bay of Pigs and the Cuban Missile Crisis reveal the crucial factors involved in the process of making decisions.

While reading a book written by Arthur M. Schlesinger *A Thousand Days: John F. Kennedy in the White House*, Irving J. Janis, a psychology professor at Yale University, was particularly fascinated by a chapter about the disaster at the Bay of Pigs. Janis (1972) was puzzled: "How could bright, shrewd men like John F. Kennedy and his advisers be taken in by the CIA's stupid, patchwork plan?" He wondered if the poor decision-making of the well-trained and highly-educated men at the White House might be similar to an average person's lapse of judgment. This reading of Schlesinger's book prompted Janis to investigate other blunders he calls "fiascoes" such as Pearl Harbor, Vietnam, Watergate, and others. His book *Groupthink: Psychological Studies of Policy Decisions and Fiascoes* is an excellent analysis of defective decisions made by key members of government in the past.

The Bay of Pigs

"How could I have been so stupid to let them go ahead?" They are John F. Kennedy's own words immediately after the invasion of the Bay of Pigs that he approved became a disaster. Like Kennedy, Janis (1972) and many other authors used the phrase "how could" in questioning the peculiar and puzzling circumstances of the decision making process of the Bay of Pigs. "The Bay of Pigs" or, to give it its more accurate name "Operation Zapata," has been called "a failure of mind, of imagination, of common sense" by Karl E. Meyer & Tad Szulc (1962), "the first blemish on the magic of the Kennedy name" by Wyden

(1979), and "fiasco" by Schlesinger (1965) and Janis (1972). The disaster of the Bay of Pigs resulted not from a set of complex problems that only experts can understand, but from simple errors made by a highly competent group of people. The plot to topple Castro's government was one of many covert skirmishes between the two superpowers in the era of a continually-intensifying Cold War. Operation Zapata was originally planned under Eisenhower administration with an aim to send CIA-trained Cuban exiles to engage in guerilla warfare. The jungle of Guatemala was the chosen to training ground for this force in order to hide US involvement.

The Director of the CIA, Allen Dulles, briefed Kennedy on Cuba on July 23, 1960 at Kennedy's summer home on Cape Cod during the heated presidential campaign between Vice President Richard Nixon and John F. Kennedy. Kennedy's concern was that the invasion might take place before the November election and with this victory Nixon would win. Nixon, for the same reason, was urging the CIA to carry out an invasion before the election. The main reason Eisenhower delayed the decision to carry out the invasion was because the lack of unity among the Cuban exiles. He wanted a more comprehensive plan, created in conjunction with a leader of the Cuban exiles who would organize a new government and direct the efforts to topple the Castro's regime before the execution of any plans.

The CIA radically changed its plan from a guerilla infiltration to a full-scale amphibious invasion. This major change was ordered - without Eisenhower's input by Richard Bissell, the Deputy Director and the chief architect of the covert operation. He thought the plan of guerilla infiltrations stalled because Castro was tougher than expected, and was particularly concerned by the CIA's difficulty making contact with the underground due to the effective Cuban secret police in controlling any resistance in Cuba. The invasion plan was based on an assumption that a successful landing would set off a general uprising. The problem in changing plans from a guerilla infiltration to an all-out invasion, however, was the issue of deniability of the US involvement. Unlike a guerilla infiltration accomplished by quietly sneaking into the country, an invasion requires a considerable amount of backing, such as supplying and protecting the invaders on the beachhead. Since Castro's army and air force would be mobilized to face the invaders, there needed to be sufficient force and support to encounter the anticipated clash, making it quite difficult to cover up the US involvement in the operation. The predicament stemming from showing overwhelming force lay in international perception that would make the US no different than the Soviet Union, ruthlessly crushing its neighbor as the USSR did Hungary. If, on the other hand, the involvement was small enough to conceal, a successful landing became that much more difficult. This is the dilemma that the newly elected Kennedy and his advisors faced soon after the inauguration.

In January 1961, Allen Dulles, the Director of CIA, and Richard Bissell, the Deputy Director of the CIA gave detailed briefing about the plans to Kennedy right after the inauguration. Both men were carry-overs from the Eisenhower administration with impeccable reputations. Schlesinger (1965) described Dulles as a director with "the coolness and proficiency." Kennedy later commented about the intimidating aspect of the CIA Director after the Bay of Pigs failed: "Dulles is a legendary figure, and it's hard to operate with legendary figures." Schlesinger (1965) depicted Bissell as a "man of high character and remarkable intellectual gifts." Other descriptions include "a human computer," "a perfectionist," and "genius" (Wyden, 1979). Another critical quality Bissell possessed was a skill in salesmanship that effectively meant "many of his peers felt he was a manipulator but liked him too well or too awed by his persuasiveness to say to outright" (Wyden, 1979). According to Schlesinger (1965), whenever Bissell presented, "we all listened transfixed." His eloquent presentations were the major persuasive factor in the group's acceptance (Janis, 1982). All other members of the group were known to be shrewd thinkers, capable of objective, rational analysis, and accustomed to speaking their minds (Janis, 1972).

Kennedy sought assessment of the CIA's plans from the military experts in the Pentagon. A small committee headed by Brigadier General David W. Gray gave their opinion. This was done with input (more appropriately called friendly pressure) from six men from the CIA. They brought no papers and gave the committee nothing more than "a verbal rundown." The men from the CIA told the Pentagon committee that the task was very urgent because: 1) the rainy season was approaching, 2) Cuba was about to receive a shipment of Soviet jet fighters, along with Czech-trained pilots, 3) the Cuban tides made landings possible only on certain days. Gray's committee was told the plan was to land at dawn at

A Critique of Science

Trinidad, a shore city in southern Cuba with a population of 18,000. Trinidad is near the foothills of the Escambray Mountains, which could become an escape site for the invaders, allowing them to "melt into the mountains." Gray's evaluation of the operation "JCSM-57-61" had "peculiar and ambiguous" contents to the President and advisers (Schlesinger, 1965). It states that ultimate success would depend on either a sizable uprising inside the country or sizable support from outside. Then oddly, without elaboration on the two conditions for success, it concludes that the plan has a fair chance of ultimate success. Schlesinger (1965) calls this a "logical gap" between the two statements: "the plan would work if one or another condition was fulfilled and the statement that the plan would work anyway." He speculates whether the gap resulted from "sloppiness in analysis" or from a "conviction" that once the invasion were launched, either internal uprising or external support would follow. The assessment of the guerilla option by JCS relied on CIA's assurance, not by a careful study (Wyden, 1979).

The plan was viewed with deep doubts by some military experts. For instance, Admiral Arleigh Burke, Chief of Naval Operation, thought the plan conceived by a group of men who had no military command experience "weak" and "sloppy." General David M. Shoup, the Commandant of the Marine Corps, also thought that the invaders "could never expect anything but annihilation." On the side of the CIA, Sherman Kent questioned whether it was the right time for the invasion as Castro's position in Cuba was likely to grow stronger rather than weaker. Roger Hilsman of the State Department was alarmed by a plan based on an assumption that a small group of men would bring down the Castro regime.

One of the assumptions in the planning of the invasion was that the exile brigade would touch off sabotage by the Cuban underground and incite armed uprisings behind the lines effectively supporting the invaders and lead to the toppling of the Castro regime (Janis, 1972). However, the truth was that a carefully conducted poll showed the overwhelming majority of Cubans supported the Castro regime had been circulated throughout the United States government. The Cuban people had long suffered poverty and injustice under the corruptive and tyrannical Batista regime, so Castro's establishment of many social reforms such as health care and education made him enormously popular.

When General Lyman Lemnitzer, of the Joint Chiefs of Staff

(JCS), presented the Trinidad plan to a White House meeting, the President rejected it as "too spectacular." He felt an amphibious landing in such a populous area would make it difficult to hide U.S. involvement. He preferred a quiet landing at night to disguise American involvement. The landing site at Trinidad was too exposed for a small number of invaders trying to slip in unnoticed. A proposed alternative landing site was the Bay of Pigs, an isolated and uninhabited area. Unfortunately, however, the site they chose had been called a "geographical and military trap" by Máximo Gomez, the master tactician of guerilla warfare during Cuba's war for independence. It is surrounded by impassable swamps and the Escambray Mountains are 80 miles away, essentially leaving no escape route for the invaders. The change in location essentially eliminated a vital factor of the operation: the invaders could no longer "melt into the mountains" and link up with the anti-Castro rebels. This fatal oversight could have been easily corrected had anyone in the group glanced at a map of Cuba.

Furthermore, the attempt to hide American involvement had already been foiled. Tad Szulc of the New York Times had already reported about the supposedly secret details about recruiting Cuban volunteers in Miami and training in the military camp in Guatemala. Despite the media's knowledge about the plan, Kennedy was confident enough to publicly state that there would not be any intervention in Cuba by the United States "under any conditions" just five days before the invasion.

Two major changes had been made to the original plan. The first was changing its nature from a guerilla infiltration to an invasion. The latter operation obviously would have required far greater force and support than the former. This crucial change was not followed by necessary modifications required. The second critical change was switching the landing site from Trinidad to the Bay of Pigs. Again, this was done without careful study of Cuba's geography or history. If executives of modern corporations made decision in such a sloppy manner, they would rapidly find themselves bankrupt in the competitive world. So on April 17, 1961, a brigade of lightly-armed 1,400 Cuban exiles landed in the Bay of Pigs. The next day, they were surrounded by 20,000 members of Castro's well-equipped army with an additional 200,000 troops in reserve. As the news of the doomed Cuban exiles reached President Kennedy, he was stunned and angry. Kennedy administration's first major decision in foreign policy turned out to be a disaster. Kennedy had to go through the humiliation of paying a ransom of $53 million in food and drugs in order to have the captured men released.

The Unheeded Doubts
As the news of doomed mission began to become known, all members of the inner circle were troubled by their obvious lapses in judgment. Even long after the fiasco, Kennedy would sometimes wonder: "how a rational and responsible government could have become involved in so ill-starred adventure." The mission went ahead in spite of doubts by most of the group including the President Kennedy. Schlesinger (1965) recalls that Kennedy "was growing steadily more skeptical as his hard questioning exposed one problem after another in the plans." He believes that if one senior adviser opposed the invasion, Kennedy would have canceled the entire mission. Schlesinger (1965) blames himself for having kept "so silent" during discussions, admitting a reluctance to bring up his objections for fear that others might think him, a history professor, "presumptuous" in challenging the experts in intelligence and military. Robert McNamara, the Secretary of Defense, stated several years later that he still felt responsible for having misadvised the President. The Secretary of State, Dean Rusk gave "gentle warnings" about possible excesses despite the fact that he thought "the chances of succeeding were zero."

When Rusk went to the SEATO (Southeast Asia Treaty Organization) conference, Chester Bowles, the Undersecretary Secretary of State, sat in his place. Bowles was appalled by what he heard but did not voice his objections at that time. He thought the plan to invade Cuba was "profoundly disturbing" and "a grave mistake." Bowles argued his objection to the plan and gave Rusk a strong memorandum opposing the plan and asked if it could be carried to the President. Rusk assured him that the project was being whittled down into a guerilla infiltration and filed the memorandum away.

Challenges to the proposed invasion of Cuba also came from the outside of the government. Senator James William Fulbright, who discovered the plan by reading newspapers vigorously voiced his opposition to the President's proposed course of action. Both Bowles and Fulbright were able to recognize the critical flaws of the invasion plan because they were not been part of the decision-making process. They lacked the "incremental descent" (Vaughn, 1996) that clouded the judgment of those involved from the beginning.

What Went Wrong?
In the aftermath of Bay of Pigs, Kennedy took publicly taking a responsibility for the disaster, and appointed the Cuban Study Group, an internal investigation committee, chaired by General Maxwell Taylor along with Dulles, Admiral Burke, and his brother Attorney General Robert Kennedy. The committee concluded that the CIA and the JCS shared the blame for the failure. Janis (1972) also lists four factors given by insiders (Schlesinger, Sorenson, Salinger, Hilsman, and others) as integral to the plan's ultimate failure. The first factor was political calculation - during the Cold War era, on politicians were preoccupied with not being seen as soft on communism. A second factor was the problem of a new administration bottled in an old bureaucracy. Kennedy's administration faced the problem without the opportunity to get to know each other and fully develop decision-making procedures. The transition period from one administration to another is a critical period susceptible to various forms of problems. The third factor was the obsession for secrecy that excluded the chance input from the intelligence branch of the CIA or the State Department. Ironically, this happened at a time when many newspapers had reported the supposedly clandestine operation. Even Castro was ready for an imminent invasion, alerted by U.S. presses. Fourth, and finally, was the individuals' concern for reputation and status. Anyone considering voicing objection to the plan felt his personal status and effectiveness might be at risk; this dilemma is called the "effectiveness trap."

Janis (1972), however, argues that these explanations are not sufficient. According to him, even if all four factors were operating simultaneously, such a faulty decision would not necessarily have happened. He introduces the "groupthink hypothesis" to account for the mishaps of the Bay of Pigs and other blunders. The groupthink hypothesis is the tendency of members of any small cohesive group to maintain *esprit de corps* by unconsciously developing a number of shared illusions and related norms that interfere with critical thinking and reality testing. Janis lists six factors occurring in groupthink:

1. *The illusion of invulnerability*: Members tend to have overoptimistic views of themselves paired with perceived weakness of their opponents. The idea of an invasion of 1,400 exiles against Castro's 200,000 troops is obviously a colossal blunder.
2. *The illusion of unanimity*: When individual perceives unanimity in a group of people who respect each other, each member is likely to feel that this united belief must be true. This stifles one's critical thinking and reality testing.
3. *Suppression of personal doubts*: Most members of the group had doubts, yet they were kept silent.
4. *Self-appointed mindguards*: Just as bodyguards aim to protect a target from any perceived threats, members of a cohesive group try to protect the dominant belief of the group by putting pressure on members with deviational viewpoints. An example in this case occurred when, in response to Schlesinger's memorandum opposing the plan, Robert Kennedy urged him: "Don't push it any further."
5. *Docility fostered by suave leadership*: Explicit or implicit support of one view by a charming leader like Kennedy can undermine the critical appraisal.
6. *The taboo against antagonizing valuable new members*: From the onset, Bissell and Dulles received preferential treatment by the group and avoiding any critical comments to the CIA's plan became an informal norm.

Kennedy's Adjustments after the Bay of Pigs
When a person experiences a major setback that could have been prevented with minimal care, he or she experiences a set of mixed emotions like regret, anger, guilty, blame, and other negative reactions. Often, these emotions dominate one's thinking and behaviors for some time. An important thing to realize, however, is that the transformations one goes through after setbacks has the potential to be either negative or positive. An example of extremely negative transformation occurred in President Richard Nixon in the wake of the fiasco of the Watergate cover-up. Many decisions Nixon made during the period after the Watergate break-in can be characterized as a "panic-like state of hypervigilance." It was the "opposite pole from the manifestly calm state

of overoptimism" of the decisions that transpired before the Watergate break-in (Janis, 1972).

Quite often, the changes made after major setbacks largely depend on the mindset of the affected person. In the aftermath of the Bay of Pigs, Kennedy was determined that he would ensure nothing like that happened to him again. His order for an internal investigation of the disaster to discover exactly what had gone wrong was a clear indication of this frame of mind. After evaluating the investigation's results, Kennedy implemented a set of changes in his decision-making procedures:

1. The first lesson was never to rely on experts (Schlesinger, 1965). Committee members were given much broader role; instead of confining one's role to particular specialty and avoiding arguing about issues on which others are expert, each member was expected to function as a critical generalist. In other words, members were encouraged to look at the big picture to comment any flaws that might exist in decisions regardless of one's expertise.
2. McGeorge Bundy was given new authority as a coordinator. He invited members of the White House, State, Defense, and CIA to lessen the sprawling mystery of government.
3. Kennedy made one more step in preventing errors. He appointed his brother Robert Kennedy and Theodore Sorensen to function as intellectual watchdogs so that they could relentlessly pursue flaws of decision from every possible direction at the expense of becoming unpopular with the group.
4. Rather than the typical rules of protocol, meetings were conducted in an atmosphere which promoted frank and freewheeling discussions without imposing formal agenda. The members were constrained as Schlesinger (1965) recalls: "The Bay of Pigs gave us a license for the impolite inquiry and the rude comment."
5. Outside experts were brought in to broaden the scope of information available.
6. Recognizing the tendency for newcomers to remain silent, members deliberately asked them to give their views.

A Critique of Science

7. To promote critical thinking, members were sometimes broken up into subgroups.
8. President Kennedy deliberately did not attend some of the meetings, particularly during the preliminary phase. His absence allowed all options to be discussed to avoid exerting undue influence.

As we shall see later, the decision-making procedure that Kennedy developed could apply to virtually all other areas of inquiry, including science. In fact, if these same decision-making procedures were fully utilized by science, I would argue that a great many problems should be resolved. All eight procedures may be more or less summed up with a single statement that all ideas and options should be equally, carefully, and thoroughly considered regardless who puts them forward.

The Cuban Missile Crisis

> That war did not come in 1962 was decided by the slimmest of margins....Kennedy and Khrushchev, men from very different cultures and backgrounds, came to the "brink" and then stepped back. That mankind and our planet are not ash and cinders is the lasting benefit of their actions.
> -Norman Polmar & John D. Gresham

Soon after photographs taken by U-2 spy planes revealed the presence of nuclear missiles in Cuba, the first meeting of the group which became known as the Executive Committee (or ExComm) was held on 11:45 A.M. October 16, 1962. They discussed a wide range of topics, such as military analysis, international situation, the status of the U.S. Jupiter nuclear missiles in Turkey, and possible response options. Kennedy's initial inclination was a swift surgical air strike on the missile sites. However, the first ExComm meeting ended with an agreement of the members *not to rule out any possible solution* for resolving the crisis. The mindset of the group can be illustrated by Robert Kennedy's urge for adding more alternatives: "Surely, there was some course in between bombing and doing nothing." Schlesinger (1965) remembers: "Every alternative was laid on the table for examination...In effect, the members walked around the problem, inspecting it first from this angle, then from that, viewing it in a variety of perspectives. In the course of the long hours of thinking aloud, hearing new arguments, entertaining new consideration, they almost all found themselves moving from one position to another." Using such an open-minded approach, the group listed possible courses of action:

1) Do nothing.
2) Exert diplomatic pressure on the Soviet Union by appealing to the United Nations.
3) Arrange for direct communication between Kennedy and Khrushchev.
4) Secretly approach Castro to warn him of drastic United States action.
5) Institute a naval blockade to prevent Russian ships from bringing any more missiles to Cuba.
6) Launch an air strike to render the missiles inoperable.
7) Carry out a limited surgical air strike with advance warnings to allow Cuban and Soviet personnel to escape the bombing.
8) Carry out a limited surgical air strike without advance warning.
9) Carry out a massive air strike against all military targets in Cuba.
10) Launch an all-out invasion.

Even though the President rejected the first two from the beginning, the group discussed the pros and cons of each alternative. That was a clear indication that no one, including the President, could hinder members from presenting their views freely. Sorensen (1966) remembers the atmosphere of "complete equality." Furthermore, recognizing lower-ranking officials' reluctance to openly disagree with their superiors, Kennedy deliberately summoned men in subordinate position for their views. In addition, outsiders and retired officials were brought in for their opinions. Not only did each member freely express their views, they also remained flexible in their judgment so that just about all of them changed their position at least once—some more than once. At the end, the ExComm chose McNamara's idea of blockade. The decision, however, was not without strong opposition; the hardliners, the generals of the Pentagon, were against it all the way to the end in favoring military action. On October 22, President Kennedy addressed to the world the decision to initiate "a strict quarantine on all offensive military equipment under shipment to Cuba." Two hours before the speech, President Kennedy met with congressional leaders to inform the situation and the decision. Senator Richard Russell of Georgia disagreed; he thought the only solution was invasion. Kennedy later told Schlesinger: "The trouble is that, when you get a group of senators together, they are always dominated by the man who takes the boldest and strongest line. That is what happened the other day. After Russell spoke, no one wanted to take issue with him. When you can talk them individually, they are reasonable."

Just before delivering the speech, he also authorized the military to go DEFCON-3. Astonishingly, two days later, General Thomas Power, the Commander-in-Chief Strategic Air Command (SAC), upgraded to DEFCON-2 (the highest state of military readiness short of war) on his own authority as the naval quarantine of Cuba became effective. This is the first and only time that the U.S. armed forces reached DEFCON-2. Furthermore, the DEFCON-2 order was transmitted unencrypted over SAC's worldwide broadcast system, which undoubtedly had been intercepted by the Soviets. Polmar & Gresham (2006) speculates: "That President Kennedy might not want to elevate SAC beyond DEFCON-3 or have the Soviets know of the command's heightened alert status never seems to have occurred to Power." Needless to say, President Kennedy was outraged when he discovered the elevation to DEFCON-2. General Power never offered an explanation the reasoning behind raising the level of readiness to DEFCON-2. Perhaps his remark to Professor William Kaufmann at a conference in December 1960 might give a glimpse of the general's frame of mind: "Restraint! Why are you so concerned with saving their lives? The whole idea is to kill the bastards! Look. At the end of the war, if there are two Americans and one Russian, we win!" Kaufmann replied: "Well, you'd better make sure that they're a man and a woman." His predecessor, General Curtis LeMay, might further reflect some of the seasoned military men's mindset at the height of the Cold War. General LeMay, brilliant, innovative, aggressive, and charismatic bomber commander during World War II, was quoted as saying: "There are only two things in the world, SAC bases and SAC target." LeMay left SAC to become the Chief-of-Staff of the Air Force. General Taylor's remark on LeMay was: "But a good bomber commander doesn't automatically make a good chief of staff, and appointing LeMay as chief of staff on the Air Force was a big mistake." It makes one wonder what would have happened if the hardliners with itchy trigger fingers had been in charge of the country during the Cuban Missile Crisis.

Of course, the conflict between hardliners and moderates also applies to the Soviet Union. Two seemingly conflicting responses came from the Soviet Union, a soft one that seemed to imply a peaceful settlement and a hard-line message that followed the next day, likely indicating a state of dissension between moderates and hardliners in the Kremlin. As seen in the French Revolution, extremists can hijack and plunge the nation into chaos. In dealing with these contradictory statements, Robert Kennedy came up with a solution - they ought to ignore the hard-line message as if it didn't exist. Kennedy's conciliatory letter responding to the peace-making message was also followed by a tougher message. Robert Kennedy delivered an oral message to tell Soviet Ambassador Dobrynin insisting that the USSR should remove the missiles, followed with the assurance that if they did not, the U.S. would remove them. The crisis was resolved with an exchange of withdrawals: the US would remove the Jupiter missile in Turkey and promised not to invade Cuba in return for the removal of the missiles in Cuba. On October 28, the "most dangerous moment" in the history of mankind abated and men in both Washington and Moscow, exhausted from days of stress and sleeplessness, could finally relax.

Joseph Kennedy, the president's father, told JFK that the Bay of Pigs was not a misfortune but a benefit. Similarly, Schlesinger (1965) agrees: "no one can doubt that failure in Cuba in 1961 contributed to success in Cuba in 1962." Kennedy's resolution that he would not make the same mistake twice and the implementation of a set of decision-making procedures were crucial in dealing with the next emergency, the Cuban missile crisis. At no other point in the nuclear age was the global nuclear war so close to occurring than during the critical two weeks from October 16 to 28, 1962. History might have been very different if Kennedy's administration was faced with the Cuban Missile Crisis without the lesson learned from the horrendous experience of the Bay of Pigs. The critical moments during the Bay of Pigs and the Cuban Missile Crisis are the reminders of the magnitude of the importance of a good and rational decision-making process. To reiterate, what is remarkable about the decision-making process in the Cuban Missile Crisis is that the group consisted of many of the same members who participated in the

A Critique of Science

Bay of Pigs debacle.

Symptoms of Groupthink and Their Significance in Science

If we compare the decision-making process of the Bay of Pigs with that of the Cuban Missile Crisis, the difference boils down to a simple thing. In the Bay of Pigs, there was lack of thoughtful consideration of all options, whereas in the Cuban Missile Crisis, committee members carefully examined all options. As simple as this may sound in theory, however, it is extremely difficult to execute in practice as the President Kennedy and the inner circle realized after the Bay of Pigs. The factors contributing to a failure to consider all options can come in various forms. After examining many other blunders and successes, Janis (1972) categorizes symptoms of groupthink, some of which we have discussed earlier:

Type I: Overestimation of the group—its power and morality

1. An illusion of invulnerability, shared by most or all the members, which creates excessive optimum and encourages taking extreme risks.
2. An unquestioned belief in the group's inherent morality, allowing the members to ignore the ethical or moral consequences of their decisions.

Type II: Closed-mindedness

3. Collective efforts at rationalization in order to discount warnings or other information that might lead the members to reconsider their assumptions before they recommit themselves to their past policy decisions.
4. Stereotyped views of enemy leaders as too deviant to warrant against genuine attempts to negotiate, or as too weak and stupid to counter risky attempts made at defeating their purposes.

Type III: Pressure toward uniformity.

5. Self-censorship of deviations from the apparent group consensus, reflecting each member's inclination to minimize to himself the importance of his doubts and counterarguments.
6. A shared illusion of unanimity concerning judgments conforming to the majority view (partly resulting from self-censorship of deviations, augmented by the false assumption that silence means consent).
7. Direct pressure on any member who express strong arguments against any of the group's stereotypes, illusions, or commitments, making clear that this type of dissent is contrary to what is expected of all loyal members.
8. The emergence of self-appointed mindguards—members who protect the group from adverse information that might shatter their shared complacency about the effectiveness and morality of their decisions.

Since Janis proposed the groupthink hypothesis more than three decades ago, it has received a great deal of attention and has been applied to other episodes like the disasters of Challenger and Columbia, hostage rescue attempts in Iran, Iran-Contra, and other serious situations. The realization that human's decision-making process within an organization can often be flawed should be a wake-up call to everyone. Any social assembly, big or small, government or private, should put forth an effort to monitor and scrutinize its decision-making process.

When we study the past blunders in military, government, corporate, or any other organizations, we tend to pass judgment as uninvolved third parties. However, I feel we should try to become a first person participant in learning from past mistakes. In other words, we should try to put ourselves in their shoes. These historical lessons are not just for learning the errors of others; more importantly, they are useful in educating ourselves, allowing us to realize our own decision-making systems have a flawed side constantly lurking in our mind. There is no unsinkable human being. Awareness of the imperfect aspect of human capacity is something everyone should strive to understand and be on guard against in order to minimize the impediments to making good decisions. Any organization, especially ones in which a simple mistake potentially brings disastrous consequences, should have personnel and resources to identify and counteract human's built-in flaws. In hindsight,

A Critique of Science

it might be easy to scoff at past lapses of judgment. However, how many of us can be certain that our judgments and actions would be any different from them if we were in their shoes? If you were a member of NASA's upper management, would you have noticed the impending disaster and stopped the Challenger launch? If you were a part of Kennedy's group that made the decisions in the Bay of Pigs, would you be certain that you could have recognized the flaws in the invasion plans and had the courage to speak out against them?

Historical lessons demand that we ask these questions of science: Is it possible that some major lapse of judgment is taking place in an organization you are a part of? What if there is some crucial matter that no one recognizes? Or what if there is someone recognizes a crucial matter yet he or she is shunned by his or her fellow workers? Suppose you are sympathetic with the outcast - would you be willing to risk being rejected by your coworkers by voicing your dissension? If you realized that a convenient way to become popular was to ridicule the loner, would you eagerly join the lynch mob? These types of questions should be aimed at the way science is practiced.

Orogee Dolsenhe

Chapter Thirteen
Preventing Maladaptive Attitudes and Practices

> His [George Owell's] genius centered on seeing how language, not physical force, would be used to manipulate minds. In fact the growing evidence in the behavioral sciences is that a smiling Big Brother has greater power to influence thought and decision-making than a visibly threatening person.
> -Margaret Singer

> Nothing in the nature of majority democracy prevents the *demos* from playing the tyrant.
> -Eli Sagan

This final chapter proposes two methods that can be used to prevent the current systemic and cultural maladaptiveness seen in science: everyone should (1) constantly remind themselves about their own impulses and flaws while making decisions and (2) be mindful of the leaders of the would-be leaders who are themselves driven by impulses and flawed assumptions as regards their decision-making procedures. If they do so, it will become much harder for them to fall victim to the phenomenon known as "social hijacking," which has been defined as some people being able to persuade others because of their prestige and communication skills. One manifestation of social hijacking has appeared within the scientific community: the development of a sophisticated set of rules that both justify and rationalize its discriminatory practices. I contend that such an attitude hinders the progression of knowledge because it obstructs people and ideas from entering the neutral space.

The most important task of the scientific community is to examine itself. The fundamental systemic flaw comes from its practitioners' efforts to make science appear superior to other domains within the academy. The rise of science's caste system in the 19th century is a direct consequence of the self-promotion of science. It legitimizes the suppression of ideas and mistreatment of fellow knowledge-seekers. A phrase like "self-correction" leads scientists to embrace the delusion that science should be exempted from any oversight and/or monitoring process. Campbell (1979) describes the scientific community as "self-perpetuating mutual admiration societies whose social systems prevent reality testing, stifle innovation as heresy, and suppress disconfirming evidence."

Fischer (2008) argues that studies of science have not yet made "human deficiencies" as causes of malfunction. This essay addresses the factors of human deficiencies in the maladaptive culture of science. Although originally intended as a critique of science, however, the impulses and biases found in science can easily be found within other domains of society as well as within myself. Thus I draw the following inference: the human decision-making process is highly fragile and can crack under even a very slight degree of pressure. A few incoherent leaders with a knack for persuasion and self-promotion can transmit his or her dysfunctional views to the masses. As a result, the maladaptive culture formulate by such people can last for decades or even millennia. In this chapter, I offer a list of impulses that we should try to overcome so that we can make better decisions.

Signs of Trouble and the Critical Period

> There is a higher court than courts of justice and that is the court of conscience. It supersedes all other courts.
>
> -Gandhi

> All that is necessary for the triumph of evil is that good men to do nothing.
>
> -Edmund Burke

> No one can terrorize a whole nation, unless we are all his accomplices.
>
> -Ed R. Murrow

In 2004, the disturbing images of Iraqi prisoners being abused and tortured at Abu Ghraib made headlines around the world. Some reporters compared them to the cold-blooded 1968 massacre of approximately 500 unarmed Vietnamese civilians – women, children, infants, and elderly people – in the small village of My Lai by American soldiers. These two appalling incidents, separated by almost four decades, share some common features, such as the ringleaders (Lt. William Calley

A Critique of Science

in My Lai and Specialist Charles Graner in Abu Ghraib) and the heroes who tried to stop the brutality. In My Lai, helicopter pilot Hugh Thompson did what he could to end the senseless killings; in Abu Ghraib, Specialist Joseph M. Darby blew the whistle. History often sees the Thompsons and the Darbys as people who are willing to risk their own wellbeing to challenge injustice and cruelty. During the Nazi era, renowned psychologist Wolfgang Kohler openly opposed the dismissal of his Jewish colleagues. One of the main measurements of humanity's moral progress may be the presence of the Thompsons, the Darbys, and the Kohlers, for they are the ones who oppose the Calleys and the Graners. The vast majority of people, who fall between these two endpoints, can be persuaded to join either side. In fact, there is a critical period that determines which side a group will tilt toward. Like the falling tree, once it gains a momentum it is extremely difficult to reverse. This fact makes early detection and undertaking prompt corrective action critically important.

The first and foremost sign of trouble is unanimity. If no one disagrees or offers an alternative view, there is likely to be some hidden pressure to conform. Perhaps one is afraid of the leader. In Abu Ghraib, Darby spent three weeks agonizing over his decision to blow the whistle because he was so afraid of Graner's retribution that he slept with a loaded weapon. A more common reason might be that people have so much respect for the leader that they lose the ability to critically assess the position advocated by a leader with a talent for communication. We must realize that no matter how skillful or charming a salesperson may be, an old beat-up Volkswagen should never become a brand new Mercedes. As unbelievable as it may sound, the domination of Darwinism within the academy and within society at large is a good example of an old beat-up Volkswagen becoming a brand new Mercedes.

One recurring problem of history is that a suave leader may be able to convince people that a given individual or group of people should be deprived of their basic rights and even a minimum degree of respect. It may be difficult to accept that what happened in Abu Ghraib, My Lai, and Nazi Germany have something in common with the science community's decision to purge science of anything having to do with metaphysics. But in fact, all of these groups have infringed upon the basic rights and minimum respect of others. The direction in which an organization will tilt largely depends upon which side engulfs more people. Here, I am reminded of the story of an old Cherokee who told his grandson that a fight is going on inside of him between two wolves. One is evil (viz., anger, envy, sorrow, regret, greed, arrogance, guilt, inferiority, lies, false pride, superiority, and ego), and the other is good (viz., joy, peace, love, hope, serenity, humility, kindness, benevolence, empathy, generosity, truth, compassion, and faith). He said that the same fight is going on inside his grandson. When the grandson asked his grandfather which wolf will win, the old Cherokee replied: "The one you feed." It is the same in society: the group that is fed more wins.

Impulse Control and the General Checklist

> ...under conditions of true complexity—where the knowledge required exceeds that of any individual and unpredictability reigns—efforts to dictate every step from the center will fail. People need room to act and adapt. Yet they cannot succeed as isolated individuals, either—that is anarchy. Instead, they require a seemingly contradictory mix of freedom and expectation— expectation to coordinate, for example, and also to measure progress toward common goals.
>
> -Atul Gawande

Atul Gawande (2010), a physician at Brigham and Women's Hospital and a professor at Harvard Medical School, argues that we can use "the lowly checklist" to deal with the plague of failures at various levels. His *The Checklist Manifesto* starts with a 1935 flight competition held at Wright Air Field in Dayton, OH. The Boeing Corporation's prototype "Model 299," which would become the famous B-17 (nicknamed the "flying fortress"), was competing against the bombers of Martin and Douglas. Boeing's bomber was expected to win easily because its performance was superior to that of its competitor: This four-engine bomber could carry five times as many bombs as the Army required, and it could fly faster and almost twice as far. Unfortunately, for some reason it crashed during the test flight and the Army declared Douglas' smaller bomber the winner. Boeing almost went bankrupt.

During their analysis of this event, the investigators determined that the crash had been caused by pilot, as opposed to mechanical, error. The veteran test pilot, Major Ployer P. Hill, had forgotten to release a new locking mechanism on the elevator and the rudder controls. Despite the crash, a group of test pilots did not lose their faith in the bomber and continued to test it. In an effort to avoid the same type of error, they devised a pilot's checklist that consisted of step-by-step checks on an index card for takeoff, flight, landing, and taxiing. The checklist essentially consists of "dumb stuff" (Gawande, 2010): the brakes are released, the instruments are set, the elevator controls are unlocked, etc. And yet this checklist of "dumb stuff" helped them fly a total of 1.8 million miles without an accident. Boeing went on to produce 12,731 of these bombers. Since then, a pilot checklist has become standard for flying any airplane. Gawande (2010) also discusses an emergency checklist that is responsible for saving a person who is drowning as well as a cleanliness checklist in intensive care units that has virtually eliminated a hospital infection.

Such checklists are essentially technical and procedural guides to be used in specific situations. I would like to propose a more general checklist, one that is geared toward wide-ranging settings, that would remind us to both check our own impulses and errors and to pay attention to our own mental process when we are making decisions. Thinking about our own thought should help us identify our inner demons and the impulses that make us magnetically drawn toward specific errors. In return, it should also help us identify someone else's misguided behavior so that we will not be persuaded to follow it.

The first item on the general checklist is that our decision-making process can be hampered by value judgments, which results in overvaluing and undervaluing. We might neglect the intrinsic merit of an idea or a person by focusing on their supposed goodness or badness. The prestige, or lack of it, attached to author or an institution should not affect the assessment of the work itself. Just as the sun's light outshines the dim light of the star, our value judgments can easily overshadow our logical and rational mental processes. This is not to say that we should dispense with making value judgments, but that we must be able to separate the two types of judgment. For example the mystique of CIA and military experts intimidated the Kennedy administration into relinquishing its own judgment when presented with the proposed invasion of Cuba at the Bay of Pigs. As a result, this undertaking was a spectacular failure. Intimidation can also occur due to numerical superiority, the mystique of mathematics, past achievements, institutional affiliation, and other factors. The opposite of this is overconfidence, which can cause us to be hasty and careless not only in our own judgment but also in dismissing the views of others. Thus we should be aware of swinging between these two extremes of insecurity and overconfidence.

The second is directly related to the first: we tend to think less of others when we achieve or possess something positive that somehow places us above them. Known as "the zero-sum impulse," this is the main reason for all types of caste systems, whether they exist in India or within the academy. Such stratified systems are prone to ill-treat those who are not at the top, who are in the minority, and who are weak.

The third is that we have an innate tendency to neglect the middle position in favor of either extreme. Once our value judgment leans toward a specific side, we are magnetically drawn to it. Referred to as the "two valued-impulse," this also applies to certainty judgment. We tend to say absolutely "yes" or "no" in evaluating any possibility. Combining an extreme stance on value judgment and certainty judgment can lead one toward extremist and fanatical positions.

The fourth is that we should be aware of the reactionary impulse. When something negative happens, we are susceptible to adopting the exact opposite position. Plato's totalitarian philosophy, which he developed after the demagogue had condemned his mentor Socrates to death by demagogue, is an example of this reactionary impulse, as are Marx's demonization of capitalism and the anti-religious agenda that arose among the intellectual community during the 18th and 19th centuries.

The fifth is the "incremental descent" (Vaughn, 1996), which we become desensitized through constant repetition resulting failure. Outsiders can play a critical role in recognizing the maladaptive culture that insiders miss or neglect. Thus we should seek their critical input, rather than seeing them as enemies.

The sixth is that we have two entirely separate systems of decision-making: one designed for producing a quick reaction and another that allows us enough time to formulate a leisurely response. Such an arrangement enables us to respond differently according to the degree of urgency that we feel. While the fast system can overtake the slow rational system, a development that has been called "emotional hijacking" (Goleman, 1995), an emotionally hijacked person can convince others to follow the ill-conceived idea that leads to "social hijacking."

The seventh is that the human mind is subject to both gullibility and rigidity when trying to reach a decision. We can easily be convinced to accept any view without much contemplation or examination of alternative possibilities. Once accepted, we can paradoxically become so rigid that no ideas and evidence that might contradict the accepted assumption can penetrate our minds. Rigidity is especially robust when one has publicly expressed his or her commitment.

The eighth is that we are more likely to be persuaded by those who can express themselves in highly emotional and absolute terms than those who are less emotional and more cautious. In addition, we tend to welcome those who use ridicule, such as "a china teapot revolving around the sun" (Russell, 1952), the "flying spaghetti monster" (Wolf, 2006), and the "dragon in my garage" (Dawkins, 2006), to reject God's existence. As Voltaire said, ridicule can overcome almost anything. One of the greatest ironies of human communication is that we are susceptible to following the least intellectually and morally sound person.

The ninth is that we have the impulse to put all of our eggs in one basket, namely, to fully commit to one idea and reject all others. In a sense, we are innately intolerant with ourselves as well as toward others. Those who consent and conform are welcomed; those who dissent are excluded. This is a truly bad way to manage both individual and societal risk.

The tenth is that we see what we expect to see. Something as simple as the perceptual judgment of a line's length can be distorted by a person's expectation (Asch, 1951). In a more ambiguous and complicated situation, expectancy plays an even greater role. One irony is that in this case, the most susceptible people express the highest confidence. In other words, people with high expectancy level have high confidence level. Then, an unfortunate paradox is that we are more likely to be persuaded by those who display high expectation.

The eleventh is the tendency to misrepresent ourselves so that others will accept us. In his *Private Truths, Public Lies,* Kuran (1995) calls this "preference falsification," a phenomenon that can develop into a trapping mechanism. What happened with communism is a good example of this. Solzhenitsyn (1974) voiced his opinion that only the people's pretended enthusiasm for communism and the government held everything together in the Soviet Union.

All of these eleven items can be summed up in just one sentence: We must allow all options to be examined. This is the exactly same principle that the Kennedy administration emphasized after the failed Bay of Pigs invasion. It is so simple but yet so hard to follow both within and among ourselves. The general checklist is the first line of defense against faulty decision-making process. Whenever it is breached, which happens quite frequently, we need another line of defense. This is the subject of the next section.

Applying Institutional Pressure

> If we value the pursuit of knowledge, we must be free to follow wherever that search may lead us. The free mind is not a barking dog, to be tethered on a ten-foot chain.
> -Adlai E. Stevenson, Jr.

> The truth is that all men having power ought to be mistrusted.
> -James Madison

The general checklist discussed above should encourage us to guard against our own and someone else's impulses when faced with having to reach a decision. In addition to impulses, we might also have to deal with a maladaptive cultural inheritance. Suppose we become aware of science's chronic maladaptiveness. How can we bring the necessary change? All cultural inheritances come in two forms: the formal and the informal. The former is seen in institutions, policies, laws, regulations, and other institutions; the latter, however, is intangible: attitudes, beliefs, and values. Consider racial segregation. Jim Crow and apartheid are *de jure* ("concerning the law") practices of segregation that have been formally and legally enacted by the government. Informal or *de facto* ("in practice but not ordained by law") segregation, on the other hand, is intangible and practiced according to the people's attitude and without the existence of any officially established decrees. Formal segregation in America ended with the Civil Rights Act of 1964 as mentioned earlier; however, the people's attitude toward it have proven far more difficult to change.

Unlike *de facto* and *de jure* segregation, the distinction between formal and informal discrimination as regards metaphysics is far less

A Critique of Science

clear. The main reason for this is that the scientific community rejected metaphysics without even engaging in a thoughtful discourse as to why it should do so. This resembles the approach of a demagogue. As this occurred over time, no single event that can be pinpointed as inaugurating this trend, as Numbers (2003) argues:

> No single event marks the transition from godly natural philosophy to naturalistic modern science, but sometime between roughly the mid-eighteenth and mid-nineteenth centuries students of nature in one discipline after another reached the conclusion that, regardless of one's personal belief, supernatural explanations had no place in the practice of science.

The flipping of the then-prevailing value system, in which metaphysical concepts were both prominent and respected, is largely the result of the anti-religious fervor that continues to sweep across the intellectual community. Naturalism, materialism, and positivism were the symbols of intellectual superiority. During the 18th and 19th centuries, the wholesale rejection of supernaturalism was considered a new way of "being cool." Thus the scientific cultural inheritance, which began without a formal declaration, gradually became the enduring and overarching principle that dominates virtually all aspects of the knowledge-seeking process. The fact is that, if the same type of discrimination was practiced in any part of the free world, it surely would be vigorously condemned.

In attempt to rectify the ensuing cultural and institutional maladaptiveness, we can start with Lawrence Kohlberg, a psychologist at the University of Chicago who enumerated the stages of moral development in his 1958 dissertation. He suggests that one's moral development consists of six distinctive stages that continue to grow throughout one's lifetime (Kohlberg, 1958):

1. Obedience and punishment orientation (How can I avoid punishment?).
2. Self-interest orientation (What's in it for me?).
3. Interpersonal accord and conformity (Social norms).
4. Authority and social-order maintaining orientation (Law and order morality).
5. Social contract orientation.

6. Universal ethical principles (Principled conscience).

One obvious implication of his theory is that people's morality does not develop at the same rate. If this is true, at which stage one should seek to induce people to respond to some desired goal? Consider the two political documents during the American Revolution. In the Declaration of Independence, the famous sentence "We hold these truths to be self-evident, that all men are created equal, that they are endowed by their Creator with certain unalienable rights, that among these are life, liberty and the pursuit of happiness." These words were designed to inspire the American colonists to revolt against British rule by appealing to the highest stage: Universal ethical principles. Whereas, the American Constitution concerns itself with finding a way to balance the dangers of too much people power and too much government power. James Madison, who is known as the godfather of the Constitution, addresses the essence of what this document, one of the most important political documents in human history, is trying to achieve:

> If men were angels, no government would be necessary. If angels were to govern men, neither external nor internal controls on government would be necessary. In framing a government which is to be administered by men over men, the great difficulty lies in this: You must first enable the government to control the governed; and in the next place, oblige it to control itself....the constant aim is to divide and arrange the several offices in such a manner as that each may be a check on the other; that the private interest of every individual, may be a sentinel over the public rights. These inventions of prudence cannot be less requisite in the distribution of the supreme powers of the state (Madison in the *Federalist Papers*, Number 51).

In adopting the caution raised by Montesquieu and Locke, Madison's intention is to prepare for the worst by inaugurating the institutional system that can prevent (or balance) from two vices: tyranny (too much government power) and anarchy (too much people power).

Like the two political documents that set examples for other nations to emulate and adopt, science may also needs two separate approaches: one to appeal universal ethical principles and another is to find ways to institutional pressure. Robert Merton (1973) put forward what he calls the "ethos of science." This ethical principle emphasizes universalism, communalism, disinterestedness, and organized skepticism. Since scientists are no angels and they have impulses and biases just like the rest of us, institutional pressure needs to be applied to them as well. At the same time, we must guard to against impinging on the academic freedom.

Establishing ways to put institutional pressure is critical because science exempted from overseeing. Most domains of modern society have intellectual watchdogs who can assess and critique what they see going on around them. Typically, media play a critical role in this endeavor. However, science is usually excluded from the media's attention given that it usually endorses virtually everything that scientists, both as individuals and as a discipline, say. Therefore, science's maladaptive cultural inheritance endures because it is free of any oversight. Aronowitz (1996) has pointed out the media's abdication of its role in this instance:

> Most science writers and journalists are the willing supplicants of a scientific establishment which passes down authoritative news and opinion about the successes of science, successfully manages its failures and, perhaps most important, marginalizes or silences alternative science both at the level of explanation and at the level of discovery.

The role of intellectual watchdog is supposedly performed by the "philosophy of science," "sociology of scientific knowledge," "science and technology studies," "history and philosophy of science," and other disciplines. However, such disciplines are often seen as the enemies, as opposed to the overseers, of science. Perhaps it's because science has been spoiled with praises and accolades from positivism that elevates science as something above and beyond other disciplines.

Prevention versus Treatment

> An ounce of prevention is worth a pound of cure.
> -Henry de Bracton

Orogee Dolsenhe

> Injustice anywhere is a threat to justice everywhere.
> -Martin Luther King, Jr.

> Our civilization is still in a middle stage, scarcely beast, in that it is no longer guided by instinct, scarcely human in that it is not yet wholly guided by reason.
> -Theodore Dreisle

Humanity's first global conflict, World War I, involved some 30 countries with 65 million men mobilized across five continents. President Woodrow Willson coined the phrase "the war to end all wars" in an attempt to justify America's entrance into the essentially European conflict. Ironically, however, it has been credited by some with paving the way to yet another global war. The Treaty of Versailles required Germany to accept much of responsibility for the conflict and to make restitution by ceding territory and paying an enormous amount of money for the next almost 70 years. The ensuing resentment became the rallying point for mobilizing the German masses and fueling Hitler's rise to power. Even more than the Treaty of Versailles, Hitler presented the Jews as a scapegoat for the nation's defeat and his effective hate speech enabled the nation to become the instrument of genocide. Its main victims were 6 million Jews, about two-thirds of Europe's total Jewish population. In addition, 5 to 11 million other people—Gypsies, Poles, Soviet prisoners of war, people with disabilities, political and religious opponents—also perished. The lesson here is that one person's hatred and insanity can be transmitted to the masses so successfully that the entire society essentially becomes an extension of a deranged mind. The famous phrase "never again," coined by the Soviet officer in command of the troops who liberated Auschwitz, symbolizes the resolution that the world should never allow such a horror happen again. And yet we continue to witness such horrors over and over again: Cambodia, Rwanda, Kosovo, and Darfur, to name just a few. Behind these ever-recurring cataclysms are those who can devise yet another justification for denying the basic rights of an individual or a group of people.

The challenge to dealing with humanity's destructive potential is similar to confronting the problems of healthcare. The former deals with the world's communal welfare; the latter deals with an individual's health. Both of them also share to major approaches: treatment and prevention. The healthcare profession suffers from narrow-mindedness and short-sightedness, which results in the avoidance of any preventative measures being devised and implemented. The vast majority of the world's healthcare costs can be attributed to treatment after the person becomes sick. For instance, America spends only about 5% or less of its healthcare budget on prevention, despite the profession's almost unanimous agreement that prevention is the most cost-effective approach.

Prevention may also be the most efficient way to avert man-made cataclysms. However, for some reason this approach is virtually nonexistent in such instances. But could there even be a preventive measure when events seem to be heading in this direction? My suggestion is quite simple: Knowledge must be accompanied by metaknowledge, which can serve as an operating manual during the knowledge-seeking process. It should restrain our impulses and point out our blind spots that we can avoid making flawed decisions and justifying discriminatory practices. This manual should remind us of the fact that all options must be thoughtfully and thoroughly examined both when seeking knowledge and when formulating decisions. It should remind us to remain open-minded so that all options can be given fair and equal hearing. If we conduct our own decision-making process with these points in mind, such basic social issues as fairness, equality, and mutual respect would automatically follow.

The scientific process is not fundamentally different from the decision-making process; rather, it is a matter of being open-minded enough so that one can judge all ideas fairly and equally based upon their intrinsic merits. Unfortunately, however, discrimination and intolerance among scientists are not only allowed – they are actually encouraged, especially in various crucial fields. Given that the scientific community has developed an elaborate system to justify this inequality and injustice, science paradoxically may be one of the most backward, if not *the* most backward, communities when it comes to the issue of basic rights.

Science can and must lead the world in providing a neutral space in which all kinds of ideas can compete with each other. It must set an example for the rest of us to follow. But it can only fulfill this responsibility by conducting a thorough and bias-free examination of itself. The goal of such self-examination is not just to benefit science, but to benefit all other fields as well. Moreover, the insights gained from this procedure should be taught as early as possible. Metaknowledge should be a required course, along with math and science, beginning with elementary school. The emphasis on metaknowledge in education may help us to identify inequality and injustice that can curve bullying that can occur in school, in neighborhoods, in workplace, in politics, and even in the international community. Such an approach just might be the most effective way to prevent the world's slide into yet another man-made disaster. A clear red flag that we are headed down this same self-destructive path yet again is that someone may find a way to justify the violation the basic rights and minimum amount of respect to which human being is entitled.

In addition, this approach should help us develop better relationships with our friends, families, coworkers, and neighbors by reducing the tensions that arise from impulsive decisions and misunderstanding. Since metaknowledge helps us make better decisions, society should use it to prevent future horrors from arising. When we start to emphasize metaknowledge, humanity will finally escape from its sophomoric phase and enter a more mature stage as regards both seeking knowledge and building social relationship. Only then will we truly become "Homo sapiens" – the "wise man," rather than "sophomore" – "wise fool."

References

Abra, J. (1998). *Should psychology be a science?: pros and cons.* Westport, Conn: Praeger.

Ahnadhuabm V. (1966). Artificial reestablishment of the lichen *Cladonia cristalla. Science*, 151, 199-201.

Al-Fasi, Muhammad ibn Ahmad (1990). Muhammad al-Murad. ed (in Arabic). *Dhayl al-Taqyid.* 3 (1 ed.). Mecca, Saudi Arabia: Umm al-Qura University.

Altemeyer, B. (1996). *The authoritarian specter.* Cambridge, MA: Harvard University Press.

Arieti, J. A., & Wilson, P. A. (2003). *The scientific and divine: Conflict and reconciliation from ancient Greece to the present.* Lanham, MD: Rowman & Littlefield.

Arjomand, S. A. (1999). The law, agency, and policy in medieval Islamic society: Development of the institutions of learning from the tenth to the fifteenth century. *Comparative Studies in Society and History*, 41, 263-293.

Aronowitz, S. (1996). The politics of the science wars. In A. Ross (ed.), *Science wars*, Durham: Duke University Press.

Aronson, E. (1999). *Social Animals.* New York: Worth Publishers.

Asch, S. E. (1951). Effects of group pressure upon the modification and distortion of judgment. In H. Guetzkow (ed.) *Groups, leadership, and men.* Pittsburgh, PA: Carnegie Press.

Augros, R., & Stanciu, G. (1987). *The new biology: Discovering the wisdom in nature.* Boston: New Science Library.

Baars, B. J. (1986). *The cognitive revolution in psychology.* New York: Guilford Press.

Baars, B. J. (2003). The Double Life of B.F. Skinner: Inner Conflict, Dissociation and the scientific taboo against consciousness. *Journal of Consciousness Studies*, 10, 5-25.

Baars, B. J. (2003). Reply to commentators. *Journal of Consciousness Studies*, 10, 79-94.

Bacon, F. (1960/1620). *The new organon and related writings.* New York: Liberal Arts Press.

Bala, A. (2006). *The dialogue of civilizations in the birth of modern science.* New York: Palgrave Macmillan.

Balashov, Y., & Rosenberg, A. (2002). *Philosophy of science:*

Contemporary readings. London: Routledge.
Barber, B. (1961). Resistance by scientists to scientific discovery. *Science,* 134, 596-602.
Barlett, D. L., & Steele, J. B. (1979). *Empire: The life, legend, and madness of Howard Hughes.* New York: Norton.
Barr, A. P). (1997). *The major prose of Thomas Huxley.* Athens, GA: University of Georgia Press.
Barry, D. H. (1998). Interpretive essay. In F. W. Thackery, & J. E. Findling (eds.), *Events that changed the world in the eighteenth century.* Westport, CT: The Greenwood Press.
Bartolomeo, P. (2007). Visual neglect. *Current Opinion in Neurology,* 20, 381-386.
Barzun, J. (1941). *Darwin, Marx, Wagner: critique of a heritage.* New York: Doubleday.
Beddall, B. G. (1968). Wallace, Darwin and the theory of natural selection. *Journal of the History of Biology,* 1, 261-324.
Behe, M. J. (1996). *Darwin's black box: the biochemical challenge to evolution.* New York: Free Press.
Behe, M. J. (2004). Irreducible complexity: Obstacle to Darwinian evolution. In W. A. Dembski, & M. Ruse (eds.), *Debating design: From Darwin to DNA.* Cambridge, UK: Cambridge University Press.
Bem, D. J. & McConnell, H. K. (1970). Testing the self-perception explanation of dissonance phenomena: On the salience of premanipulation attitude. *Journal of Personality and Social Psychology,* 14, 31.
Ben-David, J. (1971). *The scientist's role in society: A comparative study.* Englewood Cliffs, NJ: Prentice-Hall.
Berk, R. (1974). A gaming approach to crowd behavior. *American Sociological Review,* 39, 355-373.
Bertalanffy, L. von (1969). *General systems theory.* New York: George Braziller.
Béteille, A. (1969). *Caste, class, and power: Changing patterns of stratification in a Tanjore Village.* Berkeley, CA: University of California Press.
Bethell, T. (1976). Darwin's Mistake. *Harper's Magazine,* February, 70-75.
Bird, R. J. (2003). *Chaos and life: Complexity and order in evolution and*

thought. New York: Columbia University Press.
Blomberg, S. P., & Garland, Jr. T. (2002). Tempo and mode in evolution: phylogenetic inertia, adaptation and comparative method. *Journal of Evolutionary Biology*, 15, 899-910.
Blum, H. (1951). *Time's arrow and evolution*. Princeton: Princeton University Press.
Bodde, D. (1991). *Chinese thought, society, and science*. Honolulu: University of Hawaii Press.
Bossomaier, T. & Green, D. (1998). *Patterns in the sand*. Reading, MA: Helix Books.
Bouglé, C. (1908). *Essais sur le régime des castes*. Paris: Alcan.
Bowler, P. (1983). *The eclipse of Darwinism: Anti-Darwinian evolution theories in the decades around 1900*. Baltimore: Johns Hopkins University Press.
Bowler, P. (1988). *The non-Darwinian revolution: Reinterpreting a historical myth*. Baltimore, MD: Johns Hopkins University Press.
Bowler, P. (1989). *The Mendelian revolution: The emergence of hereditarian concepts in modern science and society*. Baltimore, MD: The Johns Hopkins University Press.
Bowler, P. J. (1997). Thomas Henry Huxley and the reconstruction of life's ancestry. In A. P. Barr (ed.), *Thomas Huxley's place in science and letters: Centenary essays*. Athens, GA: The University of Georgia Press.
Boyce W.T. (2004). Social stratification, health, and violence in the very young. *Annals of the New York Academy of Sciences*, 1036:47–68.
Brackman, A. C. (1980). *A delicate arrangement: The strange case of Charles Darwin and Alfred Russel Wallace*. New York: Times Book.
Breder, Jr. C. M. (1944). Ocular anatomy and light sensitivity studies on the blind fish from Cueva de los Sabinos, Mexico. *Zoologica*, 29, 131-143.
Brewer, W. H. (1873). Anticipations of natural philosophy: Maupertuis. *Nature*, 7, 402.
Brewster, D. (1855). *Memoir of the life, writings, and discoveries of Sir Isaac Newton*, Edinburgh.
Brinton, C. (1965). *The anatomy of revolution*. New York: Vintage Books.

Broman, T. H. (2003). Matter, force, and the Christian worldview in the Enlightenment. In Lindberg, D. C., & Numbers, R. L. (eds.), *When science & Christianity meet*. Chicago: University of Chicago Press.

Brooke, J. H. (1991). *Science and religion: Some historical perspectives*. Cambridge, UK: University of Cambridge Press.

Brooks, D. R. (1983). What's going on in evolution? A brief guide to some new ideas in evolutionary theory. *Canadian Journal of Zoology*, 12, 32-47.

Brooks, J. L. (1984). *Just before the origin: Alfred Russel Wallace's theory of evolution*. New York: Columbia University Press.

Browne, E. Janet (1995), *Charles Darwin: vol. 1 Voyaging*. London: Jonathan Cape.

Bruner, J. S., & Postman, L. (1949). On the perception of incongruity: a paradigm. *Journal of Personality*, 18, 106-223.

Buckley, M. J. (1987). *At the origins of modern atheism*. New Haven, CT: Yale University Press.

Bullock, A. (1964). *Hitler: A study in tyranny*. London: Harper & Row.

Burian, R. (1988). Challenges to the evolutionary synthesis. *Evolutionary Biology*, 23, 247-269.

Burke, E. (1973). *Reflections on the revolution in France*. Garden City, NY: Anchor Press (Original work published in 1790).

Burkhardt, R. W. (1977). *The spirit of system: Lamarck and evolutionary biology*. Cambridge, MA: Harvard University Press.

Burkholder, P. R. (1965). Cooperation and conflict among primitive organisms. In Edward J. Kormondy (ed.), *Reading in ecology*. Englewood Cliffs, NJ: Prentice-Hall.

Burt, D. B. (2001). Evolutionary stasis, constraints and other terminology describing evolutionary patterns. *Biological Journal of Linnean Society*, 72, 509-517.

Butler, S. (1924/1887). *Luck, or cunning, as the main means of organic modification?* New York: Ames Press.

Burtt, E. A. (1924). *The metaphysical foundations of modern physical science: A historical and critical essay*. London: Routledge and Kegan Paul.

Busky, D. F. (2002). *Communism in history and theory: From utopian socialism to the fall of the Soviet Union*. Westport, CT: Praeger.

Byrne, J. M., & Gates, R. D. (1987). Single-case study of left cerebral hemispherectomy: Development in the first five years of life. *Journal of Clinical and Experimental Psychology*, 9, 423-434.

Cairns, J., Overbaugh, J., & Miller, S. (1988). The origin of mutants. *Nature*, 335, 142-145.

Campanini, M. (1996) *'Al-Ghazzali'*, In S.H. Nasr and O. Leaman (eds), *History of Islamic Philosophy*. London: Routledge.

Campbell, D. T. (1979). A tribal model of the social system vehicle carrying scientific knowledge. *Knowledge: Creation, Diffusion, Utilization.* 1: 181-201.

Campbell, J. H. (1985). An organizational interpretation of evolution. In D. J. Depew & B. H. Weber (eds.) *Evolution at crossroads: The new biology and the new philosophy of science*. Cambridge, MA: A Bradford Book.

Caporale, L. H. (2006). An overview of the implicit genome." In Caporale, L. H. (ed.), *The implicit genome*. New York: Oxford University Press.

Carlyle, T. (1888). *On heroes, hero-worship and the heroic in history*. New York: Fredrick A. Stokes & Brother.

Castle, W. E. (1905). The mutation theory of organic evolution from the stand point of animal breeding. *Science*, 21, 522.

Cattan, H. (1955). The law of the Waqf. In Law in the Middle East (Vol. 1), *Origin and development of Islamic law*, Ed Majid, Herbert J. Liebesny Khadduri (Vol. 1). Washington D.D.: The Middle East Institute.

Caudill, E. (1997). *Darwinian myths: the legends and misuses of a theory*. Knoxville, TN: University of Tennessee Press.

Ceccarelli, L. (2001). *Shaping science with rhetoric*. Chicago: University of Chicago Press.

Chaitin, G. J. (1975). Randomness and mathematical proof. *Scientific American*, 232, 47-52.

Chomsky, N. (1959). A review of B. F. Skinner's verbal behavior. *Language*, 35, 26-58.

Chomsky, N. (1965). *Aspects of the theory of syntax*. Cambridge, MA: MIT Press.

Cicchetti, D. V. (1982). On peer review: "We have met the enemy and he is us" *Behavioral and Brain Sciences*, 5, 205.

Collins, H. & Pinch, T. (1993). *The Golem: what you should know about science*. Cambridge, UK: Cambridge University Press.

Colp, Jr. R. (1980). 'I was born a naturalist': Charles Darwin's 1838 notes about himself. *Journal of the History of Medicine and Allied Sciences*, 35, 8-39.

Cope, E. D. (1896). *The primary factors of organic evolution*. Chicago: Open Court.

Costall, A. (2006). 'Introspectionism' and the mythical origins of scientific psychology. *Consciousness and Cognition*, 15, 634-654.

Cowan, N. (2001). The magical number 4 in short-term memory: A reconsideration of mental storage capacity. *Behavioral and Brain Sciences*, 24, pp. 1–185.

Cox, O. C. (1970). *Caste, class, & race; a study in social dynamics*. New York: Monthly Review Press.

Crick, F. (1966). *Of molecules and men*. Seattle, WA: University of Washington Press.

Crick, F. (1981). *Life itself: Its origin and nature*. New York: Simon & Schuster.

Crick, F. (1994). *The astonishing hypothesis: the scientific search for the soul*. New York: Scribner.

Cullis, C. A. (1988). Control of variation in higher plants. In M. W. Ho, & S. Fox (eds.), *Evolutionary processes and metaphors*. New York: John Wiley & Sons Ltd.

Cummins D. D. How the social environment shaped the evolution of mind. *Synthese*. 2000;122:3–28.

Cuvier, G. (1835). Éloge de M. Lamarck, *Mémoires de l'Académie Royale des Sciences de l'Institut de France*, 13, i-xxxi.

Damasio, A. R. (1994). *Descartes' error: emotion, reason, and the human brain*. New York: Avon Books.

Damasio, A. (2003). *Looking for Spinoza: Joy, sorrow, and the feeling brain*. Orlando, FL: Harcourt.

Darlington, C. D. (1959). *Darwin's place in history*. Oxford: Basil Blackwell.

Darwin, C. (1876). *The origin of species* (6th edition). London: John Murray.

Darwin, F. (1892). *The life and letters of Charles Darwin*. New York: D. Appleton and Company.

Daston, L. J. (1982). The theory of will versus the science of mind. In W. R. Woodward, & M. G. Ash (eds.) *The problematic science:*

Psychology in nineteenth-century thought. New York: Praeger.
Davidson, D. (1985). Deception and division. In E. Lepore & B. McLaughlin (eds.), *Actions and events*. New York: Basil Blackwell.
Davies, E. B. (2003). *Science in the looking glass: What do scientists really know?* Oxford: Oxford University Press.
Davies, P. (1988). *The cosmic blueprint: New discoveries in nature's creative ability to order the universe*. New York: Simon and Schuster.
Davies, R. (2008). *The Darwin Conspiracy: Origins of a Scientific Crime*. London: Golden Square Books.
Davis, K. & Moore, W. E. (1945). Some principles of stratification. *American Sociological Review*, 10, 242-249.
Dawkins, R. (1976). *The selfish gene*. Oxford: Oxford University Press.
Dawkins, R. (2001). Ignorance is no crime. *Free Inquiry*, 21, no.3.
Dawkins, R. (2006). *The God delusion*. Boston: Houghton Mifflin Co.
Day, P., & Ulatowska, H. (1979). Perceptual, cognitive, and linguistic development after early hemispherectomy: Two case studies. *Brain and Language*, 7, 17-33.
Deason, G. B. (1986). Reformation theology and the mechanistic conception of nature. In D. C. Lindberg, & R. L. Numbers (eds.), *God and nature: Historical essays on the encounter between Christianity and science*. Berkeley, CA: University of California Press.
Dennett, D. (1995). *Darwin's Dangerous Idea: evolution and the meaning of life*. New York: Simon & Schuster.
Dennis, M., & Kohn, B. (1975). Comprehension of syntax in infantile hemiplegics after cerebral hemidecortication: Left hemisphere superiority. *Brain and Language*, 2, 475-486.
Denton, M. (1986). *Evolution: A theory in crisis*. Bethesda, Maryland: Adler & Adler.
Depew, D. J., & Weber, B. H. (1985). Introduction. In D. J. Depew & B. H. Weber (eds.) *Evolution at crossroads: The new biology and the new philosophy of science*. Cambridge, Mass.: A Bradford Book.
Depew, D. J., & Weber, B. H. (1988). Consequences of nonequilibrium thermodynamics for the Darwinian tradition. In B. H. Weber, D. J. Depew, & J. D. Smith (eds.), *Entropy, information, and*

evolution: Perspectives on physical and biological evolution. Cambridge, MA: MIT Press.

Descartes, R. (1637/1960). *Discourse on method and meditations* (trans. Laurence J. Lafleur). New York: Liberal Arts Press.

Desmond, A. (1982). *Archetypes and ancestors: Paleontology in Victorian London 1850-1875.* Chicago: University of Chicago Press.

D'Espagnat, B.(1979). The quantum theory and reality. *Scientific American*, November, 158-181.

Dhanani, A. (2002). The Copernican revolution. In Gary B. Ferngren, (ed.), *Science and religion*, Baltimore, MD: Johns Hopkins University Press.

Diamond, J. (1997). *Guns, germs, and steel: The fates of human societies.* New York: W.W. Norton & Co.

Diamond, J. (2005). *Collapse: how societies choose to fail or succeed.* New York: Viking.

Diderot, D. (1777). *Conversations with a Christian Lady.*

Dobzhansky, T. (1968). Adaptedness and fitness. In R. C. Lewontin (ed.), *Population biology and evolution.* New York: Syracuse University Press.

Dobzhansky, T. (1973). Nothing in biology makes sense except in the light of evolution. *American Biology Teacher*, 35, 125-129.

Dobzhansky, T. (1937). *Genetics and the origin of species.* New York: Columbia University Press.

Doise, W. (1969). Intergroup relations and polarization of individual and collective judgments. *Journal of Personality and Social Psychology*, 12, 136-143.

Dolittle, W. F. (2005). If the tree of life fell, would we recognize the sound? In J. Sapp (Ed.) *Microbial phylogeny and evolution: Concepts and controversies.* Oxford: Oxford University Press.

Dolsenhe, O. (2000). *Coherent nature: The structure and process of consciousness.* Baltimore: American Literary Press.

Dolsenhe, O. (2005). *A genuine theory of everything: Explaining consciousness through a unification of psychology, physics, and biology.* Houston: Ebookstand.

Douady, S. & Couder, Y. (1992). Phyllotaxis as a physical self-organized growth process. *Physical Review Letters*, 68, 2098-2101.

Douglas, H. E. (2009). *Science, policy, and the value-free ideal.*

Pittsburgh, PA: University of Pittsburgh Press.
Douglas, M. (1966). *Purity and danger*. New York: Routledge.
Doyle, J., Csete, M., & Caporale, L. (2006). An engineering perspective: The implicit protocols. In Caporale, L. H. (ed.), *The implicit genome*. New York: Oxford University Press.
Drake, S. (1957). *Discoveries and opinions of Galileo*. Garden City, NY: Doubleday.
Drosnin, M. (1985). *Citizen Hughes*. New York: Holt, Rinehart and Winston.
Dugatkin, L. A. (2006). *The altruistic equation: Seven scientists search for the origin of goodness*. Princeton, NJ: Princeton University Press.
Duhem, P. (1996). *Essays in the history and philosophy of science* (trans. Roger Ariew & Peter Barker). Indianapolis: Hackett Publication.
Dumont, L. (1966). *Homo Hierarchicus: The caste system and its implications*. Chicago: University of Chiago Press.
Dumontheil, I., Burgess, P. W. & Blakemore, S. J. (2008). Development of rostral prefrontal cortex and cognitive and behavioural disorders. *Developmental Medicine and Child Neurology*, 50, 168-181.
Dunn, S. (1999). *Sister revolutions: French lightning, American light*. New York: Faber and Faber.
Edgerton RB (1992) Sick societies: Challenging the myth of primitive harmony. New York: Free Press.
Ebenstein, W. (1951). *Great political thinkers: Plato to the present*. Fort Worth, TX: Holt, Rinehart and Winston.
Eddington, A. (1928/1958). *The nature of the physical world*. Ann Arbor, MI: University of Michigan Press.
Edis, T. (2006). *Science and nonbelief*. Westport, CN: Greenwood Press.
Eigen, M. (1992). *Step toward life*. Oxford: Oxford University Press.
Einstein, E. (1934). Autobiographical notes. In P. A. Schilpp (ed.), *Albert Einstein: Philosopher-scientist*. New York: Tudor.
Eiseley L. (1959). Charles Darwin, Edward Blyth, and the theory of natural selection. *Proceedings of the American Philosophical Society*, 103:94–114.
Eiseley, L. (1961). *Darwin's century: Evolution and the men who discovered it*. New York: Doubleday.
Eiseley, L. (1967). *Thomas Henry Huxley, On a piece of chalk*. New York:

Oriole Editions.
Eldridge, N. (1995). *Reinventing Darwin: The great debate at the high table of evolutionary theory*. New York: Wiley.
Eldridge, N. & Gould, S. J. (1972). Punctuated equilibria: An alternative to phyletic gradualism. In T. J. M. Schopf (ed.), *Models in paleobiology*. San Francisco: Freeman, Cooper and Co.
Elster, J. (1999). *The alchemies of the mind: Rationality and the emotions*. Cambridge, UK: Cambridge University Press.
Emery, A. E. (1988). Pierre Louis Moreau de Maupertuis (1698-1759). *Journal of Medical Genetics*, 24, 561-564.
Evans, R. (1993). *Deng Xiaoping and the making of modern China*. London: Penguin Books.
Fairbanks, J. K., & Goldman, M. (2006). *China: A new history*. Cambridge, MA: Belknap Press.
Fakhry, M. (2004). *A history of Islamic philosophy*. New York: Columbia University Press.
Farley, J. (1977). *The spontaneous generation controversy from Descartes to Oparin*. Baltimore: The Johns Hopkins University Press.
Feinberg, J. (1985). *Offense to others: The moral limits of the criminal law*. Oxford: Oxford University Press.
Festinger, L. (1954). A theory of social comparison processes. *Human Relations*, 7, 117-140.
Feyerabend, P. (1988). *Against method*. New York: Verso.
Feynman, R. (1999). *The pleasure of finding things out: The best short works of Richard P. Feynman* (edited transcript of an interview with Feynman made for BBC television program Horizon in 1981). Cambridge, MA: Perseus Books.
Fick, R. (1897). *Die social Gliederung im Nordöstlichen Indien zu Budda's Zeit mit besonder Berücksichtigung der Kastenfrage*. Kiel, (trans. M. Mitra, Calcutta, 1920).
Fischer, K. (2008). Science and its malfunctions. *Human Architecture: Journal of the Sociology of Self-Knowledge*, 6, 1-22.
Fisher, M. J. (1952). Communist doctrine and the free world: The ideology of Communism according to Marx, Engels, Lenin, and Stalin. New York: Greenwood Press.
Fisher, R. A. (1934). Adaptations and mutations. *The School Science Review*, 15, 294-301.

Fisher, R. A. (1958). *The genetical theory of natural selection* (second revised edition). New York: Dover Publication.
Flavell, J. (1976). Metacogntive aspects of problem-solving. In L. Resnick (Ed.), *The nature of intelligence*. Hillsdale, NJ: Erlbaum Assoc.
Flexner, S. & Flexner, D. (1993). *Wise words and wives' tales: The origins, meanings and time-honored wisdom of proverbs and folk saying olde and new*. New York: Avon Books.
Forrest, B., & Gross, P. R. (2004). *Creationism's Trojan horse: the wedge Of intelligent design*. New York: Oxford University Press.
Freeman, C. (2004). *The closing of the western mind: The rise of faith and the fall of reason*. New York: Alfred A. Knopf.
Freeman, D. H. (2010). *Wrong: Why experts keep failing us—and how to know when not to trust them*. New York: Little, Brown and Co.
Fu, Z. (1996). *China's legalists: The earliest totalitarians and their art of ruling*. Artmonk, N.Y.: M.E. Sharpe.
Fuller, S. (2004). *Kuhn vs. Popper: the struggle for the soul of science*. New York: Columbia University Press.
Fuller, S. (2007). *Science vs religion?: Intelligent design and the problem of evolution*. Cambridge, UK: Polity.
Galileo Galileo, (1989). Letters to the Grand Duchess Christina. In M. Finocchiaro, *The Galileo affair: A documentary history*. Berkeley and Los Angeles: University of California Press.
Gardner, H. (1985). *The mind's new science*. New York: Basic Books.
Gaukronger, S. (2006). *The Emergence of a Scientific Culture: Science and the Shaping of Modernity 1210-1685*. Oxford University Press. 2
Gawande, A. (2010). *The checklist manifesto: How to get things right*. New York: Metropolitan Books.
Gay, P. (1977). The Enlightenment: An interpretation, vol. 1, *The rise of modern paganism*. New York: Norton.
Ghirardi, G. (2005). *Sneaking a look at God's cards: Unraveling the mysteries of quantum mechanics* (trans. By Gerald Malsbary). Princeton, NJ: Princeton University Press.
Ghurye, G. S. (1969). *Caste and race in India*. Bombay: Popular Prakashan.
Gibbon, E. (1761). *Essai sur l'étude de la literature*. London.
Gigerenzer, G., Hertwig, R., van den Broek, E., Fasolo, B.,

Katsikopoulos, K. V. (2005). "A 30% chance of rain tomorrow": How does the public understand probabilistic weather forecasts? *Risk Analysis*, 25, 623-620.

Gilham, N. W. (2001). Evolution by jumps: Francis Galton and William Bateson and the mechanism of evolutionary change. *Genetics*, 159, 1383-1392.

Glass, B. (1959). Maupertuis, pioneer of genetics and evolution. In Glass B., Temkin, O. & Strauss, W. L. Jr. (eds.), *Forerunners of Darwin, 1745-1859*. Baltimore, MD: Johns Hopkins Press.

Goldschmidt, R. (1940). *The material basis of evolution*. New Haven: Yale University Press.

Goldschmidt, R. B. (1960). *In and out of the ivory tower*. Seattle: University of Washington Press.

Goldsmith, E. (1990). Evolution, neo-Darwinism and the paradigm of science. *Ecologist*, 20, 67-73.

Goldstein, M., & Goldstein, I. F. (1981). *How we know: an exploration of the scientific process*. New York: Plenum Press.

Goldstein, R. (2006). *Betraying Spinoza: The renegade Jew who gave us modernity*. New York: Nextbook.

Goldziher, I. (1981). The attitude of orthodox Islam toward the ancient sciences. In M. Swartz (ed.), *Studies in Islam*. New York: Oxford University Press.

Goleman, D. (1995). *Emotional intelligence*. New York: Bantam Books.

Gonzalez, G., & Richards, J. W. (2004). *The privileged planet: how our place in the cosmos is designed for discovery*. Washington DC: Regnery Publication.

Goodwin, B. (1976). *Analytical physiology of cells and developing organisms*. London: Academic Press.

Goodwin, B. (1994). *How the leopard changed its spots*. New York: Charles Scribner's Sons.

Goonatilake, S. (1998). *Toward a global science: mining civilizational knowledge*. Bloomington, IN: Indiana University Press.

Gould, S. J. (1977). *Ever since Darwin: reflexion on natural history*. New York: W.W. Norton Co.

Gould, S. J. (1980). Is a new and general theory of evolution emerging? *Paleobiology*, 6, 119-130.

Gould, S. J. (1981). But not Wright enough: Reply to Orzack'. *Paleobiology*, 7, 131-139.

Gould, S. J. (1982). Darwinism and the expansion of evolutionary theory. *Science*, 216, 380-387.
Gould, S. J. (1991). *Bully for brontosaurus: reflections in natural history*. New York: Norton.
Graham, A. C. (1978). *Later Mohist logic, ethics and science*. Hong Kong: Chinese University Press.
Grant, E. (1986). Science and theology in the Middle Ages. In D. C. Lindberg, & R. L. Numbers (eds.), *God and nature: Historical essays on the encounter between Christianity and science*. Berkeley, CA: University of California Press.
Grant, E. (1996). *The foundations of modern science in the Middle Ages: Their religious, institutional, and intellectual contexts*. Cambridge, UK: Cambridge University Press.
Grant, E. (2002). Aristotle and Aristotelianism. In G. B. Ferngren (ed.), *Science and religion: A historical introduction*. Baltimore: Johns Hopkins University Press.
Greenberg, D. S. (2001). *Science, money, and politics: Political triumph and ethical erosion*. Chicago: University of Chicago Press.
Greenwald, G. (2007). *A tragic legacy: How a good vs. evil mentality destroyed the Bush presidency*. New York: Crown Publishers.
Greenwood, P. H. (1981). Species-flocks and explosive radiation. In P. H. Greenwood (Ed.), *The evolving biosphere*. Cambridge: Cambridge University Press.
Gregory, M. E. (2007). *Diderot and the metamorphosis of species*. New York: Routledge.
Grene, M., & Depew, D. (2004). *The philosophy of biology: An episodic history*. Cambridge, UK: Cambridge University Press.
Grossman, R. H. S. (1937). *Plato today*. New York: Oxford University Press.
Gruber, J. W. (1960). *A conscience in conflict: The life of St. George Jackson Mivart*. Westport, CT: Greenwood Press.
Guyénot, E. (1941). *Les sciences de la vie aux XVIIe et XVIIIe siècles: l'idée d'évolution, Albin Michel, Paris*.
Haber, F. C. (1959). Fossils and the idea of a process of time in a natural history. In B. Glass, O. Temkin, & W. L. Strauss, Jr. (eds.), *Forerunners of Darwin: 1745-1859*. Baltimore, MD: Johns Hopkins Press.
Haggbloom, S. J., Harnick, R., Warnick, J. E., Jones, V. K., Yarbrough, G.

L., Russell, T. M. (2002). The 100 most eminent psychologists of the 20th century. *Review of General Psychology*, 6, 139-152.
Haldane, J. B. S. (1935). Darwinism under revision. *Rationalist Annual*.
Hall, B. G. (1988). Adaptive evolution that requires multiple spontaneous mutations. I. Mutations involving an insertion sequence. *Genetics*, 120, 887-897.
Hall, B. G. (1991). Adaptive evolution that requires multiple spontaneous mutations: mutations involving base substitutions. *Proceedings of the National Academy of Sciences, USA*, 88, 5882-5886.
Hall, J. L. (2003). Columbia and Challenger: organizational failure at NASA. *Space Policy*, 19, 239–247.
Hall, R. (1962). "General Introduction" to Marie boas Hall's *The Scientific Renaissance: 1450-1630*. London: Collins.
Hanawalt, P. C. (1989). Preferential repair of damage in actively transcribed DNA sequences *in vivo*. *Genome*, 31, 605-611.
Hardy, A. (1965). *The living stream*. London: Collins.
Harman, W. W. & Sahtouris, E. (1998). *Biology revisioned*. Berkeley: North Atlantic Books.
Hart, H. H. (1952). Masochism, passivity, and radicalism. *Psychoanalytic Review*, October, 1952.
Hartle, A. (2003). *Michel de Montaigne: accidental philosopher*. New York: Cambridge University Press.
Hartley, D. (1749). *Observations on man*. London.
Haskins, C. H. (1927). *The Renaissance of the twelfth century*. Cambridge, MA: Harvard University Press.
Haught, J. (2000). *God after Darwin: A theology of evolution*. Boulder, CO: Westview Press.
Hayakawa, S. I. (1941). *Language in action*. New York: Harcourt, Brance and Company.
Hayakawa, S. I., & Hayakawa, A. R. (1990). *Language in thought and action*. San Diego, CA: Harcourt Brace Jovanovich.
Highkin, H. R. (1958). Temperature-induced variability in peas. *American Journal of Botany*, 45, 626-631.
Hyatt, A. (1882). Transformation of Planorbis at Steinheim, with remarks on the effects of shell and animals. *American Naturalist*, 16, 441-453.
Hellman, H. (1998). *Great feuds in science: Ten of the liveliest disputes ever*. New York: John Wiley and Sons.

Henderson, B. (2005). *Open Letters to Kansas School Board.*
Heuss, T. (1932). *Hitlers Weg: Eine historisch-politische Studies über den Nationalsozialismus.* Berlin: Stuttgart.
Himmelfarb, G. (1959). *Darwin and Darwinian revolution.* New York: W.W. Norton.
Hinson, E. G. (1996). *The early church: Origins to the dawn of the Middle Ages.* Nashville, TN: Abingdon Press.
Ho, M. W. (1988). On not holding nature still: Evolution by process, not by consequences. In M. W. Ho, & S. W. Fox (Eds.), *Evolutionary metaphors.* John Wiley and Sons.
Ho, M. W. & Saunders, P. T. (1984). *Beyond neo-Darwinism: An introduction to the new evolutionary paradigm.* London: Academic Press.
Ho Peng-ti, (1967). *The Ladder of Success: Aspects of mobility in China, 1368-1911.* Rev. ed. New York: Columbia University Press.
Hobson, J. A. (1902). *Imperialism, A study.* George Allen & Unwin.
Hofsteded, G. (1980). *Culture's consequences.* Beverly Hills, CA: Sage.
Holbach, Paul Henri Thirty d." *Le Christianisme dévoilé.* London.
Hood, L., Campbell, J. H. & Elgin, S. C. R. (1973). The organization, expression and evolution of antibody genes and other multigene families. *Annual Review of Genetics,* 9, 305-353.
Horgan, J. (1997). *The end of science: Facing the limits of knowledge in the twilight of the scientific age.* New York: Broadway Books.
Horn, J. (1988). *Why we sleep: The functions of sleep in human and other animals.* Oxford: Oxford University Press.
Hudelson, R. (1993). *The rise and fall of communism.* Boulder, CO: Westview Press.
Huff, T. E. (2003). *The rise of early modern science: Islam, China, and the West.* Cambridge, UK: Cambridge University Press.
Hume, D. (1984/1739-40). *A treatise of human nature.* London: Penguin.
Hunter, C. G. (2007). *Science's blind spot: the unseen religion of scientific naturalism.* Grand Rapids, MI: Brazon Press.
Hutton, J. H. (1946). *Caste in India.* Cambridge, UK: Oxford University Press.
Hutton, R. H. (1870). Pope Huxley. *The Spectator* (January 29, 1870).
Huxley, L. (1901). *Life and letters of Thomas H. Huxley.* London: Macmillain.
Huxley, L. (1916). *Life and letters of Thomas Huxley,* 2 vols. New York:

D. Appleton.
Huxley, T. (1876). *Darwiniana* (Collected Essays, II), 187-226, Appleton, New York.
Huxley, T. (1893). *Collected essays*. London: George Olms.
Huxley, T. (1900). *Life and letters of Thomas Huxley*. London: Gregg International.
Huxley, T. H. (1882). Charles Darwin. *Nature*, 25, 597.
Hyatt, A. (1882). Transformation of Planorbis at Steinheim, with remarks on the effects of shell and animals. *American Naturalist*, 16, 441-453.
Irons, P. (2007). *God on trial: dispatches from America's religious battlefields*. New York: Vikings.
Irvine, W. (1955). *Apes, angels, and Victorians: A joint biography of Darwin and Huxley*. New York: Weidenfeld and Nicolson.
Israel, J. I. (2001). *Radical enlightenment: Philosophy and the making of modernity, 1650-1750*. New York: Oxford University Press.
Ito, T. A., Larsen, J. T., Smith, N. K., & Cacioppo, J. T. (1998). Negative information weighs more heavily on the brain: The negativity bias in evaluative categorizations. *Journal of Personality and Social Psychology*, 75, 887-900.
Jablonka, E. & Lamb, M. (1995). *Epigenetic inheritance and evolution: Epigenetic inheritance and evolution: The Lamarckian dimension*. Oxford: Oxford University Press.
Jablonka, E. & Raz, G. (2009). Transgenerational epigenetic inheritance: Prevalance, mechanisms, and implications for the study of heredity and evolution. *Quarterly Review of Biology*, 84, 131-176.
Jacob, F. (1982). *The possible and the actual*. New York: Pantheon Books.
James, J. W. (1962). Conflict centripetal and directional selection. *Heredity*, 17, 487-499.
James, W. (1892). A Plea for Psychology as a 'Natural Science', *Philosophical Review*, Vol. 1, 146-153.
Janis, I. L. (1972). *Groupthink: psychological studies of policy decision and fiascoes*. Boston, MA: Houghton Mifflin.
Jefferys, W. H. (2005). Review: The privileged planet. *Reports of the National Center for Science Education*, 25, 47-49.
Jensen, J. V. (1991). *Thomas Henry Huxley: Communicating for science*.

Newark: University of Delaware Press.
Jeon, K. W. (1972). Development of cellular dependence on infective organisms: micrugical studies in Amoebas. *Science*, 176, 1122-1123.
Jeon, K. W. & Danielli, J. F. (1971). Microsurgical studies with large free living amoebas. *International Review of Cytology*, 30, 49-89.
Jones, J., & Wilson, W. (1995). *An incomplete education.* New York: Ballantine.
Josephson, B. D. (2003). We think that we think clearly, but that's only because we don't think clearly: Brian Josephson on mathematics, mind and the human world. In P. C. Hogan and L. Pandit, (eds.), *Rabindranath Tagore: University and Tradition*, London: Associated University Press.
Judson, H. (1979). *The eighth day of creation.* New York: Simon & Schuster.
Karn, M. A., & Penrose, L. S. (1951). Birth weight and gestation time in relation to maternal age, parity, and infant survival. *Ann. Eugen.*, 16, 147-164.
Karve, I. (1961). *Hindu society – An interpretation.* Poona, India: Deshmukh Prakashan.
Keller, E. F. (2000). *The century of the gene.* Cambridge, MA: Harvard University Press.
Kilroy-Silk, R. (1973). *Socialism since Marx.* New York: Taplinger.
Kim, Y. H., Yu, R, Kulik, S. P., Shih, Y., & Scully, M. O. (2000). A delayed choice quantum eraser. *Physical Review Letters*, 84, 105.
Kimura, M. (1968). Evolutionary rate at the molecular level. *Nature*, 217, 624-646.
Kimura, M. & Ohta, T. (1971). *Theoretical aspects of population genetics.* Princeton: Princeton University Press.
King, D. (1996). An interview with professor Brian Goodwin. *GenEthics News*, 11, 6-8.
King, J. L. & Jukes, T. H. (1969). Non-Darwinian evolution: random fixation of selectively neutral mutations. *Science*, 164, 788-798.
Kitano, H. (2009). Systems biology: Toward system-level understanding of biological systems. In H. Kitano (ed.), *Foundations of systems biology.* Cambridge, MA: MIT Press.
KNMI. (2002). Het weer nader verlaard: Neerslagkans [The weather explained: Precipitation chance].

http://www.knmi.nl/voorl/nader/neerslagkans.htm.
Knowles, D. (1962). *The evolution of medieval thought*. New York: Vintage Books.
Kohlberg, Lawrence (1958). *The Development of Modes of Thinking and Choices in Years 10 to 16*. Ph. D. Dissertation, University of Chicago.
Kohler, W. (1931). *The mentality of apes* (trans. Ella Winter). New York: Harcourt, Brace. Edited by Frederick Burkhardt, Duncan M. Porter, Sheila Ann Dean, Paul S. White, Sarah Wilmot. (Cambridge University Press 2001).
Kohn, D., Murrell, G., Parker, J., & Whitehorn, M. (2005). What Henslow taught Darwin: How a herbarium helped to lay the foundations of evolutionary thinking. *Nature*, 436, 643-645.
Koestler, A. (1959). *The sleepwalkers: a history of man's changing vision of the universe*. New York: Grosset & Dunlap.
Koestler, A. (1968). *The ghost in the machine*. New York: Macmillan.
Kölliker, R. A., von (1864). Letter. In F. Burkhardt, D. M. Porter, S. A. Dean, P. S. White, & S. Wilmot (eds.), *The correspondence of Charles Darwin*, (vol.12). Cambridge: Cambridge University Press.
Korzybski, A. (1933). *Science and sanity: an introduction to non-Aristotelian systems and general semantics*. Lakeville, Conn: International Non-Aristotelian Library Publishing Co.
Kosambi, D. D. (1964). *The culture and civilization of ancient India in historical outline*. India: Vikas Publishing House.
Kramer, J. (1992). Letter from Europe. *New Yorker*, May 25, p. 43.
Krantz, D. L., & Wiggins, L. (1973). Personal and impersonal channels of recruitment in the growth of theory. *Human Development*, 16, 133-156.
Krebs, R. E. (1999). *Scientific development and misconceptions through the ages: A reference guide*. Westport, CT: Greenwood Press.
Krug, K. S., & Weaver, C. A., III. (2005). Eyewitness memory and metamemory in product identification: Evidence for familiarity biases. *Journal of General Psychology*, 132, 429-445.
Kuhn, T. (1962). *The structure of scientific revolution*. Chicago: University of Chicago Press.
Kuran, T. (1995). *Private truths, public lies: The social consequences of preference falsification*. Cambridge, MA: Harvard University

Press.
Kuran, T. (2010). *The long divergence: How Islamic law held back the Middle East*. Princeton, NJ: Princeton University Press.
Labinger, J. A., & Collins, H. (2001). *The one culture?: A conversation about science*. Chicago: University of Chicago Press.
Lacey, H. (1999). *Is science value free?: Values and scientific understanding*. London: Routledge.
Langer, U. (2005). *The Cambridge Companion to Montaigne*. Cambridge, UK: Cambridge University Press.
Lashley, K. S. (1950). In search of the engram. In *Symposium of the Society for the Experimental Biology* (Vol. 4). New York: Cambridge University Press.
Leahey, T. H. (1987). *A history of psychology: Main currents in psychological thought*. Englewood Cliffs, NJ: Prentice-Hall.
Leahey, T. H. (1991). *History of modern psychology*. Englewood Cliffs, NJ: Prentice Hall).
Leahey, T. (1992). The mythical revolutions of American psychology. *American Psychologist*, 47, 308-318.
Leaman, O. (1999). *A Brief Introduction to Islamic Philosophy* Polity Press.
LeDoux, J. E. (1994). Emotion, memory and the brain [Review]. *Scientific American*, 270 (6), 32-39.
LeDoux, J. E. (1996). *Emotional brain*. New York: Simon & Schuster.
Leiser, G. (1986). Notes on the madrasa in medieval Islamic society. *The Muslim World*, 76, 16-23.
Lennox, J. G. (1993). Darwin was a teleogist. *Biology and Philosophy*, 8, 409-421.
Lim, R. (1995). *Public disputation: Power and social order in late antiquity*. Berkeley.
Lima-de-Faria, A. (1988). *Evolution without selection: Forms and function by autoevolution*. Amsterdam: Elsevier.
Lindley, D. (1993). *The end of physics: The myth of a unified theory*. New York: Basic Books.
Lindberg, D. C. (1976). *Theories of vision from al-Kindi to Kepler*. Chicago: University of Chicago Press.
Lindberg, D. C. (2007). *The beginnings of Western science: The European scientific tradition in philosophical, religious, and institutional context, prehistory to A.D. 1450*. Chicago: The

University of Chicago Press.
Linholm, C. (1990). *Charisma*. Cambridge, MA: Basil Blackwell.
Lipset, S. M. & Bence, G. (1994). Anticipation of failure of communism. *Theory Soc.* 23(1), 169-210.
Lloyd, G. E. R. (2002). *The ambitions of curiosity: Understanding the world in ancient Greece and China*. Cambridge, UK: Cambridge University Press.
Loftus, E. F., & Schooler, J. W. (1985). Information-processing conceptualizations of human cognition: past, present, and future. In B. D. Ruben (ed.), *Information and behavior*, Vol. 1. New Brunswick: Transaction Books.
Lovelock, J. (1988). *The age of Gaia: A biography of our living earth*. New York: W.W. Norton & Co.
Løvtrup, S. (1987). *Darwinism: the refutation of a myth*. New York: Croom Helm.
Lynch, M. (2001). Situated knowledge and common enemies: Therapy for the science wars. In Labinger, J. A., & Collins, H. (eds.). *The one culture?: A conversation about science*. Chicago: University of Chicago Press.
Macbeth, N. (1971). *Darwin retried: an appeal for reason*. Boston, MA: Gambit.
MacMullen, R. (1997). *Christianity and paganism in the fourth to eighth centuries*. New Haven.
Mahoney, M. J. (1976). *Scientist as subject: The psychological imperative*. Cambridge, MA: Ballinger.
Mahoney, M. J. (1979). Psychology of the Scientist: An Evaluative Review, *Social Studies of Science*, 9, 349-375.
Malik, K. (2002). *Man, beast, and zombie: What science can and cannot tell us about human culture*. New Brunswick, NJ: Rutgers University Press.
Mannan, S. K., Mort, D. J., Hodgson, T. L., Driver, J., Kennard, C., & Husain, M. (2005). Revisiting previously searched locations in visual neglect: role of right parietal and frontal lesions in misjudging old locations as new. *Journal of Cognitive Neuroscience*, 17, 340-354.
Manschreck, C. L. (1985). *A history of Christianity in the world*. Englewood Cliffs, NJ: Prentice-Hall.
Margulis, L. (1970). *Origin of eukaryotic cells*. New Haven, CT: Yale

University Press.
Margulis, L. (1981). *Symbiosis in cell evolution*. San Francisco: Freeman.
Margulis, L. & Sagan, D. (2002). *Acquiring genomes: A theory of the origin of species*. New York: Basic Books.
Martin, B. (1993). Stamping out dissent: Too often, unconventional or unpopular scientific views are simply suppressed. *Newsweek*, April 26, 1993, 49-50.
Marx, K. (1872). Amsterdam Speech (September 8, 1872). In Robert Tucker, (ed.), *The Marx-Engels Reader*. (New York: Norton, 1978).
Maslow, A. H. (1962). *Toward a psychology of being*. Princeton, NJ; Van Nostrand.
Mattick, J. S. (2004). The hidden genetic program of complex organisms. *Scientific American*, 291, October, 60-67.
Maupertuis, P. (1745). *Vénus Physique, contenant deux dissertations, l'une sur l'origine des hommes et des animaux; et l'autre sur l'origine des noirs*. La Haye.
Maurer, A. (1962). *Medieval philosophy*. New York: Copleston.
Maynard Smith, J. (1964). Group selection and kin selection. *Nature*, 201, 1145-1147.
Mayr, E. (1976). *Evolution and the diversity of life*. Cambridge, MA: Belknap Press.
Mayr, E. (1980). Prologue: Some thoughts on the history of the evolutionary synthesis. In E. Mayr & W. B. Provine (eds.), *The evolutionary synthesis: Perspectives on the unification of biology*. Cambridge, MA: Harvard University Press.
Mayr, E. (1982). *The growth of biological thought: Diversity, evolution, and inheritance*. Cambridge, MA: Belknap Press.
Mayr, E. (1984). The unity of the genotype. In Robert N. Brandon & Richard M. Burian (eds.), *Genes, organisms, populations: Controversies over the units of selection*. Cambridge, MA: The MIT Press.
Mayr, E. (1985). How biology differs from the physical sciences. In D. J. Depew, & B. H. Weber (eds.),*Evolution at a crossroads: The new biology and the new philosophy of science*. Cambridge, MA: A Bradford Book.
Mayr, E. (1988). *Toward a new philosophy of biology: Observations of an evolutionist*. Cambridge, MA: Belknap Press.

Mayr, E. (1993). What was the evolutionary synthesis? *Trends in Ecology and Evolution*, 8, 31-34.
Mayr, E. (1996). The modern evolutionary theory. *Journal of Mammalogy*, 77, 1-7.
Mayr, E. (2004). *What makes biology unique?: consideration on the autonomy of a scientific discipline*. New York: Cambridge University Press.
Mayr, E. & Provine, W. B.(1980), *The evolutionary synthesis: Perspectives on the unification of biology*. Cambridge, MA: Harvard University Press.
McClintock, B. (1984). The significance of responses of the genome to challenge. *Science*, 226, 792-801.
McClellan, III. J. E. & Dorn, H. (2006). *Science and technology in world history: An introduction*. Baltimore, MD: Johns Hopkins University Press.
McComas, W. F. (1997). The discovery & nature of evolution by natural selection: Misconceptions & lessons from the history of science. *American Biology Teacher*, 59 (8), 492-500.
McCourt, M. E. (2001). Performance consistency of normal observers in forced-choice tachistocope visual line bisection. *Neuropsychologia*, 39, 1065-1076.
McCourt, M. E., Garlinghouse, M., & Slater, J. (2000). Centripetal versus centrifugal bias in visual line bisection: Focusing attention on two hypotheses. *Frontiers in Bioscience*, 5, 58-71.
McDermott, R. (1998). *Risk-taking in international politics: Prospect theory in American foreign policy*. Ann Arbor, Michigan: University of Michigan Press.
McLellan, D. (1973). *Karl Marx: A biography*. New York: Palgrave Macmillan.
McMullin, E. (2000). Values in science. In W. H. Newton-Smith (ed.), *A companion to the philosophy of science*. Malden, MA: Blackwell.
Medawar, P. B. (1961). Critical notice. *Mind*, 277, 99-106.
Megill, A. (2002). *Karl Marx: The burden of reason (Why Marx rejected politics and the market)*. Lanham, MD: Rowman & Littlefield.
Mele, A. (1997). Real self-deception. *Behavioral and Brain Sciences*, 20(1), 91-102.
Mele, A. (2001). *Self-deception unmasked*. Princeton, NJ: Princeton

University Press.
Meinhardt, H. (1995). *The algorhythmic beauty of sea shells*. Berlin: Springer-Verlag.
Melehy, H. (1997). *Writing cogito: Montaigne, Descartes, and the institution of the modern subject*. New York: State University of New York Press.
Mendoza, E. R. (2009). Systems biology: Its past, present and potential. *Philippine Science Letters*, 2, 16-34.
Merton, R. K. (1968). The Matthew effect in science. *Science*, 159, 56-63.
Merton, R. K. (1973). The neglect of the sociology of science. In N. Storer (ed.), *The sociology of science: Theoretical and empirical investigations*. Chicago: University of Chicago Press. (originally published as foreward to Bernard Barber, Science and the Social Order, New York: The Free Press).
Meslier, J. (1729/2009). *Testament: memoir of the thoughts and sentiments of Jean Meslier*. Amherst, NY: Prometheus Books.
Mesulam, M. (1985). Attention, confusional states, and neglect. In M. Mesualm (ed.), *Principles of behavioral neurology*. New York: Oxford University Press.
Methvin, E. H. (1973). *The rise of radicalism: The social psychology of messianic extremism*. New Rochelle, N.Y." Arlington House.
Meyer, A. G. (1954). *Marxism, The unity of theory and practice*. Cambridge, MA: Harvard University Press.
Meyer, K. E., & Szulc, T. (1962). *The Cuban invasion: The chronicle of a disaster*. New York: Frederick A. Praeger.
Meyer, S. C. (2003). DNA and the origin of life: Information, specification, and explanation. In J. C. Campbell, & S. C. Meyer (eds.), *Darwinism, design, and public education*. East Lansing, MI: Michigan State University Press.
Michelet, J. (1967). *History of the French Revolution*. Chicago: University of Chicago Press.
Milgram, S. (1974), *Obedience to Authority; An Experimental View*. New York: Harper & Row.
Mill, J. S. (1843). *System of logic, Ratiocinative and inductive*. London: John W. Parker and Son.
Miller, A. (1983). *For your own good*. New York: Farrar, Strauss & Giroux.
Mills, C. W. (1956). *The power elite*. Oxford University Press.

Miller, G. A. (1956). The magical number seven, plus or minus two: Some limits on our capacity for processing information. *Psychological Review*, 63, 81-97.

Miller, K. R. (2004). The flagellum unspun: The collapse of "irreducible complexity." In W. A. Dembski, & M. Ruse (eds.), *Debating design: From Darwin to DNA*. Cambridge, UK: Cambridge University Press.

Milner, M., Jr. (1994). *Status and sacredness: A general theory of status relations and an analysis of Indian culture*. New York: Oxford University Press.

Milosz, C. (1953). *The captive mind*. New York: Vintage International.

Mivart, St. G. (1871). *On the genesis of species*. London: Macmillan.

Mivart, St. G. (1897). *Some reminiscences of Thomas Henry Huxley, Nineteenth Century*, XLII, 994-995.

Miyazaki, I. (1976). *China's examination hell: The civil service examinations of imperial China* (trans. Schirokauer, Conrad). New York: Weatherhill.

Moldoveanu, M. & Stevenson, H. (2001). The self as a problem: the intra-personal coordination of conflicting desires. *Journal of Socio-Economics*, 30, 295-330.

Monod, J. (1971). *Chance and necessity: an essay on the natural philosophy of modern biology*. New York: Vintage Books.

Morowitz, H. J. (1968). *Energy flow in biology: biological organization as a problem in thermal physics*. New York: Academic Press.

Morowitz, H. J. (1984). Rediscovering the mind. In D. R. Hofstadter, & D. C. Dennett (Eds.), *The mind's I: Fantasies and reflections on self and soul*. New York: Basic Books (originally published in *Psychology Today*, August, 1980).

Morowitz, H. J. (1992). *Beginning of cellular life: Metabolism recapitulates biogenesis*. New Haven: Yale University Press.

Morowitz, H. J. (2002). *The emergence of everything: How the world became complex*. New York: Oxford University Press.

Moscovich, S., & Zavalloni, M. (1969). The group at a polarizer of attitudes. *Journal of Personality and Social Psychology*, 12, 125-135.

Mukherjee, B. (1963). Caste-ranking among Rajbanshis in North Bengal (Bala ratnam, ed.) *Anthropology on the March*.

Muller, H. J. (1927). Artificial transmutation of the gene. *Science*, 66, 84-

87.

Murphy, A. H., Lichtenstein, S., Fischhoff, B., & Winkler, R. L. (1980). Misinterpretations of precipitation probability forecasts. *Bulletin of the American Meteorological Society*, 61, 695-701.

Nagel, T. (2008). Public education and intelligent design. *Philosophy & Public Affairs*, 36, 187-205.

Narasimhachary, M. (2002). The caste system: An overview. *IK Foundation Lecture Series, Indian Culture in the Modern World*, 23rd Oct. 2002, London.

Needham, J. (1942). *Biochemistry and morphogenesis*. Cambridge: Cambridge University Press.

Needham, J. (1970). *Clerks and craftsmen in China and the West*. Cambridge: Cambridge University Press.

Nehru, Jawaharlal (1960). *Proceedings of the National Institute of Science of India* 27.

Nesfield, J. C. (1885). *Brief view of the caste system of the North-Western provinces and Oudh: together with an examination of the names and figures shown in the census report, 1882, being an attempt to classify on a functional basis all the main castes of the United Provinces, and to explain their gradations of rank and the process of their formation*. Allahabad: North-Western Provinces and Oudh Government Press.

Newell, A., & Simon, H. A. (1972). *Human problem solving*. Englewood Cliffs, NJ: Prentice Hall.

Nicholls, E. R., Hadgraft, N. T., Chapman, H. L., Loftus, A. M., Robertson, J., Bradshaw, J. L. (2010). A hit-and-miss investigation of asymmetries in wheelchair navigation. *Attention, Perception, & Psychophysics*, 72, 1576-1590.

Nordenskioeld, E. (1928). *The history of biology*. New York: Knopf.

Nurse, P. (2008). Life, logic and information. *Nature*, 454, 424-426.

Numbers, R. L. (2003). Science without God: Natural laws and Christian beliefs. In Lindberg, D. C., & Numbers, R. L. (eds.), *When science & Christianity meet*. Chicago: University of Chicago Press.

Oatley, K., & Johnson-Laird, P. N. (1987). Towards a cognitive theory of emotions. *Cognition and Emotion*, 1, 29-50.

Oldenberg, H. (1897). *Buddha: sein Leben, Seine Lehre, seine Gemeinde*. Berlin.

Onfray, M (2006). Jean Meslier and "The gentle inclination of nature," *New Politics*, 10,

Osborn, H. F. (1893). The rise of the mammalia in North America. *American Journal of Science*, 46, 373-392, 448-466.

Osborn, H. F. (1894). *From Greeks to Darwin*. London: Macmillan.

Owen, R. (1860). Darwin on the origin of species. *The Edinburgh Review* 3, 487-532.

Oyserman, D., Coon, H. M., & Kemmelmeier, M. (2002). Rethinking individualism and collectivism: Evaluation of theoretical assumptions and meta-analysis. *Psychological Bulletin*, 128, 3-72.

Palermo, D. S. (1971). Is a scientific revolution taking place in psychology? *Science Studies*, 1, 135-155.

Paley, W. (1802/2006). *Natural theology*. Cambridge, UK: Oxford University Press.

Parsons, T. (1957). The distribution of power in American society. *World Politics*, 10, 139.

Pascal, B. (1977). *Pensées*, trans. With an introduction by A. J. Krailsheimer. London: Cox and Wyman.

Pears, D. (1984). *Motivated irrationality*. New York: Oxford University Press.

Pearson, H. (2006). What is a Gene? *Nature*, 441, 399-401.

Peirce, C. (1896). Lessons from the history of science. *Collected Papers*, 1: 19-49.:

Peng, X. (1987). Demographic consequence of the Great Leap Forward in China's provinces. *Population and Development Review*, 13 (4), 639-670.

Perrie, M. (2000). The Russian revolution. In D. Parker (ed.), *Revolutions and the revolutionary tradition: In the West 1560-1991*. London: Routledge.

Peters, D., & Ceci, S. (1982). Peer-review practices of psychological journals: The fate of submitted articles, submitted again. *Behavioral and Brain Sciences*, 5, 187-255.

Peters, F. E. (1968). *Aristotle and the Arabs: The Aristotelian tradition in Islam*. New York: New York University Press.

Pfeifer, E. J. (1965). The genesis of American neo-Lamarckism. *ISIS*, 56(2), 156-167.

Pinker, S. (2004). Why nature & nurture won't go away. *Daedalus*, Fall,

1-13.
Polanyi, M. (1968). Life's irreducible structure. *Science*, 160, 1308-1312.
Polmar, N., & Clancy, J. D. (2006). *DEFCON-2: Standing on the brink of nuclear war during the Cuban missile crisis*. Hoboken, NJ: John Wiley & Sons.
Popkins, R. H., & Neto, J. R. M. (2007). *Skepticism: An anthology*. New York: Prometheus Books.
Popper, K. R. (1972). Objective knowledge: An evolutionary approach. Oxford, UK: Clarendon Press.
Pratto, F., & John, O. P. (2005). Automatic vigilance: The attention-grabbing power of negative social information. In Hamilton, D. L. (ed.), *Social cognition: Key readings*. New York: Psychology Press.
Provine, W. (1971). *The origins of theoretical population genetics*. Chicago: University of Chicago Press.
Provine, W. B. (1992). Progress in evolution and meaning in life. In C. K. Waters, & A. Van Halden (eds.), *Julian Huxley, biologist and statesman of science*. Houston: Rice University Press.
Provine, W. B. (2001). *The origins of theoretical population genetics*. Chicago:University of Chicago Press (originally published in 1971).
Quastler, H. (1964). *The emergence of biological organization*. New Haven, CT: Yale University Press.
Rabow, J., Fowler, F. J., Jr. (1966). The role of social norms and leadership in risk taking. *Sociometry*, 29, 16-27.
Rawls, J. B. (1971). *A theory of justice*. Cambridge, MA: Belknap Press of Harvard University Press.
Redlich, F. C. (1999). *Hitler: diagnosis of a destructive prophet*. New York: Oxford University Press.
Redner, H. (1986). *The ends of philosophy: An essay in the sociology of philosophy and rationality*. Totowa: NJ: Rowman & Allenheld.
Redner, H. (1987). *The ends of science: An essay in scientific authority*. Boulder, CO: Westview Press.
Reich, D., Thangaraj, K., Patterson, N., Price, A. L., & Singh, L. (2009). Reconstructing Indian population history. *Nature*, 461, 489-494.
Reid, R. G. B. (2007). *Biological emergences: Evolution by natural experiment*. Cambridge, MA: Bradford Book.
Riedl, R. (1978). *Order in living organisms: a system analysis of*

evolution. New York: Wiley.

Rendel, J. M. (1943). Variations in the weight of hatched and unhatched ducks' eggs. *Biometrica*, 33, 48-58.

Renn, J., & Schemmel, M. (2003). *Mechanics in the Mohist canon and its European counterpart*. A presentation given at the 3rd International Symposium on Ancient Chinese Books and Records of Science and Technology in Tuebingen from March 31 to April 3, 2003.

Richards, R. J. (1983). Why Darwin delayed, or interesting problems and models in the history of science. *Journal of the History of the Behavioral Science*, 19, 45-53.

Riddiford, A., & Penny, D. (1984). The scientific status of modern evolutionary theory. In Pollard, J. W. (Ed.), *Evolutionary theory: paths into the future*. Chichester: John Wiley & Sons.

Rilling, J. K. (2006). Human and nonhuman primate brains: Are they allometrically scaled versions of the same design? *Evolutionary Anthropology*, 15, 65-77.

Robertson, I. H., & Heutink, J. (2002). Rehabilitation of unilateral neglect. In W. H. Brouwer, A. H. van Zomeren, I. J. Berg, J. M. Bouma, & E. H. F. de Haan (Eds.), *Neuropsychological rehabilitation: a cognitive approach*. Amsterdam: Boom.

Roediger, H. L. (1980). Memory metaphors in cognitive psychology. *Memory & Cognition*, 8, 231-246.

Rose, S. (1997). *Lifelines: biology beyond determinism*. Oxford: Oxford University Press.

Ross, G. M. (1990). Science and philosophy. In R. C. Olby, G. N. Cantor, J. R. R. Christie, & M. J. S. Hodge (eds.), *Companion to the history of modern science*. London: Routledge.

Ross, H. (2001). *The creator and the cosmos*. Colorado Springs, Co: NavPress.

Rothbard, M. (1990). Concepts of the role of intellectuals in social change toward laissez faire, *The Journal of Libertarian Studies*, 9(2), Fall.

Rowland, W. (2003). *Galileo's mistake: A new look at the epic confrontation between Galileo and the Church*. New York: Arcade Publishing.

Ruse, M. (1979). *The Darwinian revolution: Science red in tooth and claw*. Chicago: University of Chicago Press.

Ruse, M. (1997). Thomas Henry Huxley and the status of evolution as science. In A. P. Barr (ed.), *Thomas Huxley's place in science and letters: Centenary essays*. Athens, GA: The University of Georgia Press.
Russell, B. (1927). *An outline of philosophy*. London: George Allen & Unwin.
Russell, B. (1952). *The impact of science on society*. London: George Allen & Unwin.
Sabra, A. I. (1987). The appropriation and subsequent naturalization of Greek science in Medieval Islam: A preliminary statement. *History of Science*, 25, 223-243.
Safina, C. (2010). Darwinism must die so that evolution may live. *New York Times*, April 30.
Sagan, C. (1992). *The demon-haunted world: Science as a candle in the dark*. New York: Random House.
Saliba, G. (2007). *Islamic science and the making of the European Renaissance*. Cambridge, Mass.: MIT Press.
Schilthuizen, M. (2001). *Frogs, flies, and dandelions: Speciation—The evolution of new species*. Oxford, UK: Oxford University Press.
Schlesinger, A. M., Jr. (1965). *A thousand days: John F. Kennedy in the White House*. Cambridge, MA: Houghton, Mifflin.
Schlichting, C., & Pigliucci, M. (1998). *Phenotypic evolution: A reaction norm perspective*. Sunderland, MA: Sinauer.
Schmutzer, E. & Schiltz, W. (1983). *Galileo Galilei, Biographien hervorragender Naturwissenschaftler, Techniker und Mediziner 19*, B. G. Teubner, Verlagsgesellschaft, Leipzig, Germany.
Schoenblum, J. A. (1999). The role of legal doctrine in the decline of Islamic Waqf: A comparison with the Trust (Inheritance System) (Symposium: International Trust, Part 2), *Vanderbilt Journal of Transnational Law*, 32, 1191.
Schroeder, G. (1997). *The science of God*. New York: Free Pres.
Schroeder, G. L. (2001). *The hidden face of God: How science reveals the ultimate truth*. New York: Free Press.
Schwaab, E. H. (1992). *Hitler's mind: A plunge into madness*. New York: Praeger.
Schwartz, J. H. (1999). *Sudden origins: fossils, genes, and the emergence of species*. New York: Wiley.

Searle, J. R. (1993). The problem of consciousness. In Ciba Foundation Symposium, E*xperimental and theoretical studies of consciousness*. Chichester: John Wiley & Sons.

Selfridge, O. G. (1959). Pandemonium: A paradigm for learning. In D. V. Blake and A. M. Utreley (eds.), *Proceedings of the symposium on the mechanisation of thought processes*. London: H.M. Stationer Office.

Semendferi, K., Lu, A., Schenker, N., & Damasio, H. (2002). Humans and great apes share a large frontal cortex. *Nature Neuroscience*, 5, 272-276.

Shapiro, A. M. (1976). Seasonal polymorphism. In T. Dobzhansky, M. R. Hecht, & W. C. Steere (Eds.), *Evolutionary Biology*. New York: Appleton-Century-Crofts.

Shapiro, J. A. (1992). Natural genetic engineering. *Genetica*, 86, 99-111.

Shapiro, R. (1986). *Origins: a skeptic's guide to the creation of life on earth*. New York: Summit Books.

Shea, W. R., & Artigas, M. (2003). *Galileo in Rome: the rise and fall of a troubled genius*. New York: Oxford University Press.

Simon, H. A. (1967). Motivational and emotional control of cognition. *Psychological Review*, 74, 29-39.

Simpson, G. G. (1949). *The meaning of evolution*. New Haven: Yale University Press.

Simpson, G. G. (1953). *The major features of evolution*. London: Columbia University Press.

Singh, V. P. (1976). *Caste, Class and Democracy: Changes in a stratification system*. Cambridge, MA: Schenkman Publishing.

Sinnot, E. S. (1950). *Cell and psyche: The biology of purpose*. Chapel Hill: The University of North Carolina Press.

Skinner, B. F. (1953). *Science and human behavior*. New York: Macmillan Company.

Skinner, B. F. (1957). *Verbal behavior*. New York: Appleton-Century-Crofts.

Slack, G. (2007). *The battle over the meaning of everything: Evolution, intelligent design, and a school board in Dover, PA*. San Francisco, CA: John Wiley & Sons.

Smith, B. (1967). *Adolf Hitler: His family, childhood, and youth*. Stanford, CA: Hoover Institution on War/Stanford University Press.

Smocovitis, V. B. (1996). *Unifying biology: The evolutionary synthesis and evolutionary biology*. Princeton, NJ: Princeton University Press.

Snsolabehere, S. & Iyengar, S. (1995). *Going negative: How attack ads shrink and polarize the electorate*. New York: Free Press.

Sokal, A. D. (2001). What the *Social Text* affair does and does not prove. In K. M. Ashman, & P. S. Baringer (eds.), *After science wars*. London: Routledge.

Solé, R. V. & Goodwin, B. (2000). *Signs of life: how complexity pervades biology*. New York: Basic Books.

Solzhenitsyn, A. I. (1974). *A letter to the Soviet leaders*. New York: Harper & Row.

Sorensen, T. C. (1966). *Kennedy*. New York: Bantam.

Spencer, H. (1896). *The study of sociology*. Appleton.

Spetner, L. (1996). *Not by chance*. New York: Judaica Press.

Staats, A. W. (1983). *Psychology's crisis of disunity: Philosophy and method for a unified science*. New York: Praeger.

Staats, A. W. (1985). Unified positivism: Philosophy for unionomic psychology. In W. J. Baker, M. E. Hyland, H. Van Rappard, & A.W. Staats (Eds.), *Current issues in theoretical Psychology*. Amsterdam: North-Holland.

Standen, A. (1950). *Science is a sacred cow*. New York: E.P. Dutton.

Stanley, S. M. (1979). *Macroevolution: Pattern and process*. San Francisco: W. H. Freeman.

Steffens, B. (2006). *Ibn al-Haytham: First scientist*. Greensboro, NC: Morgan Reynolds Publishing.

St. Augustine (1982). *The literal meaning of Genesis*, trans. Jon Hammond Taylor. New York: Newman Press.

Steele, A. (1868). *The law and custom of Hindu castes: Within the Dekhun provinces subject to the presidency of Bombay*.

Sterelny, K. (2007). Snafus: an evolutionary perspective. *Biological Theory*, 2, 1-11.

Stewart, M. (1997). *The truth about everything: An irreverent history of philosophy*. New York: Prometheus Books.

Stoner, J. A. F. (1961). *A comparison of individual and group decisions involving risk*. Unpublished master's thesis. Massachusetts Institute of Technology, Cambridge, MA.

Strauss, L. (1965). *Spinoza's critique of religion*. New York: Schocken

Books.
Strober, G. S., & Strober, D. H. (1993). *Let us begin anew: An oral history of the Kennedy presidency*. New York: HarperCollins.
Stroun, M., Anker, P. & Auderset, G. (1970). Natural release of nucleic acids from bacteria into cells. *Nature*, 227, 607-608.
Sunstein, C. R. (2009). *Going to the extremes: How like minds unite and divide*. New York: Oxford University Press.
Taleb, N. N. (2007). *The black swan*. New York: Penguin.
Taylor, F. J. R. (1974). Implications and extensions of the serial endosymbiosis theory of the origin of eukaryotes. *Taxonomy*, 23, 229-258.
Teilhard de Chardin, P. (1975). *The phenomenon of man*. New York: Harper Colophon.
Templeton, A. R., & Giddings, L. V. (1981). Macroevolution conference. *Science*, 211, 770-773.
Terrall, M. (2002). *The man who flattened the earth: Maupertuis and the sciences in the Enlightenment*. Chicago: University of Chicago Press.
Thompson, D. (1942). *On growth and form*. London: Cambridge University Press.
Thorndike, E. L. (1920). A constant error in psychological ratings. *Journal of Applied Psychology*, 4, 25-29.
Thrower, J. (1971). *A short history of western atheism*. London: Pemberton Books.
Toates, F. (2004). Skinner's double life as both perpetrator and innocent victim: A reply to Baars. *Journal of Consciousness Studies*, 11, 57-63.
Tolman, E. (1951). *Behavior and psychological man*. Berkeley, CA: University of California Press.
Tolman, E. C. (1932). *Purposive behavior in animals and men*. New York: Century.
Tolman, E. C. (1951). *Behavior and psychological man: essays in motivation and learning*. Berkeley, CA: University of California Press.
Triandis, H. C. (1995). *Individualism & collectivism*. Boulder, CO: Westview Press.
Trimble, V. (2008). Universe of multiverse? *Classical and Quantum*

Gravity, 25, 229.
Tulving, E. (1994). Forward. In J. Metcalfe & A. Shimamura (Eds.), *Metacognition; Knowing about knowing.* Cambridge, MA: Bradford.
Turner, J. S. (2007). *The tinkerer's accomplice: How design emerges from life itself.* Cambridge, MA: Harvard University Press.
Tyler, P. E. (1997). Obituary: Deng Xiaoping: A political wizard who put China on the capitalist road. *New York Times,* February 20, 1997.
Ulanowicz, R. E. (1997). *Ecology, the ascendant perspective.* New York: Columbia University Press.
Uttal, W. R. (2000). *The war between mentalism and behaviorism: on the accessibility of mental processes.* Mahwah, NJ: Erlbaum Associates.
Van den Daele, W. (1977). The social construction of science: Institutionalization and definition of positive science in the latter half of the seventeenth century. In E. Mendelsohn, P. Weingart, & R. Whitley (eds.), *The social production of scientific knowledge.* Dordrecht, Holland: D. Reidel Publishing.
Van Regenmortel, M. H. V. (2004). Reductionism and complexity in molecular biology. *Journal of Molecular Recognition,* 17, 145-148.
Vaughan D. (1996). *Normalization of deviance. The Challenger launch decision: risky technology, culture and deviance at NASA.* Chicago: The University of Chicago Press.
Versluis, A. (2006). *The new inquisitions: Heretic-hunting and the intellectual origins of modern totalitarianism.* New York: Oxford University Press.
Viereck, P. (1965). *The root of the Nazi mind.* New York: Knopf.
Von Bertalanffy, L. (1960). Principles and theory of growth. In W. W. Nowinski (Ed.), *Fundamental aspects of normal and malignant growth.* Amsterdam: Elsevier.
Von Stein, L. (1842). *Socialismus und Communismus Des Heutigen Frankreichs. Ein Beitrag Zur Zeitgeschichte.* Leipzig: Otto Wigand.
Vorzimmer, P. J. (1977). The Darwin reading notebooks (1838-1860). *Journal of the History of Biology,* 10, 107-153.
Waddington, C. H. (1960). Evolutionary adaptations. In S. Tax (ed.), *Evolution after Darwin* (vol.1). Chicago: University of Chicago

Press.

Waddington, C. H. (1961). *Mathematical challenges to the Neo-Darwinian interpretation of evolution* (eds. P. S. Moorhead & M. M. Kaplan). Philadelphia: Wistar Institute Press.

Waite, R. G. (1977). *The psychopathic god*. New York: Basic Books.

Wald, G. (1954). The origin of life. *Scientific American*, 191 (August), 44-53.

Wallace, A. R. (1958). *On the tendency of varieties to depart indefinitely from the original type*.

Wallace, A. R. (1905). *My Life: A records of events and opinions*. New York.

Wang, Chong (1962). *Lunheng* (trans. Forke, Alfred). New York: Paragon Book Gallery.

Warren, H. C. (1917). Numerical effects of natural selection acting on Mendenlian characters. *Genetics*, 2, 305-312.

Watkins, M. J. (1990). Mediationism and the obfuscation of memory. *American Psychologist*, 45, 328-335.

Watson, B. (1968). *The complete works of Chuang Tzu* (trans.). New York: Columbia University Press.

Watson, D. L. (1938). *Scientists are human*. London: Watts & Co.

Watson, D. M. S., (1926). Croonian Lecture. The evolution and origin of the Amphibia. *Philosophical Transactions of the Royal Society*, 214 B, 189-257.

Watson, J. B. (1913). Psychology as the behaviorist views it. *Psychological Review*, 20, 158-177.

Watson, J. B. (1925). *Behaviorism*. New York: Norton.

Watson, J. D. (1965). *The molecular biology of the gene*. New York: W. A. Benjamin.

Watson, J. D. (1970). *Molecular biology of the gene*. New York: Benjamin.

Watt, W. M. (1963). *Muslim Intellectual: A Study of al-Ghazali*, Edinburgh.

Weber, M. (1930). *The Protestant ethic and the spirit of capitalism*. New York: Routledge.

Weimer, W. B. & Palermo, D. S. (1973). Paradigms and normal science in psychology, *Science Studies*, 3, 211-244.

Weinberg, S. (1977). *The first three minutes*. New York: Basic Books.

Weinberg, S. (1992). *Dreams of a final theory*. New York: Pantheon

Books.
Weinberg, S. (2001). *Facing up: Science and its cultural adversaries*. Cambridge, MA: Harvard University Press.
Weinberg, J., Diller, L., Gordon, W. A., Gerstman, L. J., Lieberman, A., Lakin, P. (1977). Visual scanning training effect on reading-related tasks in acquired right brain damage. *Archives of Physical Medicine and Rehabilitation*, 58, 479-486.
Weizsäcker, C. F. von. (1952). *The world view of physics* (trans. Marjorie Greene). Chicago: University of Chicago Press.
Wells, G. L., Small, M., Penrod, S., Malpass, R. S., Fulero, S. M., & Brimacombe, C. A. F. (1998). Eyewitness identification procedures: Recommendations for lineups and photospreads. *Law and Human Behavior*, 22, 603-647.
Wells, J. (2006). The *politically incorrect guide to Darwinism and intelligent design*. Washington D.D.: Regnery Publishing.
Wesson, R. (1991). *Beyond natural selection*. Cambridge, MA: The MIT Press.
White, P. (2003). *Thomas Huxley: Making the "man of science."* New York: Cambridge University Press.
Whitehead, A. N. (1960). *Process and reality: an essay in cosmology*. New York: Harper.
Whitehead, C. (2004). Everything I believe might be delusion. Whoa! *Journal of Consciousness Studies*, 11, 68-88.
Whyte, L. L. (1949). *The unitary principle in physics and biology*. London: Cresset Press.
Wigner, E. P. (1962). In I. J. Good (ed.), *The scientist speculate*. Kindswood, England: The Windmill Press.
Williams, G. C. (1966). *Adaptation and natural selection*. Princeton, NJ: Princeton University Press.
Williams, L. P. (1973). Kant, Naturophilosophie and scientific method. In R. N. Giere, & R. S. Westfall (eds.), *Foundations of scientific method: the nineteenth century*. Bloomington, IN: Indiana University Press.
Williams, N. (1997). Biologists cut reductionist approach down to size. *Science*, 277, 476-477.
Williamson, P. G. (1981). Palaeontological documentation of speciation in Cenozoic molluscs from Turkana Basin. *Nature*, 293, 437-443.

Wilson, D. S. (1999). Tasty slice—but where is the rest of the pie? *Evolution and Human Behaviour*, 20, 279-287.
Wilson, E. O. (1975). *Sociobiology*. Belkup: Press of Havard University.
Winfree, A. T. (1987). *The timing of biological clocks*. New York: Scientific American Library.
Winter, M. (1992). *Egyptian society under Ottoman rule*. London: Routledge.
Wozniak, R. H. (1997). Commentary on "Psychology as the behaviorist views it" John B. Watson (1913). In W. G. Bringman, R. Miller, & C. E. Early (Eds.). *A pictorial history of psychology*. Chicago: Quintessence.
Woodhead, L. (2004). *An introduction to Christianity*. Cambridge, UK: Cambridge University Press.
Wyden, P. (1979). *Bay of Pigs: The untold story*. New York: Simon and Schuster.
Yamamoto, M. (1979). Compensating capacity: Interchromosomal effects of heterochromatic delections on recombination in *Drosophila melanogaster*. *Genetics*, 93, 437-448.
Yang, K. (1988). Will societal modernization eventually eliminate cross-cultural psychological differences? In M. Bond (ed.), *The cross-cultural challenge to social psychology*. Newbury Park, CA: Sage.
Yule, G. U. (1902). Mendel's laws and their probable relations to intra-racial heredity. *New Phytologist I*, 193-207, 222-238.
Zajonc, R. (1980). Feeling and thinking (emotion versus ration): Preferences need no inferences. *American Psychologist*, 35, 151-175.
Ziman, J. (1982). Bias, incompetence, or bad management? *Behavioral and Brain Sciences*, 5, 245.
Zink, C. F., Tong, Y., Chen, Q., Bassett, D. S., Stein, J. L., & Meyer-Lindenberg, A. (2008). Know your place: Neural processing of social hierarchy in humans. *Neuron*, 58, 273-283.
Zorzi, M., Priftis, K., Meneghello, F. (2002). The spatial representation of numerical and nonnumerical sequences: evidence from neglect. *Neuropsychologia*, 44, 1061-1067.

INDEX

Abra, J. 34
Abu Ghraib 94, 356, 357
Al-Fasi, M. 140
Al-Ghazali 142, 143, 144, 148
Alhazen (Ibn al-Haytham) 137, 138, 140, 177, 292, 308
Al-Kindi 133
Al-Jahiz 136
American Constitution 109, 270, 282, 321, 364
Altemeyer, B. 112
Amygdala 83, 84, 86, 87
Anthropic principle 263
Aquinas 154, 155, 163, 213
Arendt, H. 111
Arieti, J. A. 142, 143, 144
Aristotle 22, 82, 90, 93, 132, 133, 137, 141, 142, 148, 149, 154, 155, 157, 261, 275, 327
Arjomand, S. A. 139, 140, 144
Aronowitz, S. 365
Aronson, E. 113
Aspect, A. 329
Athens Academy 68, 149, 167, 322
Aryans 53, 54
Asch, S. E. 40, 41, 330, 361
Augros, R. 212, 245, 260, 266
Avicenna (Ibn Sinna) 137, 142, 143, 308
Axial age 134
B-17 358
Baars, B. 32, 37, 38, 44, 80
Bacon, F. 79, 122, 241, 319, 334
Bala, A. 134

Balashov, Y. 179
Banausoi 135
Barber, B. 224
Barlett, D. L. 60, 61
Barr, A. P. 207
Barry, D. H. 109
Bartolomeo, P. 96
Barzun, J. 183, 197
Bayle, P. 167
Bay of Pigs 338-347
Beddall, B. 201
Behaviorism 32-47
Behe, M. 247, 250, 251, 252, 254, 272, 280, 281
Belousov-Zhabotinsky Reaction (BZ Reaction) 299
Bem, D. J. 329
Ben-David, J. 152, 162, 178, 288
Berlin Wall 66, 76, 99, 105
Béteille, A. 52
Bethell, T. 188
Biometrician 226, 227, 234
Bird, R. J. 243, 262
Bissell, R. 101, 102, 339, 340, 345
Black swan event 312, 330, 331
Blomberg, S. P. 222, 295
Blue collar workers 125, 136, 137, 308
Blyth, E. 195, 199, 213
Bodde, D. 131, 135, 136
Bohm, D. 165
Bohr, N. 284
Boorstin, D. J. 100
Bossomaier, T. 315
Bouglé, C. 51, 52, 63

Boveri, T. 227
Bowler, P. 197, 202, 211, 214, 216, 218, 221, 222, 224, 227, 228
Boyce, W. T. 27
Boyle, R. 174, 179
Brackman, A. C. 183, 198, 199, 201
Brahmans 49, 50, 51, 53, 54, 64
Breder, Jr. C. M. 298
Brewster, D. 177
Brinton, C. 108
Brodmann Area 10 (BA 10) 85, 86
Broman, T. H. 219
Brooks, D. R. 232, 303
Brooks, J. L. 201
Browne, E. 189
Bruner, J. 40, 330
Buckley, W. F. Jr. 66
Buddha 100, 124, 329
Bullock, A. 118
Burian, R. M. 227, 237, 302
Burke, E. 106, 341
Burkhardt, R. W. 196
Burt, D. B. 294
Burton, R. A. 103
Burtt, E. A. 177, 178
Busky, D. F. 69, 72
Butler, S. 193, 194, 195, 199
Butler Act 266, 267, 271
Byrne, J. M. 264
Caccini, T. 158, 159
Cairns, J. 295
Calley, W. 356
Campanini, M. 143, 144
Campbell, D. T. 355
Campbell, J. H. 296, 297

Cardinal Bellarmine 156, 159
Caporale, L. H. 255, 296
Castro, F. 72, 278, 339, 341, 344, 348
Castro, R. 278
Caudill, E. 274
Cautious shift 94
Ceccarelli, L. 232, 238, 239
Ceci, S. 26, 28, 103, 225
CERN 288
Certainty judgment 103, 313, 360
Chain, E. B. 259
Chaitin, G. J. 262
Challenger 18, 352, 353
Chambers, R. 189, 196, 206, 213, 214
Chomsky, N. 43, 44, 97, 98
Chuang Tzu 124, 127, 318
Cicchetti, D. V. 28
Civil Rights Movement 76, 362
Cognitive revolution 43, 44, 48, 304, 334
Collins, F. 246, 247, 256
Collins, H. 232
Colp, R. Jr. 185
Colombe, L. de 158
Columbia 18, 352
Communism 35, 36, 65-77, 79, 87, 88, 89, 91, 92, 94, 99, 105, 108, 118, 253, 279, 312, 344, 361
Comte, A. 176, 177, 309
Confucianism 124, 125, 126, 129
Constantine 146, 147
Cope, E. D. 221, 222, 294
Copernicus 134, 144, 158, 159, 161, 263
Correns, C 224

A Critique of Science

Costall, A. 35
Cox, O. C. 50
Counter-Reformation 21
Crick, F. 48, 80, 244, 261,
Critical period 344, 356, 357
Crude stupidities 314
Cuba 66, 99, 278, 279, 328, 339-353
Cuban Missile Crisis 347-353
Cultural Revolution 75, 204
Cummins, D. D. 27
Cuvier, G. 196, 253
Czechoslovakia 105
Dalton, J. 285
Damasio, A. R. 82, 83
Darby, J. M. 357
Darlington, C. D. 184, 193, 194, 195, 196, 197
Darwin, C. 103, 180, 181, 183-226, 236, 243, 245, 248, 250, 251, 266, 269, 274, 275, 276, 281, 292, 297, 302, 303, 312, 320
Daston, L. J. 34
Davies, E. B. 167
Davies, P. 263, 288
Davies, R. 199
Davidson, D. 39
Davis, K. 56
Dawkins, R. 184, 190, 253, 255, 276, 302, 361
Day, P. 264
Deason, G. B. 175
De Bracton, H. 365
De Broglie, L. 248
De Chardin, T. 263
Deism 213, 327
Democritus 174

Deng Xioping 71, 76
Dennett, D. 176, 184, 253, 302
Dennett, E. 223
Dennis, M. 264
Denton, M. 249, 250, 303
Depew, 232, 303
Descartes, R. 37, 81, 170, 175, 179, 300, 317, 319
Desmond, A. 206, 207, 215, 218
Derham, W. 212
De Sousa, R. 82
Devil effect 27
De Vries, H. 207, 211, 224, 227, 228, 233
Dhanani, A. 138, 143, 144
Diderot, D. 167, 168, 169, 170, 316,
D'Holbach, P. H. 167, 170, 180
Diamond, J. 16, 17, 18, 121, 122, 124
Dumontheil, I. 85
Dembski, W. A. 280, 281
D'Espagnat, B. 287
Dingle, H. 268
Diocletian 146
Discovery Institute 270, 271, 272, 283
DNA testing 102
Dobzhansky, T. 188, 233, 238, 239, 277, 278
Doise, W. 94
Dolittle, W. F. 297
Dolsenhe, O. 79, 265, 299, 300, 304
Donovan, W. J. 114
Dostoevsky 244
Double-slit experiment 286
Douglas, H. E. 26,

407

Douglas, M. 49, 59, 308
Dover trial 251, 267-273
Doyle, J. 256
Drake, S. 160
Dravidians 53
Dreisle, T. 366
Drosnin, M. 62
Dugatkin, L. A. 190
Duhem, P. 134
Dulles, A. 101, 339, 340, 344, 345
Dumont, L. 49, 50, 52, 59, 63, 64
Dunn, S. 109
Dunning-Kruger effect 324
Easter Island 16-19
East Germany 105
Ebenstein, W. 70
Eckart, D. 115
Ecuador 107
Eddington, A. 263, 286, 287
Edict of Milan 146
Edis, T. 327
Edwards v. Aguillard 271
Egypt 107
Eichmann, A. 111, 112, 114
Eigen, M. 257
Einstein, A. 22, 46, 98, 162, 184, 263, 285, 286, 302, 307, 312, 328, 329
Eiseley, L. 195, 202, 214, 224
Eisenhower, D. 44, 101, 102, 339, 340
Elan vital 127
Eldridge, N. 237, 302, 303
Elster, J. 37, 82
Emery, A. E. 207
Endosymbiosis 295
Engels, F. 66, 70, 71, 72

EPR paradox 328
Evans, R. 75
Everett, H. 284
Fairbanks, J. K. 125
Fakhry, M. 138
Farley, J. 261
Falsificationism 311-314
Fatwa 140
Feinberg, J. 252
Festinger, L. 27, 329
Feyerabend, P. 315
Feynman, R. 98, 323, 324
Fick, R. 51, 63
Fischer, K, 178, 356
Fischoff, S. 183, 204
Fisher, M. J. 69, 71, 72, 73, 91
Fisher, R. A. 233, 234, 235, 235, 236
Flavell, J. 320
Flight checklist 359
Flemming, W. 227
Flew, A. 272
Flexner, S. 93
French Revolution 106, 108, 109, 110, 111, 253, 350
fMRI 47, 85
Forrest, B. 272
Freeman, C. 146, 147, 149
Freeman, D. H. 315
Fu, Z. 129, 130
Galileo, G. 46, 133, 134, 136, 144, 155-163, 169, 172, 230, 267, 292, 308, 317
Galton, F. 16, 225, 226
Game theory 57
Gamow, G. 256
Gardner, H. 44
Gaukroger, S. 148, 149, 152

Gawande, A. 358, 359
Gay, B. 61
Gay, P. 174
General checklist 358-362
Ghurye, G. S. 51, 52, 63, 64
Gibbon, E. 152, 220
Gigerenzer, G. 101
Gilham, N. W. 225
Glass, B. 207-211
God 23, 63, 91, 141, 142, 143, 145, 148, 149, 154, 155, 162, 163, 166, 167, 169, 170, 171, 173, 174, 175, 180, 203, 211, 212, 213, 220, 221, 246, 247, 259, 268, 270, 275-278, 288, 327, 332, 361
Goffman, E. 33
Golden Age (Athens) 67
Golden Age (Islam) 132, 145
Goldman, M. 125
Goldschmidt, R. 279
Goldstein, M. 328
Goldstein, H. 162
Goldziher, I. 140
Goleman, D. 84, 87, 118, 360
Gonzalez, G. 263, 264, 281, 282
Goodwin, B. 255, 296, 299, 300
Goonatilake, S. 24, 134
Gould, S. J. 203, 217, 218, 233, 37, 238, 274, 302
Grafman, J. 85
Graham, A. C. 86
Graner, C. 357
Grant, E. 141, 152, 155
Grasse, P-P. 198
Gray, A. 192, 193

Great Leap Forward 75-77
Greenberg, D. S. 289
Greenwald, G. 92
Gregory, M. E. 180
Gross, D. 316
Grossman, R. H. S. 68
Group polarization 94, 95, 104
Group selection 190
Groupthink 344, 351, 352
Guatemala 107, 333, 339, 342
Gullibility 331, 361
Guyénot, E. 211
Haack, S. 307
Haber, F. C. 180
Haeckel, E. 202, 222
Haggbloom, S. J. 38
Haldane, J. B. S. 188, 234, 235, 236
Hall, B. G. 295
Hall, R. 133
Halo effect 26, 27
Hamilton, A. 66
Hamilton, W. D. 190, 236
Hanawalt, P. C. 293
Han Fei 129
Harman, W. W. 301
Hart, H. H. 21
Hartle, A. 319
Haskins, C. H. 150
Haught, J. 277
Hayakawa, S. I. 90, 91, 92, 98, 117
Henderson, B. 276
Henslow, J. S. 184, 185
Heuss, T. 118
Himmelfarb, G. 215
Hinson, E. G. 145-149
Hitler, A. 29, 67, 88, 90, 99, 111,

113-119
HMS Beagle 185, 189, 199
Ho, M. W. 131, 262
Hobbes 110, 161, 319
Hobson, J. A. 73
Hoffer, E. 100
Hofstede, G. 107, 108
Hood, L. 293
Hooker, J. 189, 197, 199, 200, 201, 202
Horgan, J. 48
Horn, J. 266
Household responsibility system 76
Hoyle, F. 262
Hudelson, R. 71
Huff, T. E. 122, 131, 133, 140, 151
Hughes, H. 60-64
Hull, D. 183
Human genome project 256
Hume, D. 25, 82, 104, 171, 309
Hus, J. 21
Hunter, C. G. 279
Hut, P. 325
Hutton, J. H. 64
Hutton, R. H. 214
Huxley, J. 237, 238
Huxley, T. H. 189, 192, 202-207, 210, 211-220, 226, 233, 252, 274
Hyatt, A. 221, 294
Ibn al-Salah 140
Indian caste system 31, 49-65, 105, 178, 307, 308, 313, 326
Intellectual laziness 39, 73, 243, 265, 301, 329

Intelligent design 247, 249, 251, 252, 269, 270, 271, 272, 273, 280, 281
Internal representative dialogue (IRD) 330
Iron curtain 105
Irons, P. 268, 272
Irreducible complexity 247, 250, 253, 254, 280
Irvine, W. 195
Israel, J. I. 169
Ito, T. A. 29
Iyengaar, S. 29
Jablonka, E. 297, 298
Jacob, F. 327
Jacobins 109
James, J. W. 226
James, W. 33, 36, 284
Janis, I. L. 337, 338, 339, 340, 341, 344, 346, 351
Japan 107
Jati 50, 51
Jeans, J. 246
Jefferys, W. H. 282
Jensen, J. V. 214, 215
Jeon, K. W. 295, 297
Johannsen, W. 186
John, O. P. 29
Johnson, L. 76
Johnson-Laird, P. N. 82
Jones, J. 132
Josephson, B. 324
Judson, H. 301
Kafka, 38
Kant, I. 33, 81, 174, 175, 211, 229, 230
Karoui, N. E. 230
Karve, I. 54

Keen, E. 33
Keller, E. F. 253, 256
Kennedy, J. F. 31, 192, 275, 328, 337-353
Kennedy, R. F. 35, 108
Kepler, J. 46, 134, 156, 157
Khrushchev, N. 101, 102, 347, 348
Kihlstrom, J. F. 47
Kilroy-Silk, R. 72, 73
Kim, Y. H. 286
Kim Il Sung 72, 204
Kimura, M. 293, 302
King, M. L. Jr. 28, 76, 79, 109, 125, 156, 308, 366
Kin selection 190
Kitano, H. 253
Knowles, D. 149
Kohn, D. 185
Koestler, A. 157, 158, 159, 161, 302
Kohlberg, L. 363
Kohler, W. 314, 315, 357
Kölliker, R. A. von 192
König, S. 209, 210
Korzybski, A. 90, 91, 311
Kosambi, D. D. 54
Kosovo 113, 366
Kramer, J. 105
Krantz, D. L. 251
Krebs, R. E. 19
Krug, K. S. 102
Kshatriyas 50, 51, 54
Kubizek, A. 116
Kuhn, T. 176, 241, 303, 304, 314, 322, 323, 335
Kuran, T. 49, 93, 97, 104, 105, 139, 361

Labinger, J. A. 7
Lacey, H. 25, 26
Lamarck, J. B. 196, 197, 198, 210, 220, 253
Lamarckism 221, 222, 223, 228
Langer, U. 320
Langer, W. C. 114
Language acquisition device (LAD) 97
Lao Tzu 126, 318
Lashley, K. 43
Lavoisier, A. 179, 285
Lawrence, W. 194, 195, 213
Leahey, T. H. 34, 35, 229, 309, 310
Leaman, O. 143
LeDoux, J. E. 84
Legalism 124, 128, 129
Leibniz 174, 209
Leiser, G. 140
Lele 59
LeMay, C. 349
Lenin, V. 68, 71, 72, 73, 74, 130
Lennox, J. G. 192
Leucippus 174, 285, 291
Li, D. 230, 231
Li Si 129
Lim, R. 149
Lama-de-Faria, A. 262, 303
Lincoln, A. 156, 165
Lindberg, D. C. 123, 137, 154
Lindley, D. 300
Linholm, C. 118
Lloyd, G. E. R. 124
Locke, J. 110, 311, 319, 364
Loftus, E. F. 80
Lord Kelvin 241, 285, 312
Lovelock, J. 266

Løvtrup, S. 183, 184, 189, 195, 197, 198, 201, 202, 232, 234, 235, 236
Lun-heng 128
Lyell, C. 180, 189, 197, 199, 200, 201, 202
Lynch, M. 289
Macbeth, N. 188
MacMullen, R. 149
Madison, J. 56, 274, 282, 321, 362, 364
Madrasas 138, 139, 140, 141, 144, 145, 151, 152, 153, 321
Mahoney, M. J. 7, 39, 230, 313
Malik, K. 178
Manichean worldview 91, 108, 215
Mann, T. 259
Mannan, S. K. 96
Manschreck, C. L. 147
Manu 51, 54, 59, 64
Mao 37, 67, 68, 72, 75, 76, 204
Margulis, L. 187, 223, 261, 270, 295
Martin, B. 283
Marx, K. 66, 69-75, 77, 87-89, 91, 92, 99, 106, 108, 118, 119, 312, 360,
Maslow, A. 117
Mason, F. 246
Matthew, P. 193, 194, 195
Mattick, J. S. 256
Maupertuis, P. 207-211, 253
Maurer, A. 142
Maynard Smith, J. 190
Mayr, E. 179, 211, 221, 223, 226, 229, 236, 237, 238, 244, 266, 292, 293, 303

McCarthy, J. 35, 36, 42, 43, 158, 253
McCain, J. 219
McClintock, B. 296
McComas, W. F. 185, 205
McCourt, M. E. 96, 97
McCulloch, W. 43
McDermott, R. 101, 102
McLean v. Arkansas Board of Education 271
McLellan, D. 88
McMullin, E. 25
Mead, G. H. 316
Megill, A. 71
Meinhardt, H. 300
Mele, A. 39
Mendel, G. 207, 208, 224-227
Mendelians 207, 223, 226, 227, 232, 233, 234
Mendelism 225, 226, 227, 232, 233, 234
Mendoza, E. R. 253
Merton, R. K. 27, 176, 365
Meslier, J. 167, 168, 169
Mesulam, M-M. 96
Metacognition 320-325
Metaphysics 8, 65, 143, 149, 178, 309, 310, 312, 316, 357, 362, 363
Methvin, E. 87, 88, 89, 91, 92, 110, 116
Meyer, A. G. 89
Meyer, K. E. 338
Meyer, S. C. 258, 259
Michelet, J. 109
Milgram, S. 111, 112
Mill, J. S. 222

Miller, A. 114
Miller, G. A. 43, 44
Miller, K. 250, 251, 272
Miller, S. L. 259, 260
Mills, C. W. 326
Milner, M. 50, 51
Milosz, C. 105
Mivart, S. 215-219
Miyazaki, I. 131
Mohism 124-126, 128
Moldoveanu, M. 329
Monod, J. 244, 260, 292, 293, 301
Montaigne, M. de 38, 173, 316-320
Montesquieu 110, 364
Moore, W. E. 56
Morgan, T. H. 228
Morowitz, H. J. 260, 261, 299
Mo Tzu 125, 126
Mukherjee, B. 52
Muller, H. J. 228
Murrow, E. R. 36, 37, 253, 356
Mutationism 207
Mutawillis 139
Mutual dependence principle 325, 326
My Lai 356, 357
Nagel, T. 224, 273
Narasimhachary, M. 54
Natural Selection 181, 183-208, 211, 216, 217, 220, 221, 223, 225, 245, 248, 250, 260, 264, 265, 271, 275, 276, 278, 292, 294, 300, 302, 304, 312
Natural theology 213, 248, 249, 277

Nazism 87-93, 99, 118
Nehru, J. 19
Needham, J. 121, 124, 134, 136, 264, 302
Neo-Darwinism 184, 224-239, 249, 261, 262, 270, 277, 279, 292, 297, 298, 302, 303
Nero 146
Nesfield, J. C. 51, 52
Neufeld, P. 102
Newton, I. 46, 126, 133, 136, 144, 170, 173, 174, 175, 177, 178, 184, 205, 209, 210, 227, 245, 285, 292, 300, 302, 308, 311, 312, 313, 315, 317
Nicholls, E. R. 97
Nigeria 107
Nixon, R. 192, 275, 229, 345
North Korea 66, 92, 99
"Not me" myth 331-335
Nurse, P. 254
Oatley, K. 82
Obsessive compulsive disorder (OCD) 66, 62, 63, 64
October Revolution 72
Oldenberg, H. 64
On-deck memory 86
One Hundred Schools Period 124
Onfray, M. 168
Oparin, A. 259
Orwell, G. 31
Osborn, H. F. 197, 210, 211, 212, 294
Owen, R. 188, 189, 205, 206, 215, 216, 217,

220, 248
Oyserman, D. 106
Palace Examination System 130-131
Palermo, D. S. 304
Paley, W. 212, 213, 248
Panama 107
Parsons, T. 56
Pascal, B. 213
Pasteur, L. 261
Pears, D. 39
Pearson, K. 206
Pearson, H. 226, 255
Pagels, H. 287
Peirce, C. 19
Peng, X. 75
Perrie, M. 71, 74
Persuasive argument 95, 171
Peter, L. J. 203
Peters, D. 26, 28, 103, 225
Peters, E. F. 143
Physics envy 34, 244, 245
Pigliucci, M. 232
Pinker, S. 16
Plato 66-68, 81, 132, 133, 134, 149, 318, 329, 360
Plotinus 284
Poincare, H. 331
Polanyi, M. 258, 281
Pol Pot 72
Polmar, N. 347
Pope Urban VIII 160
Popkins, R. H. 319
Popper, K. 49, 187, 203, 241, 291, 311-314
Postman, L. 40, 330
Powers, F. G. 101
Prana 127

Pratto, F. 29
Priestly, J. 213
Protestant Reformation 21
Provine, W. 223, 226, 233, 234, 237, 303
Pseudoneglect 97
Qadi 139
Qin dynasty 124
Qin Shi Huangdi 68, 124, 129
Quastler, H. 257
Rabow, J. 94
Raspai, F. V. 275
Rawls, J. B. 93
Ray, J. 212
Reagan, R. 92
Redlich, F. 115
Redner, H. 178
Reich, D. 53
Rees, M. 327
Reid, R. G. G. 223, 233, 237, 303
Reign of Terror 109
Rendel, J. M. 226
Renn, 126
Reno, J. 102
Richards, J. 263, 281
Richards, R. J. 189
Riddiford, A. 188, 303
Riedl, R. 291
Rigidity 116, 117, 119, 330, 331-335, 361
Rilling, J. K. 85
Risky shift 94
Robespierre, M. 110, 111
Robertson, F. W. 226
Robertson, I. H. 96, 97
Roediger, H. 46
Roggeveen, J. 16, 17, 19
Romer, O. 262

Rose, S. 179
Rosenberg, A. 115
Ross, G. M. 178, 179
Ross, H. 263
Ross, G. M. 276
Rothbard, M. 127
Rousseau 110, 123,
Rowland, W. I. 160, 161
Ruse, M. 202, 206, 207
Rusk, D. 343
Russell, B. 38, 276, 320, 327, 331, 349, 361
Rutherford, E. 46, 179, 285
Rwanda 94, 113, 366
Sabra, A. I. 133, 140
Safina, C. 204
Sagan, C. 19, 244, 246, 276
Sagan, D. 187
Sagan, E. 109, 355
Saliba, G. 141
Salmon, F. 230
Saltationism 223, 225, 226, 233, 278
Sapp, J. 278
Sarawak Law 199
Saunders, P. T. 301, 303
Savonarola, G. 21
Saxe, J. G. 23
Schlesinger, A. M. Jr. 338-341, 343-347,
349, 350
Schmutzer, E. 156
Schooler, J. 80
Scheck, B. 102
Schemmel, M. 126
Schoenblum, J. A. 139
Schroeder, G. 258, 259
Schrodinger, E. 283

Schrodinger's cat 284, 285
Schurz, C. 89
Schwaab, E. F. 113, 114-118
Schwartz, J. H. 198, 234-236, 301
Scopes trial 266, 267
Searle, J. R. 33
Selfridge, O. G. 80
Semendferi, K. 85
Shang Dynasty 124
Shapiro, J. A. 260, 296, 298
Shea, W. R. 159, 161
Simpson, G. G. 221, 224, 237, 294, 300
Short-term memory 86
Sierra Leone 107
Simon, H. A. 82, 85
Singer, M. 331, 355
Singh, V. P. 51
Skinner, B. F. 37-44, 334, 335
Slack, G. 268, 270
Smith, A. 39, 70, 106
Smith, B. F. 115
Smocovitis, B. 8, 178, 228, 245, 302
Smolin, L. 264
Snsolabehere, S. 29
Social comparison 27, 95
Sockman, R. W. 280
Socrates 68, 124, 133, 167, 318, 319, 360
Sole, R. V. 300
Sokal, A. D. 324
Solzhenitsyn, A. 105
Sorensen, T. C. 346, 348
South Korea 107
Spencer, H. 123, 187, 198, 220, 222
Spetner, L. 293

Spring and Autumn Period 124
Sputnik 65, 271
St. Augustine 149, 211, 212, 220, 274, 276, 278, 317
Staats, A. W. 34, 45
Stalin, J. 66-68, 71, 72, 120, 204
Standen, A. 19, 21
Stanley, S. M. 186
Steele, A. 51
Steele, J. B. 60, 61
Steffens, B. 138
Stevenson, A. E. Jr. 362
Stewart, M. 229
Stoner, J. A. F. 94
Strauss, L. 163
Streetlight effect 315
Stroun, M. 297
Sudan 113
Sudras 50, 51, 54
Sunstein, C. R. 94
Supreme court 73, 102, 267, 271
Sutherland, S. 331
Synthetic theory 221, 224, 232, 239
Systems biology 253, 254
Szent-Gyorgi, A. 291
Szulc, T. 338, 342
Taiwan 107
Taleb, N. N. 330
Taoism 124, 126, 127
Taylor, F. J. R. 295
Teleology 192, 193
Templeton, J. 100, 237
Terrall, M. 171, 208, 209, 210
Tertullian 148, 180, 319
Thales 128, 166, 318
Theistic evolution 219, 221, 274, 277

Thompson, D. 299, 300, 302
Thompson, H. 329
Thompson, J. J. 285
Thorndike, E. L. 26
Thorpe, W. 248
Thrower, J. 167, 171
Toates, F. 304
Tolman, E. L. 42
Toynbee, A. 16, 121
Triandis, H. C. 107
Trimble, V. 284
Trotsky, L 72
Tulving, E. 48
Turner, J. S. 251, 265
Twain, M. 99, 101
Two-valued orientation 106, 108, 215
Tyler, P. 76
U-2 spy plane 101, 102, 347
Ulanowicz, R. E. 245
Unmoved Mover 141, 275
Untouchables 50, 63
Vaisyas 50, 51, 54
Value judgment 27, 98, 100, 103, 359, 360
Van den Daele, W. 179
Van Regenmortel, M. H. V. 244
Varna 49, 50, 51
Vaughn, D. 18, 344, 360
Versluis, A. 65, 148, 149
Viereck, P. 118
Visual neglect 95-97
Voltaire 168, 169, 180, 209, 210, 211, 220, 253, 361
Von Nageli, C. 224
Von Neumann, J. 43
Von Stein, L. 88
Von Tschermak, E. 224

Von Bertalanfy, K. L. 254
Von Weizsäcker, C. F. 287
Vorzimmer, P. J. 195
Wacq 139, 153
Waddington, C. H. 186, 236, 302
Waqif 139, 140
Waite, R. G. 115, 116
Wald, G. 261
Wallace, A. 37
Wallace, A. R. 191, 194, 199, 200, 215, 248
Wang Chong 127
Warring States Period 128
Watkins, M. J. 45
Watson, B. 127
Watson, J. B. 32-38, 42, 304
Watson, J. D. 179, 204, 242
Watt, W. M. 143
Weaver, C. A. III 102
Weber, M. 25, 106
Wedge document 270, 272, 275
Weimer, W. B. 304
Weinberg, J. 97
Weinberg, S. 230, 281, 284, 293, 314, 323
Weldon, W. F R. 226
Wells, G. L. 102
Wells, J. 252
Wells, W. C. 194
Wesson, R. 226
White, P. 202, 206, 214, 215
White collar workers 136, 137, 308
Whitehead, A. N. 66, 67
Whitehead, C. 304
Wigner, E. P. 284, 286
Wilberforce, Bishop 274
Williams, G. C. 190

Williams, L. P. 229
Williams, R. M. Jr. 24
Wilson, D. S. 304
Wilson, E. O. 103, 204, 221
Winter, M. 140
Woodhead, L. 147
Wollstonecraft, M. 49
Working memory 86
World War I 72, 74, 90, 115, 116, 366
World War II 43, 65, 98, 99, 114, 118, 123, 288, 349
Wozniak, R. H. 34
Wyden, P. 338, 340, 341
Xiaogang farmers 74-77
Yan Jinchang 74
Yang, K. 107
Zajonc, R. 25
Zero-sum impulse 57, 58, 59, 135, 136, 138, 176, 177, 178, 307, 309, 360
Zhou dynasty 124
Zink, C. F. 27
Zorzi, M.

A Critique of Science